D1387879

Solid-Phase Extraction

CHEMICAL ANALYSIS

A SERIES OF MONOGRAPHS ON
ANALYTICAL CHEMISTRY AND ITS APPLICATIONS

Editor
J. D. WINEFORDNER

VOLUME 147

A WILEY-INTERSCIENCE PUBLICATION

JOHN WILEY & SONS, INC.

New York / Chichester / Weinheim / Brisbane / Singapore / Toronto

Solid-Phase Extraction

Principles and Practice

E. M. THURMAN

U.S. Geological Survey
Lawrence, Kansas, USA

M. S. MILLS

Zeneca Agrochemicals
Jealott's Hill, UK

A WILEY-INTERSCIENCE PUBLICATION

JOHN WILEY & SONS, INC.

New York / Chichester / Weinheim / Brisbane / Singapore / Toronto

Learning Resources
Centre

This book is printed on acid-free paper. ∞

Library of Congress Catologing-in-Publication Data:

Thurman, E. M. (Earl Michael), 1946–
    Solid-phase extraction : principles and practice / E. M. Thurman, M. S. Mills
        p.   cm — (Chemical analysis ;   v.   147)
    "A Wiley-Interscience publication."
    Includes bibliographical references and index.
    ISBN 0-471-61422-X (cloth : alk. paper)
    1. Extraction (Chemistry)   I. Mills, M. S.   II. Title.
III. Series.
QD63.E88T58   1998
543′.0892—dc21   97–27535

Printed in the United States of America.
10 9 8 7 6 5 4 3

Dedicated to Our Parents

CONTENTS

PREFACE

The objectives of this book are, first, to provide a straightforward explanation of how solid-phase extraction (SPE) works, including information on the chemistry of the solid phase and the mechanisms of interaction; second, to provide detailed applications of SPE in clinical, environmental, and natural product chemistry. The selectivity and sophistication of SPE chemistry and sorbents have diversified so rapidly since their inception in the late 1970s that experience is required to make intelligent decisions on choices for SPE sorbents. Third, a discussion of the latest SPE technology provides insight into automation and new methods in sample preparation. Overall, it is hoped that this book will serve as an important reference tool for students, laboratory technicians, and application chemists who are current users of solid-phase extraction.

E. M. Thurman
M. S. Mills

ACKNOWLEDGMENTS

We thank our reviewers and colleagues in the field of solid-phase extraction for their ideas and helpful comments: Jennifer Field, Marie-Claire Hennion, Lynn Jordan, Craig Markell, and Martha Wells. A special thanks to Lynn Jordan for her help on preparing the chapter on automation. Much of this chapter was taken from our joint notebook from the class sponsored by the American Chemical Society on Solid-Phase Extraction in Environmental and Clinical Chemistry. The fine graphics work of Lisa Zimmerman is gratefully acknowledged. Also we appreciate the generosity of all cited chemical company manufacturers for supplying detailed information on their products, figures, and photographs as cited, and for figures from cited references. In particular, we thank Darrell Adams, Linda Alexander, Joe Arsenault, John Berg, Judie Blackwell, Dennis Blevins, Edouard Bouvier, Udo Brinkman, Michael Burke, Richard Calverley, Eric DiConte, Frank Dorman, Bardick Ellam, W. Dale Felix, Ron Goldberg, Gerad Haak, Jun Haginaka, Csaba Horvath, James Huckins, Reed Izatt, Martha and Robert Johnson, Lynn Jordan, David Kennedy, Paul Kester, Ron Majors, Craig Markell, Wesley Meyers, Jennifer Miner, Lori Morrison, A. Wesley Moyers, Gary Nixon, Lydia Nolan, Regina Patel, Marisue Paulus, S. Tinsley Preston III, Peg Raisglid, Sean Randall, Rein Reitsma, Rolf Schlake, Lisa Serfass, Robert Shirey, Nigel Simpson, Joe Stephovich, Robert Taylor, Michael Telepchak, Steven Thibodeaux, K.C. Van Horne, Andrew Wood, Xiaogen Yang, and Michael Young. Special thanks are given to Carla Fjerstad, Editor at John Wiley, and Christine Punzo, Assistant Managing Editor, for encouragement and help. Finally, we accept the responsibility for all errors and misquotations of products. The use of brand names is for identification purposes only and does not imply endorsement by the authors. We accept sole responsibility for this work.

CHEMICAL ANALYSIS

A SERIES OF MONOGRAPHS ON
ANALYTICAL CHEMISTRY AND ITS APPLICATIONS

J. D. Winefordner, *Series Editor*

xxi

Solid-Phase Extraction

CHAPTER

1

OVERVIEW OF SOLID-PHASE EXTRACTION

1.1   WHAT IS SOLID-PHASE EXTRACTION?

Solid-phase extraction (SPE) is a method of sample preparation that concentrates and purifies analytes from solution by sorption onto a disposable solid-phase cartridge, followed by elution of the analyte with a solvent appropriate for instrumental analysis. The mechanisms of retention include reversed phase, normal phase, and ion exchange. Traditionally, sample preparation consisted of sample dissolution, purification, and extraction that was carried out with liquid–liquid extraction. The disadvantages with liquid–liquid extraction include the use of large volumes of organic solvent, cumbersome glassware, and cost. Furthermore, liquid–liquid extraction often creates emulsions with aqueous samples that are difficult to extract, and liquid–liquid extraction is not easily automated. These difficulties are overcome with solid-phase extraction. Thus, solid-phase extraction was invented in the mid-1970s as an alternative approach to liquid–liquid extraction.

Initially, SPE was based on the use of polymeric sorbents, such as XAD resins (polymeric adsorbents), which were packed in small disposable columns for use on drug analysis. The early environmental applications consisted of both XAD resins and bonded-phase sorbents, such as C-18 (McDonald and Bouvier, 1995). These precolumns were used for sample trace enrichment prior to liquid chromatography and were often done on-line, which means at the same time as liquid chromatography. However, these first, steel, on-line precolumns quickly were replaced with an off-line column made of plastic in order to be both inexpensive and disposable. Eventually, the term solid-phase extraction was coined for these low-pressure extraction columns (Zief et al., 1982). Thus, solid-phase extraction is an analogous term to liquid–liquid extraction, and in fact, solid-phase extraction might also be called liquid–solid extraction. However, it is the term solid-phase extraction or the acronym SPE that has become the common name for this procedure.

Solid-phase extraction columns now are typically constructed of polypropylene or polyethylene and filled with 40-μm packing material with different functional groups. A 20-μm polypropylene frit is used to contain from 50 mg to 10 g of packing material. A liquid sample is passed through the column and analytes are concentrated and purified. The sample volume that

1

can be applied ranges from 1 mL to over 1 L. The sample may be applied to the column by positive pressure or by vacuum manifold. After quantitative sorption of the analyte, it is removed with an appropriate elution solvent.

Therefore, SPE is a form of *digital* liquid chromatography that removes the solute onto a solid-phase sorbent by various sorption mechanisms. The term digital refers to the on/off mechanism of sorption and desorption (Wells and Michael, 1987). The goal of SPE is to quantitatively remove the analyte from solution and completely recover it in an appropriate solvent. Purification consists of removing the analyte from interfering compounds and concentrating the analyte in a small volume of solvent. For example, pesticides are concentrated from a water sample by SPE into a small volume of organic solvent for analysis by gas chromatography/mass spectrometry. Interfering substances, such as humic and fulvic acids, ionic metabolites, and salts, are removed. Another example is the removal of drugs from blood or urine for analysis by liquid or gas chromatography. SPE removes interfering proteinaceous substances, salts, and urea; then, it concentrates the analytes and places them in a solvent suitable for chromatographic analysis.

Typically, SPE replaces liquid–liquid extraction as a sample preparation tool and provides a method that is simple and safe to use. The benefits of SPE include high recoveries of analytes, purified extracts, ease of automation, compatibility with chromatographic analysis, and reduction in the consumption of organic solvents. As a result of the flexibility that SPE offers, it has found application in the preparation of environmental, clinical, and pharmaceutical samples. Examples include the concentration of trace organic pollutants from water, the purification of drugs of abuse from blood and urine, and the extraction of organic compounds from food and beverages. The simplicity of the SPE procedure and the use of disposable SPE supplies have encouraged the design of automated sample preparation stations, which decrease the time and cost of sample preparation. Finally, recent advances in on-line methods of SPE allow automation of sample preparation directly to both liquid and gas chromatography (Hennion and Pichon, 1994; Barcelo and Hennion, 1995; Brinkman, 1995).

1.2. HOW TO DO SOLID-PHASE EXTRACTION

Figure 1.1 illustrates the four-step process of SPE. First the solid-phase sorbent is conditioned (step 1). This simply means that a solvent is passed through the sorbent to wet the packing material and to solvate the functional groups of the sorbent. Furthermore, the air present in the column is removed and the void spaces are filled with solvent. Typically the conditioning solvent is methanol, which is then followed with water or an aqueous buffer. The

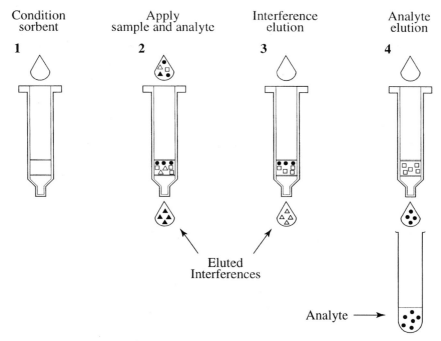

Figure 1.1. Process of solid-phase extraction. (Reproduced in somewhat modified form from the catalog pages of International Sorbent Technology, Ltd., copyright. ISOLUTE and IST are registered trademarks of International Sorbent Technology, Ltd.)

methanol followed by water or buffer activates the column in order for the sorption mechanism to work properly for aqueous samples. Care must be taken not to allow the bonded-silica packing or the polymeric sorbent to go dry. In fact, if the sorbent dries for more than several minutes under vacuum, the sorbent *must* be reconditioned. If it is not reconditioned, the mechanism of sorption will not work effectively and recoveries will be poor for the analyte.

Another cleaning step of the sorbent also may be added during conditioning, if necessary. Simply, the eluting solvent is passed through the column after the methanol wetting step to remove any impurities that may be present in the packing material. This cleaning step would then be followed by methanol and aqueous buffer, which prepares the column for sample addition.

Next the sample and analyte are applied to the column (step 2). This is the retention or loading step. Depending on the type of sample, from 1 mL to 1 L of sample may be applied to the column either by gravity feed, pumping, aspirated by vacuum, or by an automated system. It is important that the mechanism of retention holds the analyte on the column while the sample is added. The mechanisms of retention include van der Waals (also called nonpolar, hydrophobic, partitioning, or reversed-phase) interaction, hydrogen bonding, dipole–dipole forces, size exclusion, and cation and anion exchange. This retention step is the digital step, or "on/off mechanism," of solid-phase extraction. During this retention step, the analyte is concentrated on the sorbent. Some of the matrix components may also be retained and others may pass through, which gives some purification of the analyte.

Step 3 is to rinse the column of interferences and to retain the analyte (Fig. 1.1). This rinse will remove the sample matrix from the interstitial spaces of the column, while retaining the analyte. If the sample matrix was aqueous, an aqueous buffer or a water–organic-solvent mixture may be used. If the sample was dissolved in an organic solvent, the rinse solvent could be the same solvent.

Finally, step 4 is to elute the analyte from the sorbent with an appropriate solvent that is specifically chosen to disrupt the analyte–sorbent interaction, resulting in elution of the analyte. The eluting solvent should remove as little as possible of the other substances sorbed on the column. This is the basic method of solid-phase extraction. There is an alternate method where the interferences are sorbed and the analyte passes through the column. In this case, the method is terminated at step 2 of Figure 1.1. The analyte is collected and assayed directly. However, this method is rarely used.

1.3. COLUMNS AND APPARATUS FOR SOLID-PHASE EXTRACTION

The sorbents used for SPE are packaged in three basic formats. There are disks, cartridges, and syringe barrels. Figure 1.2 shows the different types of

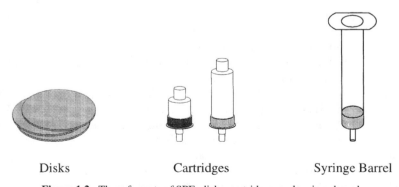

Disks   Cartridges   Syringe Barrel

Figure 1.2. Three formats of SPE: disks, cartridges, and syringe barrel.

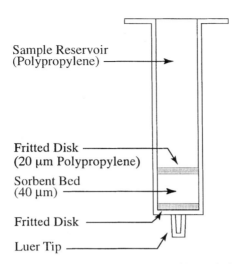

Sample Reservoir
(Polypropylene)

Fritted Disk
(20 μm Polypropylene)

Sorbent Bed
(40 μm)

Fritted Disk

Luer Tip

Figure 1.3. Anatomy of a syringe barrel. (Published with permission of J. T. Baker.)

presentations of SPE products. The disks are available in different diameters from 4 to 90 mm. The "standard" disk size is 47 mm. Cartridges vary from as little as 100 mg to 1 g or more. Syringe barrels are available in different volumes and with different masses of packing material. Syringe barrels range in size from 1 to 25 mL and packing weights from 50 mg to 10 g. These various sorbents allow for the effective treatment of different types of samples and different sample volumes.

Currently, the most commonly used format for SPE consists of a syringe barrel that contains a 20-μm polypropylene frit at the bottom of the syringe with 40-μm sorbent material and a 20-μm polypropylene frit at the top (Fig. 1.3). The syringe barrel is typically polypropylene with a male Luer tip fitting and is disposable. Some vendors do make glass syringe barrels with Teflon frits, but these configurations are used less frequently. The glass and Teflon system is used when one is interested in the analysis of plasticizers or is concerned with the potential sorption of specific analytes onto the polyethylene tube. Solvent reservoirs may be used to increase the volume of the syringe barrel. Reservoirs are typically 50 to 100 mL in volume. Coupling fittings are used to join the reservoirs and syringe barrels between the Luer fitting and the opening of the syringe barrel (Fig. 1.4).

The barrel of the syringe terminates in a male Luer tip. The male Luer tip is the standard fitting on SPE cartridges so that they are interchangeable with different SPE vacuum manifolds. The vacuum manifold is used to draw the sample and eluting solvents through the syringe barrel under negative pressure

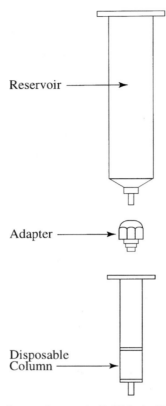

Reservoir

Adapter

Disposable
Column

Figure 1.4. Disposable column and reservoir. (Published with permission of J. T. Baker.)

by applying a vacuum to the manifold. Figure 1.5a shows a typical vacuum manifold system, which is fitted with a small vacuum pump and a waste receiver. Stopcock valves are available to control the vacuum applied to each column. Other types of sample processing that may be used include centrifugation and positive pressure, which forces the sample through the syringe barrel from above. Simple gravity flow through the syringe barrel or cartridge may also be used (Fig. 1.5b).

A second format for SPE is the cartridge design (Fig. 1.2). It consists of a polyethylene body with both a female and male Luer tip for positive pressure

Figure 1.5. Techniques for processing SPE cartridges, vacuum manifold, gravity, centrifugation, and positive pressure. (Reproduced in somewhat modified form from the catalog pages of International Sorbent Technology, Ltd., copyright. ISOLUTE and IST are registered trademarks of International Sorbent Technology, Ltd.)

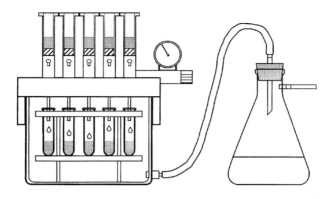

Vacuum Manifold

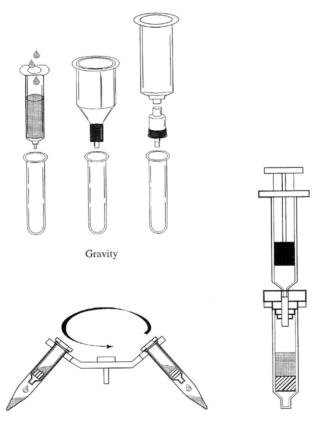

Gravity

Centrifugation

Positive Pressure

Figure 1.6. Photomicrograph of an Empore extraction disk cross section containing 12-μm particles. (Published with permission of 3M.)

from a syringe or negative pressure from a vacuum manifold. Polyethylene frits of 20 μm are placed at either end of the cartridge to hold the packing material in place. The packing material is packed and compressed to improve or optimize flow characteristics. The third type of SPE format is the disk, which is constructed in several styles by different vendors. One of the popular extraction disks, the Empore Extraction Disk (see Manufacturers Guide, Appendix), consists of 8- to 12-μm particles of packing material imbedded into an inert matrix of polytetrafluoroethylene (PTFE) fibrils (Fig. 1.6).

Because the particles are suspended in PTFE, no binder is required to give structure to the disk and the matrix is essentially inert. The disks are not coated with PTFE so that they can interact with the solvent and sample during extraction. The disks are available in membrane format as loose disks or are placed in syringe-barrel format called an extraction disk cartridge. The syringe-barrel format consists of a standard polyethylene syringe that is fitted with a Teflon 20-μm frit, Empore disk, and a prefilter of glass fiber. This arrangement allows for microscale work using the disk. Disks are conditioned and used in a similar fashion to the packed columns with flow of sample by negative pressure by vacuum.

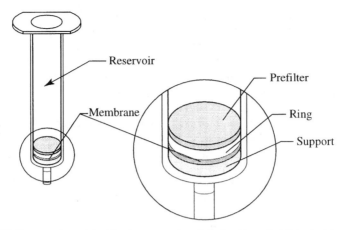

Figure 1.7. Construction of the Empore extraction disk cartridge. (Published with permission of 3M.)

A major advantage of the disk format is rapid mass transfer because of the greater surface area of the 8- to 12-μm particles, which results in high flow rates for large volume samples. This rapid flow rate is especially useful for environmental samples where 1 L of water may be processed in as little as 15 min. Rapid mass transfer because of imbedding of small particles into the disk also means that channeling is reduced and small volumes of conditioning and elution solvents may be used. For example, a 4-mm disk in syringe format, called the extraction disk cartridge, requires only 100 μL of elution solvent, and a 7-mm disk in syringe format uses only 250 μL (Fig. 1.7). This volume is small compared to the milliliter amounts applied to a 3- to 5-mL syringe barrel that contains loose packing.

Another type of disk, SPEC made by Ansys, Inc., uses a glass-fiber matrix rather than Teflon to hold the sorbent particles. This disk has a somewhat more rapid flow rate and is more rigid and thicker than the Teflon disk. There is another disk called the Speedisk made by J. T. Baker that consists of 10-μm packing material that is sandwiched between two glass-fiber filters without any type of Teflon binder. The various types of disks are discussed in detail in Chapter 11.

1.4. SORBENTS AND MODES OF INTERACTION

The sorbents used for SPE are similar to those used in liquid chromatography, including normal phase, reversed phase, size exclusion (commonly called wide pore), and ion exchange. Normal-phase sorbents consist of a stationary phase that is more polar than the solvent or sample matrix that is applied to the SPE sorbent. This means that water is not usually a solvent in normal-phase SPE because it is too polar. Normal-phase sorbents, therefore, are used in SPE when the sample is an organic solvent containing an analyte of interest. Polar interactions, such as hydrogen bonding and dipole–dipole interactions, are the primary mechanisms for solute retention.

Reversed-phase sorbents are packing materials that are more hydrophobic than the sample. Reversed-phase sorbents are commonly used in SPE when aqueous samples are involved. The mechanism of interaction is van der Waal's forces (also called nonpolar, hydrophobic, or reversed-phase interactions) and occasionally secondary interactions such as hydrogen bonding and dipole–dipole interactions. Size-exclusion sorbents utilize a separation mechanism based on the molecular size of the analyte. It is a method only recently being used in SPE, usually in conjunction with reversed phase or ion exchange. Ion-exchange sorbents isolate analytes based on the ionic state of the molecule, either cationic or anionic, where the charged analyte exchanges for another charged analyte that already is sorbed to the ion-exchange resin. Solid-phase extraction applications in this case are essentially identical to classical ion exchange.

Thus, the mechanisms of interaction include hydrogen bonding and dipole–dipole forces (polar interactions), van der Waal's forces (nonpolar or hydrophobic interactions), size exclusion, and cation and anion exchange. Some sorbents combine several interactions for greater selectivity. The extensive line of sorbent chemical structures facilitates one of the most powerful aspects of SPE, that is, selectivity. Selectivity is the degree to which an extraction technique can separate the analyte from interferences in the original sample. The number of possible interactions between the analyte and the solid phase facilitates this selectivity.

Table 1.1 lists the common sorbents that are available for SPE and their mode of action (i.e., reversed phase, normal phase, ion exchange, and size exclusion). Typically the sorbents consist of 40-μm silica gel with approximately 60-Å pore diameters. Chemically bonded to the silica gel are the phases for each mode of action. For reversed-phase sorbents, an octadecyl (C-18), octyl (C-8), ethyl (C-2), cyclohexyl, and phenyl functional groups are bonded to the silica. Typical loading of reversed-phase sorbents varies from approximately 5% for the C-2 phase to as much as 17% for the C-18 phase. The percent loading is the amount of C-2 or C-18 phase present by weight of carbon. The capacity of the sorbent in milligrams per gram (mg/g) of analyte that may be sorbed is related to both the chemistry of the phase and the loading weight of carbon. Polymeric sorbents, such as styrene divinylbenzene and carbon, also are used for reversed-phase SPE. These sorbents were some of the

Table 1.1. Common Sorbents Available for SPE

Sorbent	Structure	Typical Loading
	Reversed Phase	
Octadecyl (C-18)	$-(CH_2)_{17}CH_3$	17%C
Octyl (C-8)	$-(CH_2)_7CH_3$	14%C
Ethyl (C-2)	$-CH_2-CH_3$	4.8%C
Cyclohexyl	$-CH_2CH_2$-cyclohexyl	12%C
Phenyl	$-CH_2CH_2CH_2$-Phenyl	10.6%
Graphitized carbon	Aromatic carbon throughout	
Copolymers	Styrene–divinylbenzene	
	Normal Phase	
Cyano (CN)	$-(CH_2)_3CN$	10.5%C, 2.4%N
Amino (NH₂)	$-(CH_2)_3NH_2$	6.4%C, 2.2%N
Diol (COHCOH)	$-(CH_2)_3OCH_2CH(OH)CH_2(OH)$	8.6%C
Silica gel	$-SiOH$	—
Florisil	Mg_2SiO_3	—
Alumina	Al_2O_3	—
	Ion Exchangers	
Amino (NH₂)	$-(CH_2)_3NH_2$	1.6 meq/g
Quaternary amine	$-(CH_2)_3N^+(CH_3)_3$	0.7 meq/g
Carboxylic acid	$-(CH_2)_2COOH$	0.4 meq/g
Aromatic sulfonic acid	$-(CH_2)_3$-Phenyl-SO_3H	1.0 meq/g
	Size Exclusion	
Wide pore hydrophobic (Butyl)	$-(CH_2)_3CH_3$	5.9%C
Wide pore ion exchangers	$-COOH$	12.2%C

classical reversed-phase sorbents introduced in the 1960s. They are currently produced in purified form and are useful for the isolation of more polar solutes that have low capacities on the C-18 reversed-phase sorbents.

For normal-phase SPE, cyanopropyl (CN), aminopropyl (NH₂), and diol functional groups are chemically bonded to the silica gel. The loading on the cyano, amino, and diol columns are sufficiently large ($\sim 6-10\%$ as carbon) that they may sometimes be used for reversed-phase applications, especially for the removal of hydrophobic solutes from water or other polar solvents. These hydrophobic solutes would otherwise sorb too strongly to a more hydrophobic C-8 or C-18 sorbent and would be difficult to elute. Straight silica gel also is used for normal-phase SPE along with Florisil (magnesium silicate), and alumina (aluminum oxide in neutral, basic, and acidic forms).

Ion-exchange sorbents usually contain both weak and strong cation and anion functional groups bonded to the silica gel (Table 1.1). Strong cation-exchange sorbents contain ion-exchange sites consisting of sulfonic acid groups, and weak cation-exchange sorbents contain sites consisting of carboxylic acid groups. Strong anion-exchange sites are quaternary amines, and weak anion-exchange sites are primary, secondary, and tertiary amines. Strong and weak refers to the fact that strong sites are always present as ion-exchange sites at any pH, while weak sites are only ion-exchange sites at pH values greater or less than the pK_a, which determines whether a site contains a proton or not. The typical loading for an ion-exchange sorbent is expressed in milliequivalents per gram (meq/g) of sorbent, which is called the exchange capacity of the sorbent. The values vary from ~ 0.5 to 1.5 meq/g. These exchange capacities are somewhat less than a typical ion-exchange resin, which will have from 2 to 5 meq/g because of a higher density of ion-exchange sites. Also these SPE ion-exchange sorbents are not as rugged as the polymeric ion-exchange resins because of the silica matrix of the SPE sorbent, which is susceptible to dissolution by strong acid or base (see discussion in Chapter 6). The typical ion-exchange resin, however, consists of a crosslinked styrene–divinylbenzene polymer. Recently, several manufacturers are producing styrene–divinylbenzene-based ion-exchange SPE sorbents that are more rugged than silica-bonded sorbents because the styrene–divinylbenzene sorbents are stable from pH 1 to 13.

Size-exclusion sorbents called wide-pore sorbents (Table 1.1) use a silica-gel matrix with a large pore size, approximately 275 to 300 Å, rather than the 60-Å pores of most bonded-phase silicas. The advantage of the larger pore size is that molecules of larger molecular weight (>2000 daltons) may enter the pore of the sorbent and sorb by hydrophobic, polar, or ion-exchange interactions. Two examples are shown in Table 1.1. One is a hydrophobic sorbent of C-4 (butyl) with a carbon loading of almost 6% and the other is a weak cation sorbent using the carboxyl exchange site. More examples of these sorbents are given in Chapter 6.

Another packing material, which is not listed in Table 1.1, was recently introduced for drug analysis. It is a mixed-mode resin. This packing material contains both a bonded reversed-phase group (typically a C-8) and a cation-exchange group on a silica gel or a polymeric matrix. The combination of bonded groups is used so that both types of mechanisms retain the analyte at different times, or simultaneously, in the clean-up of complex samples of urine and blood. The principle of the mixed-mode resin is that different wash solvents may be used to remove interferences, but that the solute is always retained by one or both of the interactions. This method is especially useful for drug analysis and is discussed further in Chapter 8.

1.5. APPLICATIONS OF SOLID-PHASE EXTRACTION

Table 1.2 shows a general application guide for the use of SPE sorbents. The C-18 reversed-phase sorbent has historically been the most popular packing material and has been used most frequently. The surface of the sorbent is one of the most hydrophobic and has a large capacity. Capacity is the amount of analyte sorbed (usually expressed in milligrams/gram) before breakthrough occurs. Applications of C-18 reversed phase include isolation of hydrophobic species from aqueous solutions (such as drugs and metabolites from urine,

Table 1.2. Selected Application Guide for SPE

Sorbent	Application
C-18	*Reversed-phase application—one of the most hydrophobic phases*
	Isolation of hydrophobic species from solution
	Drugs in serum, plasma, and urine
	Desalting of peptides
	Organic acids in wine
	Pesticides in water by trace enrichment
Graphitized carbon polymeric sorbents (styrene–divinylbenzene)	*Reversed-phase application—one of the most hydrophobic phases*
	Trace enrichment of polar pesticides from water
	Isolation of polar drug metabolites
C-8	*Reversed-phase application – hydrophobic phase*
	Isolation of hydrophobic species from aqueous solution
	Drugs from serum, urine, and plasma
	Peptides in serum and plasma
Silica	*Normal-phase application—polar neutral phase*
	Isolation of low to moderate polarity species from non-aqueous solution
	Lipid classification
	Separation of plant pigments
	Removal of fat soluble vitamins
	Clean up of pesticides from soil extraction and food residue
Florisil	*Normal-phase application—polar slightly basic phase*
	Isolation of low to moderate polarity species from non-aqueous solution
	Pesticides in food and feeds
	Polychlorinated biphenyls in transformer oil
	Clean up of pesticides from soil extraction and food residue

(contd.)

Table 1.2. (*continued*)

Sorbent	Application
Alumina A	*Normal-phase application—acidic polar phase*
	Isolation of hydrophilic species in nonaqueous solution
	Low-capacity cation exchange
	Sugars and caffeine in cola beverages
	Additives in feeds
Cation exchange	*Cation-exchange phase*
	Isolation of cation analytes in aqueous or nonaqueous solutions
	Fractionation of weakly basic proteins and enzymes
Anion exchange	*Anion-exchange phase*
	Isolation of anionic analytes in aqueous or nonaqueous solutions
	Extraction of acidic and weakly acidic proteins and enzymes
	Removal of acidic pigments from wines, fruit juices, and food extracts
	Removal of organic acids from water
Mixed mode	*Reverse-phase (C_8) and cation-exchange phase*
	Isolation of basic and amphoteric drugs from serum, plasma, and urine
Aminopropyl, NH$_2$	*Normal-phase, Reverse phase, and weak cation exchange*
	Low-capacity weak anion exchanger
	Drugs and metabolites from body fluids
	Petroleum and oil fractionation
	Phenols and plant pigments
Cyanopropyl, CN	*Normal-phase and reversed phase*
	Analytes in aqueous or organic solvents
	Drugs and metabolites in physiological fluids
Diol, OH	*Normal-phase and reversed phase*
	Analytes in aqueous or organic solvents
	Drugs and metabolites in physiological fluids

serum, plasma, and other biological fluids); desalting of peptides and oligonucleotides; isolation of pigments from wine and beverages; and trace enrichment of pesticides from water for analysis by gas chromatography/mass spectrometry (GC/MS) or high-pressure liquid chromatography (HPLC).

Graphitized carbon and reversed-phase polymeric sorbents also are frequently used in environmental applications, such as trace enrichment, for soluble molecules that are not isolated by reversed-phase sorbents, such as C-18. Very water soluble analytes require a more hydrophobic sorbent with greater surface area per gram for complete retention. Carbon and polymeric sorbents also may be used for polar metabolites of drugs and pharmaceuticals that are poorly retained on C-18. Another advantage of the aromatic sorbents is their selective interaction with the aromatic rings of analytes. Because both the graphitized carbon and the styrene–divinylbenzene structures contain aromatic rings they have the ability to sorb analytes by a specific π–π interaction. This sorption mechanism may selectively isolate aromatic compounds.

The C-8 reversed-phase sorbents (Table 1.2) are often the most popular sorbents for drug analysis because of a shorter hydrocarbon chain than a C-18 sorbent. The shorter chain length makes it much easier for secondary interactions between the analyte and the silica gel that enhances retention of the analyte. This added interaction is useful in the purification of drugs and metabolites from blood and urine because they contain basic nitrogen atoms that may hydrogen bond to the silica gel.

Normal-phase sorbents such as silica and Florisil are used to isolate low to moderate polarity species from nonaqueous solutions. Examples of applications include lipid classification, plant pigment separations, and separations of fat-soluble vitamins from lipid extracts, as well as the clean-up of organic solvent concentrates obtained from a previous SPE method or liquid–liquid extraction. Alumina is used to remove polar species from nonaqueous solutions. Examples include vitamins in feeds and food and antibiotics and other additives from feed. Normal-phase chromatography has been used for a number of years, and most applications for normal-phase column chromatography may be easily transferred over to normal-phase SPE.

Cation and anion exchange is used to isolate ionic compounds from either aqueous or nonaqueous solutions. Examples of applications are isolation of weakly basic proteins, removal of acidic pigments from wines and fruit juices, and the removal of organic acids from water. Many of the applications of classical ion exchange may be used in ion-exchange SPE; however, care must be exercised in the use of strong acids and bases with SPE ion-exchange sorbents that are based on a silica matrix. Furthermore, care must be taken not to exceed the ion-exchange capacity of the sorbent.

Finally, sorbents such as aminopropyl, cyanopropyl, and diol can be used for both reversed-phase and normal-phase separations. Many manufacturers supply their sorbents in variety packs that may be used for methods development. Also quality assurance reports are commonly available for the various sorbents, which is a good indication of their reproducibility. Later chapters will show specific applications, such as environmental, drugs and pharmaceuticals, and food and natural products.

1.6. AUTOMATION OF SOLID-PHASE EXTRACTION

Automation of a manual SPE method can provide many benefits, which include safety, improved results, and cost savings. Automating a manual SPE method removes the analyst from extended contact with biological samples, such as biological fluids (i.e., blood, serum, and urine) that may contain viruses or environmental samples that may contain hazardous substances. Because automated workstations are mechanical, they can work in environments that are hostile, which include noisy production locations or a refrigerated room. Automation also reduces cost by freeing up personnel from the use of vacuum manifolds and tedious pipetting, column conditioning, and elution steps. An automated SPE instrument allows 24-h operation, with or without supervision. It provides structured and automated methods development, which makes method transfer between personnel within the laboratory or between laboratories a more routine procedure. For these reasons automation provides for better utilization of resources, such as staff, space, and instrumentation.

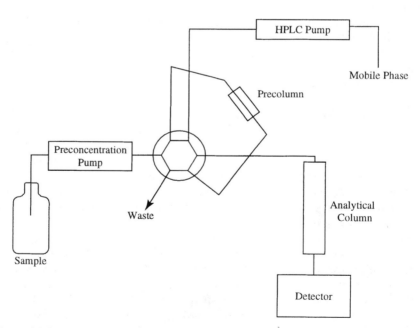

Figure 1.8. Schematic example of automated on-line SPE–HPLC system. [After Hennion and Pichon (1994), published with permission.]

The use of automation results in improved precision because of reduced operator errors compared to that of manual methods. Operator fatigue from the repetitive motions of SPE is eliminated. Just-in-time analysis may be used and automated SPE methods may be linked directly with liquid chromatography or gas chromatography for totally automated analysis. Furthermore, overnight operation is possible to make maximum use of time for sample production.

There are many types of automation equipment for SPE. They include semiautomated instruments, where some intervention is required; workstations that carry out the entire SPE operation without intervention; and customized SPE where robotic systems carry out many activities besides SPE and are custom designed for the user. Finally, there are on-line SPE–HPLC (Fig. 1.8) systems that allow the user to merely add the sample to the autosampler and analyze the sample directly. The concept of on-line SPE is that a sample is being pumped and processed onto the SPE cartridge while the liquid chromatograph or gas chromatograph is processing the preceding sample. This process saves time because the sample goes immediately to the instrument for analysis without sample storage, loss of analyte, or loss of time. The concept is that there is a second pump separate from the HPLC that processes the sample while the HPLC operates. Through computer control of switching valves, the SPE cartridge is eluted directly onto the HPLC column (Fig. 1.8). Automated on-line SPE is available for both HPLC and GC/MS. Chapter 10 will go into the details of these various technologies and discuss the latest developments.

1.7. MANUFACTURERS AND BIBLIOGRAPHIES

There are a number of companies that manufacture SPE products, and a current selection (1996) is shown in Table 1.3. At least 20 major companies produce bonded-phase sorbents, automated SPE equipment, and a variety of packing materials and accessories (see Appendix). There are also numerous automated systems currently available, which are discussed in detail in Chapter 10. Several major companies also print SPE application bibliographies: they are Waters, Varian Associates, and J. T. Baker. Waters also has a computerized version of its application bibliography that runs on both DOS and Macintosh computers.

As a result of the variety of sorbents and manufacturers, the choice of sorbents at first appears complicated and perhaps overwhelming. However, many of the sorbents are similar, and the details of these sorbents, such as endcapping and percent carbon loading, are discussed in this book so that correct choices may be easily made. For example, Chapters 3 through 6 will explore the variety of sorbents and how to make informed decisions for many applications of SPE. Many of the chapters will give detailed product information on

Table 1.3. Selected Manufacturers of SPE Products

Company	Applications
3-M	Disks, large selection
Alltech	Bonded sorbents, large selection, disks
Ansys	Disks, large selection
Interaction Chromatography	Polymeric phases
International Sorbent Technology Ltd.	Bonded sorbents, polymeric phases
J. T. Baker	Bonded sorbents, large selection, glass tubes available
J &W Scientific, Inc.	Bonded sorbents
Jones Chromatography	Bonded sorbents, large selection
Macherey-Nagel Duren	Bonded sorbents, large selection, glass tubes available
Phenomenex, Inc.	Bonded sorbents
Restek	Bonded sorbents
Supelco	Bonded sorbents, graphitized-carbon sorbents
United Chemical Technologies	Bonded sorbents
Varian Associates	Bonded sorbents, large selection
Waters	Bonded sorbents, large selection
Whatman, Inc.	Bonded sorbents
World Wide Monitoring	Mixed-mode sorbents, drug analysis specialists
Applied Separations, Inc.	Automation equipment for SPE
Gilson	On-line automation
Hamilton	Liquid handling
Hewlett-Packard	On-line automation for HPLC and GC
Jones Chromatography	On-line automation for HPLC
Spark Holland	On-line automation for HPLC and GC
Tekmar-Dohrmann	Automated SPE for environmental analysis
Zymark	Automation: clinical and environmental
Varian Associates	Applications bibliography
Waters	Applications bibliography, software bibliography

the major vendors of SPE products and automated SPE equipment. Examples from the literature can also be quite helpful in deciding the best course of action, as well as examples from the manufacturers' application bibliographies. Two manuals are available from the manufacturers of SPE products, one from Varian (1992) and the other from J. T. Baker (1995). Both are listed in the bibliography of this chapter. There is also another book on SPE, entitled *Solid Phase Extraction: Principles, Strategies, and Applications*, edited by Simpson (1997). With this information, the analyst should be well equipped to perform methods successfully with solid-phase extraction.

1.8. HISTORY AND FUTURE OF SOLID-PHASE EXTRACTION

Solid-phase extraction developed from classical chromatography, which is the use of an adsorbing medium to separate analytes according to their differing equilibrium affinities for the sorbing medium. Chromatography is historically marked from the first work of Tswett in 1906 (Berezkin, 1990) on the separation of plant pigments, using a tool he called "chromatography." Tswett used calcium carbonate to sorb various plant pigments with petroleum ether as the mobile or moving phase through the column. This separation would be called a normal-phase separation by today's nomenclature. Some of the highlights of this history are shown in Table 1.4. A detailed review of the evolution of chromatography and the use of solid sorbents in liquid chromatography is given by Ettre (1980, 1991).

As early as the 1930s, silica, alumina, Florisil, and kieselguhr (diatomaceous earth) were used as solid sorbents for normal-phase sample concentration, for what would now be called normal-phase SPE. The next major development occurred in 1941 when Martin and Synge (Nobel Prize for work in chromatography) published their work on partition chromatography. They

Table 1.4. History of SPE

1906	Term "chromatography" coined by Tswett
1930	Normal-phase applications for liquid chromatography
1941	Partition chromatography
1950	Reversed-phase chromatography; widespread use of charcoal as a sorbent
1960	Bonded sorbents synthesized
1968	Polymeric sorbents, XAD resins developed
1973	Gilpin and Burke (1973) develop chlorosilane bonded phases for HPLC
1974	XAD resins used for trace organic contaminants in water (Junk and others, 1974)
1975	C-18 reversed phase becomes popular for HPLC
1975	The term "trace enrichment" coined
1978	Sep-Pak introduced by Waters
1979	Analytichem introduces C-18 in syringe format
1980	Automation of SPE begins
1982	The term "SPE" coined by Zief and others of J. T. Baker
1985	Proliferation of manufacturers and new SPE phases, such as mixed mode
1989	3M introduces the disk format for SPE
1992	Introduction by Supleco of solid-phase microextraction (SPME)
1993	Proliferation of automation products for SPE
1995	On-line analysis by SPE–HPLC is commonplace
1996	On-line analysis by SPE–GC is becoming routine, including automated SPME

held a polar solvent (water) stationary by adsorbing it to silica gel and moved a second solvent (chloroform modified with ethanol) through the column as the mobile phase to isolate and separate acetyl derivatives of amino acids. In this system, chromatographic separation was achieved, not by adsorption of the compounds onto a solid phase as Tswett had performed but by partitioning of the solute between two liquid phases according to a distribution isotherm. Thus, the basic concept of partition chromatography was introduced.

Soon after, Howard and Martin (1950) published an account of the first use of what was to become known as reversed-phase chromatography. Instead of using a polar stationary phase, such as silica or calcium carbonate, to sorb polar compounds from a nonpolar solvent, they made the stationary phase nonpolar to sorb the nonpolar compounds from a polar solvent. They treated silica with dichlorodimethylsilane, which modified the surface of the silica to a nonpolar phase. This phase effectively held a nonpolar solvent stationary while a polar solvent was acted as a mobile phase. They separated long-chain fatty acids in an aqueous–methanol (80 : 20) mobile phase by partitioning of the solutes into the nonpolar stationary phase, which was n-octane saturated with methanol.

The basic concept of reversed-phase chromatography was changed from a liquid stationary phase to a solid stationary phase for direct adsorption of analytes onto the solid surface by Abel and others (1966). They synthesized a solid reversed phase by reacting silica with a trichlorosilane to produce a chemically modified, bonded, silica surface. This was used as solid packing in gas chromatography. This modification of the silica surfaces by bonding with organosilanes revolutionized chromatography and led to the production of a host of different modified-silica phases by varying the chemistry of the organosilane. These sorbents quickly became the basic packing materials for HPLC.

During this same time period other reversed-phase sorbents were being used for sorption chromatography. For example, charcoal (carbon) was one of the first solid phases used for drug purification in the 1950s and 1960s and for organic analysis of water samples (Jeffrey and Hood, 1958). Because of irreversible sorption of organic solutes onto charcoal, however, other sorbents were examined. For example, Rohm and Haas developed a suite of solid phases during the 1960s by the name of XAD resins. These resins were a suite of polymeric sorbents consisting of styrene–divinylbenzene and polyacrylate structures of varying pore sizes. The resins made an important contribution to the extraction of polar organic contaminants from aqueous samples (Junk et al., 1974) and for the extraction of humic substances from water (Thurman and Malcolm, 1981).

The XAD resins were packed into cartridges and used by clinical chemists for the removal of drugs from biological fluids as early as 1972. These cartridges, formally marketed by Brinkman, represent one of the first commercial applications of cartridge solid-phase extraction. In 1971, some of the first disposable cartridges were produced by BioRad in syringe-barrel form and packed with ion-exchange materials. Another producer of the syringe format disposable tube (1976) was Manhattan Instruments with the Jet-Tube, which was packed with diatomaceous earth.

Meanwhile, the use of bonded reversed-phase columns in HPLC was common, and there were applications for the concentration of organic solutes directly onto the head of the HPLC columns with the term *trace enrichment* being used for this process (Little and Fallick, 1975). However, interferences also concentrated at the head of the column, such as proteins in biological samples and humic substances in water samples; thus, column life was greatly shortened. The answer was a small precolumn positioned ahead of the analytical column to retain contaminants or to retain analytes in trace enrichment. The concept of the precolumn quickly led to the use of separate or off-line low-pressure SPE cartridges.

Waters introduced the Sep-Pak cartridge for trace enrichment of solutes by reversed-phase C-18 in 1978. Waters also produced a silica cartridge for normal-phase SPE. These cartridges were packed by a special radial compression technique for good flow characteristics. In 1979, Analytichem (now Varian, Harbor City, CA) remodeled the Jet-Tube after taking over Manhattan Instruments and marketed the cartridges as the Tox-Elut, Clin-Elut, and Chem-Elut series. All of these tubes were packed with diatomaceous earth of differing purity. Furthermore, Analytichem C-18 sorbents were also made available in 1979 in syringe-barrel format. This switch to the syringe-barrel format was extremely popular and is still widely used today. J. T. Baker followed in 1982 and a family of SPE cartridges and syringe formats became available for SPE. In fact, the term solid-phase extraction was first introduced by Zief and others (1982) of J. T. Baker, and it is the most commonly used term today for sorbent extraction.

The 1980s saw a proliferation of packing materials and formats by many vendors (see Appendix for Manufacturers' Information). In the late 1980s, mixed-mode sorbents were introduced for drug analysis. These sorbents are extremely popular for drugs of abuse because they are able to purify to a great extent the complex mixtures of proteinaceous substances present in blood and urine samples. In 1989, 3M introduced Teflon disks impregnated with 8- to 12-μm reversed-phase packing for use with large volume samples. Since then several other manufacturers have introduced various SPE disks for environmental and drug use. The disks have been made of varying sizes from microdisks of 4 mm to large disks (90 mm) for trace enrichment of water samples. The disks are available in syringe format and as free disks.

In 1992, solid-phase microextraction (SPME) was introduced by Supelco Inc. This method uses a microfiber that contains a coating similar to the reversed-phase sorbents of the gas chromatograph. The fiber is emersed directly into a water sample for sorption of the organic molecules and subsequently the fiber is desorbed by direct insertion into the inlet of the gas chromatograph. Furthermore, it is now possible to automate SPME for routine analysis of many compounds from water by either GC/MS or HPLC.

Finally, on-line SPE has become routine for HPLC and is beginning to become more commonplace for GC/MS (Barcelo and Hennion, 1995; Brinkman, 1995). The latest trends in SPE are toward the use of disks and microfibers and the automation of many of the SPE products. Finally, there continue to be new manufacturers of SPE products, as well as a continued expansion of new packing materials and even wider applications for SPE.

SUGGESTED READING

Baker, J. T., Inc., 1995. Bakerbond Application Notes, Baker, J. T., Phillipsburg, NJ, p. 27.

Ettre, L. S. 1991. A year of anniversaries in chromatography, Part 1 from Tswett to partition chromatography; *Am. Lab.*, January, 48C–48J.

Horack, J. and Majors, R. E., 1993. Perspectives from the leading edge in solid-phase extraction; *LC-GC*, **11**: 74–90.

McDonald, P. D. and Bouvier, E. S. P. 1995. *Solid Phase Extraction Applications Guide and Bibliography, A Resource for Sample Preparation Methods Development*, 6th ed, Waters, Milford, MA. This publication is also available on disk for computer searching by analyte.

Mills, M. S. and Thurman, E. M. 1993. Symposium on solid-phase extraction in environmental and clinical chemistry, *J. Chromatog.*, **629**: 1–93.

Poole, S. K., Dean, T. A., Oudsema, J. W., and Poole, C. F. 1990. Sample preparation for chromatographic separations and overview; *Anal. Chim. Acta*, **236**: 3–42.

Simpson, N. and Van Horne, K. C. 1993. *Sorbent Extraction Technology Handbook*, Varian Sample Preparation Products, Harbor City, CA.

Zief, M. and Kiser, R. 1990. Sample preparation; *Am. Lab.*, January: 70–82.

Zief, M and Kiser, R. 1988. *Sorbent Extraction for Sample Preparation*, J. T. Baker, Phillipsburg, NJ.

REFERENCES

Abel, E. W., Pollard, F. H., Uden, P. C., and Nickless, G. 1966. A new gas-liquid chromatographic phase, *J. Chromatog.*, **22**: 23–28.

Barcelo, D. and Hennion, M.-C. 1995. Online sample handling strategies for the trace level determination of pesticides and their degradation products in environmental waters, *Anal. Chim. Acta*, **318**: 1–41.

Berezkin, V. G., 1990. *Compiler, Chromatographic Adsorption Analysis: Selected Works of M. S. Tswett.* Ellis Horwood, London and New York (as cited in Ettre 1980).

Brinkman, U. A. Th. 1995. On-line monitoring of aquatic samples, *Environ. Sci. Tech.*, **29**: 79A–84A.

Ettre, L. S. 1980. Evolution of liquid chromatography: A historical overview, in *HPLC*, C. Horvath (Ed.), Vol. I, Academic Press, New York, pp. 1–74, Chapter 1.

Gilpin, R. K. and Burke, M. F., 1973. Role of trimethylsilanes in tailoring chromatographic adsorbents, *Anal. Chem.*, **45**: 1383–1389.

Hennion, M. and Pichon, V. 1994. Solid-phase extraction of polar organic pollutants from water, *Environ. Sci. and Tech.*, **28**: 576A–583A.

Howard, G. A. and Martin, A. J. P. 1950. The separation of the C12-C18 fatty acids by reversed-phase partition chromatography, *J. Biochem.*, **46**: 532–538.

J. T. Baker, Inc. 1995. *Bakerbond SPE bibliography*, J. T. Baker, Phillipsburg, NJ.

Jeffrey, L. M. and Hood, D. W., 1958. Organic matter in seawater; and evaluation of various methods for isolation, *J. Marine Res.*, **17**: 247–271.

Junk, G. A., Richard, J. J., Grieser, M. D., Witiak, D., Witiak, J. L., Arguello, M. D., Vick, R., Svec, H. J., Fritz, J. S., and Calder, G. V. 1974. Use of macroreticular resins in the analysis of water for trace organic compounds, *J. Chromatog.*, **99**: 745–762.

Little, J. N. and Fallick, G. J. J. 1975. New considerations in detector-application relationships, *J. Chromatog.*, **112**: 389–397.

Martin, A. J. P. and Synge, R. L. M. 1941. A new form of chromatogram employing two liquid phases, *J. Biochem.*, **35**: 1358–1368.

McDonald, P. D. and Bouvier, E. S. P. 1995. *Solid Phase Extraction Applications Guide and Bibliography, A Resource for Sample Preparation Methods Development,* 6th ed. Waters, Milford, MA.

Simpson, N. 1997. *Solid Phase Extraction: Principles, Strategies, and Applications,* Marcel Dekker, New York.

Thurman, E. M. and Malcolm, R. L. 1981. Preparative isolation of aquatic humic substances; *Environ. Sci. and Tech.*, **15**: 463–466.

Varian Sample Preparation Products, 1992. *Applications Bibliography,* Varian, Harbor City, CA.

Wells, M. J. M. and Michael, J. L. 1987. Reversed-phase solid-phase extraction for aqueous environmental sample preparation in herbicide residue analysis, *J. Chromatog. Sci.*, **25**: 345–350.

Zief, M., Crane, L. J., and Horvath, J. 1982. Preparation of steroid samples by solid-phase extraction, *Am. Lab.*, May, 120–130.

THEORY OF SORPTION AND ISOLATION

2.1. INTRODUCTION

There are three principle mechanisms of separation and isolation in solid-phase extraction (SPE). They are reversed phase, normal phase, and ion exchange. These are the same major separations used in high-pressure liquid chromatography (HPLC). The difference between HPLC and SPE arises in that liquid chromatography separates the compounds in a continuously flowing system of mobile phase; whereas SPE retains the solutes onto the solid phase while the sample passes through, followed by elution of the analyte with an appropriate solvent. The SPE is a simple on/off type of liquid chromatography (Wells and Michael, 1987); the secret of its use lies in one's ability to turn the retention switch on and off. The analyte is isolated from solution during the sorption phase and is eluted from the sorbent in a more concentrated and, usually, more pure form. This chapter addresses the theory of sorption with a simple explanation of the mechanisms of isolation used in SPE. It begins with the synthesis of bonded sorbents and how functional groups are attached to silica particles. Next the different mechanisms of separation are discussed: reversed phase, normal phase, and ion exchange, and how these mechanisms are manipulated to retain analytes. The concept of mixed mode is then addressed considering both the intentional and unintentional use of this interaction. Finally, the last section deals with elution conditions and releasing the analyte from the sorbent.

2.2. SYNTHESIS OF SORBENTS

2.2.1. The Silica Particle

In order to understand the theory of sorption in SPE, a simple understanding of the solid phase is required. Typically the SPE sorbent consists of a 40- to 60-μm silica particle (silica gel) onto which a liquid phase is chemically bonded. The silica gel cannot be used directly with aqueous solvent mixtures because the water deactivates the silica to such an extent that it has only weak interactions with most substances during the isolation process. So weak, in fact, that there is essentially no retention. Thus, it is necessary for the silica surface to be made hydrophobic in nature for it to be functional with aqueous solvents.

One of the early attempts at bonding a hydrocarbon phase to the silica was by Abel and co-workers (1966) who synthesized gas chromatographic phases by bonding C-18 hydrocarbons to silica gel through the use of a silica chloride. They cite earlier work by Weiss and Weiss (1954) for the ideas for this research. Other early work was that of Halasz and Sebastian (1969) who attached a hydrocarbon phase through the use of silicon–oxygen–carbon linkages. Gilpin and Burke (1973) described an alternate bonding process that involved the use of chlorosilanes. On reacting a chlorosilane with the hydroxyl groups of the silica gel, the hydrocarbon chain is attached by the stronger silicon–oxygen–silicon link. This type of linkage was to become the basis for the majority of all future bonded phases.

The organosilane stationary phase has been called a brush phase. It consists of reacting a silica surface with dimethyloctylsilyl-chlorosilane at elevated temperatures, which causes the bonding of the dimethyloctylsilyl group to the surface (Fig. 2.1). As a result, the product is a surface covered with dimethyloctylsilyl chains, like bristles of a brush. Thus, the term brush phase sometimes is used. This derivatization is monofunctional in that only one Si–O–Si link exists between the silica particle and the alkyl-bonded phase.

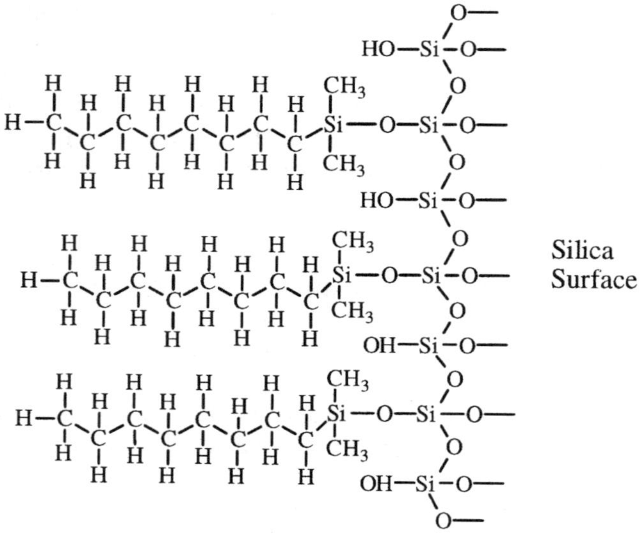

Figure 2.1. Brushlike surface of the alkyl-bonded phase of the silica gel.

This monofunctional bonded phase was originally the most popular reversed-phase sorbent with the brush-type phase commonly having 4, 8, or 18 carbon atoms attached. These phases are called C-4, C-8, and C-18 phases, respectively. More recently, other bonded phases have been developed, such as C-2, phenyl, CN, and so forth (Table 1.1). The early attempts at making bonded phases were not as reproducible as one would have liked; thus, work later described by Scott and Simpson (1992), using a fluidized-bed method for bonding, became a popular method for synthesizing the bonded phases. In spite of the technology behind the bonding of the reversed phase, there still exists substantial variation between bonded phases of different manufacturers, and sometimes even between batches of the same manufacturer's bonded phases, which has been attributed to differences in the surface composition of the silica gel itself (Nawrocki, 1991).

For example, the specific surface area of silica may range from ~100 to 850 m^2/g, with a typical range for SPE of 200 to 600 m^2/g. Generally, the particles are irregular in size with an average particle diameter of 40 μm, an average pore size of 60 Å, and an average surface area of 500 m^2/g range (Zief and Kiser, 1987). Generally, pores with a width of less than 2 Å are considered to be micropores. Pores from 2 to 50 Å are called mesopores, and these are the pores where the main part of the retention occurs (Nawrocki, 1991). Pores with a diameter larger than 50 nm are classified as macropores and are important in size-exclusion chromatography. Typically a pore size of 50 Å will show some exclusion of molecules greater than 2000 molecular weight (Aiken et al., 1979). Thus, if isolation of larger molecules is required, special sorbents may be obtained with pore sizes as large as 300 Å.

The pore diameter will affect the bonded phase by controlling the coverage density of the alkyl-bonded phase and the migration of analytes in and out of the pores during sorption. When the silica has pores less than ~10 nm diameter the possibility of "bottle-necking" occurs (Fig. 2.2). In this case, the pore cannot be easily accessed by the monochlorosilane reagent and so there are zones of the sorbent that remain underivatized.

Another consideration is that the silica particles contain unreacted silanols and siloxanes. The silanols are considered to be strong adsorption sites (Nawrocki, 1991), while the siloxanes are hydrophobic. Silanols on the surface may be single, geminal, or vicinal in form (Fig. 2.3). A single silanol is simply an isolated hydroxyl group on the surface of the silica gel. A vicinal silanol pair is two hydroxyl groups on adjacent silicon atoms, while the geminal silanols are two hydroxyl groups on the same silicon atom (Fig. 2.3). A pair of vicinal or geminal silanols can form a so-called bonded pair because of the ability for them to hydrogen bond to each other. The significance of these groups is that the three types of underivatized silanols may be reactive, with at least three different strengths for hydrogen bonding or weak cation

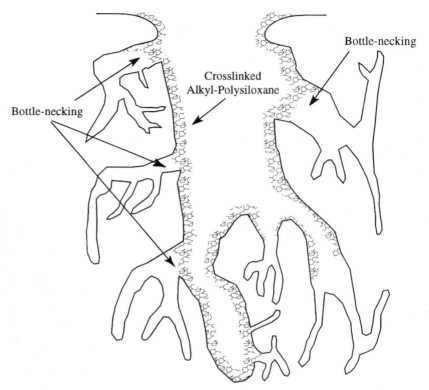

Figure 2.2. Example of bottle-necking of silica pores, which prevents bonding of the alkyl phase in the deeper pores of the silica gel. [Reproduced from the *Journal of Chromatographic Science* by permission of Preston Publications, A Division of Preston Industries, Inc. Also published with permission of the authors, Horvath and Melander (1977).]

exchange with any solute that may be bound to the bonded phase of the sorbent.

Thus, there are important interactions that can exist between weakly basic compounds and the free silanols of the silica gel. For example, many drugs contain amino-nitrogen atoms that may hydrogen bond with the silanols as well as interact hydrophobically with the C-8 or C-18 surface of a reversed-phase sorbent. This extra interaction may be quite useful in the retention of these compounds and should be kept in mind during methods development. Because the pK_a of a silanol varies from 4 to 5, they are considered weakly acidic and can also give rise to cation-exchange sites. Thus, any compounds capable of becoming positively charged as a function of pH also may interact with the silica surface of any hydrophobic sorbent.

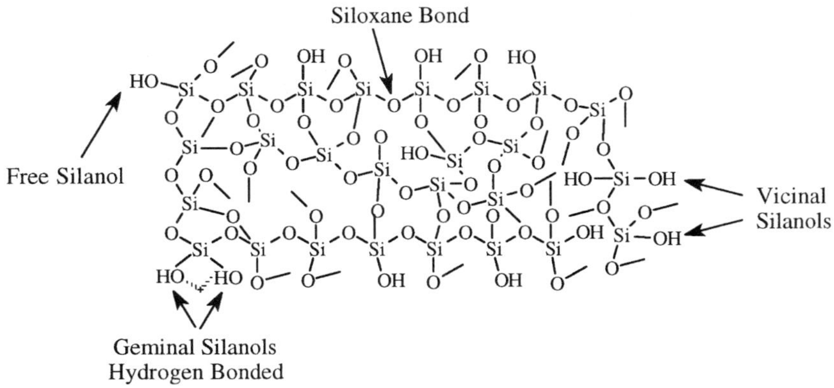

Figure 2.3. Structure of a silica gel showing siloxane bond, single free silanol, geminal silanols, and vicinal silanols.

Another characteristic of the silica particle that should be considered is its stability as a function of pH. The solubility of the silica particles increases rapidly at pHs above 7.0, with pH 7.5 considered the upper limit of usefulness. Similarly, the solubility of the silica will increase rapidly at pHs below 2.0. Thus, this simple understanding of the silica surface is useful when designing methods for SPE and affect the choices of solvents for washing and eluting the SPE sorbents.

2.2.2. Bonded Phases

Most commercially available bonded phases are of the siloxane type, having the Si–O–Si–C bond. They are prepared by the anhydrous reaction of silica-gel silanols with the monofunctional organochloro- or organoalkoxysilanes, as shown in Figure 2.4. The bonded phase contains the functional groups that are used for the type of separation, such as reversed phase, normal phase, or ion exchange. The synthesis reaction is shown in Figure 2.4 for an R group consisting of C-8, one of several typical phases used in reversed-phase SPE.

The most common bonded phase, of course, is the C-18 phase because of its widespread use in HPLC and its immediate adoption for SPE in the early 1980s. Also the C-18 phase is a versatile phase for use with aqueous samples over a wide range of polarities. The degree of surface loading by the R group of the organosilane (C-2 through C-18) will vary from 5 to 19% by weight as carbon for the packing material. The lower loading is for the short-chain alkyl groups, such as the C-2, with the highest loading for the longer chain C-18 groups. The capacity of the sorbent is generally greater as the percent loading

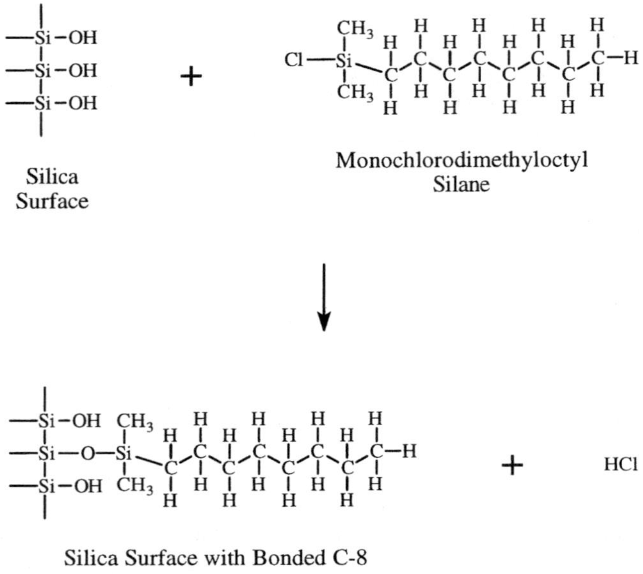

—Si—OH
—Si—OH   +
—Si—OH

Silica
Surface

CH₃ H   H   H   H
Cl—Si—C—C—C—C—C—H
CH₃ H   H   H   H

Monochlorodimethyloctyl
Silane

↓

—Si—OH CH₃ H   H   H   H
—Si—O—Si—C—C—C—C—C—H   +   HCl
—Si—OH CH₃ H   H   H   H

Silica Surface with Bonded C-8

Figure 2.4. Synthesis of a monofunctional bonded reversed-phase sorbent (C-8).

increases; thus, the C-18 sorbents have the greatest capacity expressed in milligrams of analyte sorbed per gram of sorbent.

However, the R group can contain any of the functional groups shown in Table 1.1. This includes reversed phases, such as C-2, C-4, C-8, C-18, cyclohexyl, phenyl, and even a single C-1 is available. The cyclohexyl and phenyl groups typically are bonded with a three-carbon alkyl chain connecting them to the surface of the silica gel. Likewise, the normal-phase sorbent and ion-exchange sorbents typically will contain a three-carbon linking chain to the silica-gel surface. Structures are shown in later chapters on reversed-phase and normal-phase SPE.

Because the bonded phases often use the monoalkyl silanes to bond the R group to the surface (Fig. 2.4), the R group is subject to hydrolysis by acid or base. Another derivatization reagent, trichloroalkylsilane, called a trifunctional derivative, has been introduced and is now routinely used in SPE (Fig. 2.5). This reagent yields a phase that is more stable to acid because the organosilane is attached to the silica surface at several locations. However, the trifunctional phases may sometimes have less capacity for some polar analytes that may bind by secondary interactions to the hydroxyl groups of the silica gel, which are now less available for retention of the analyte.

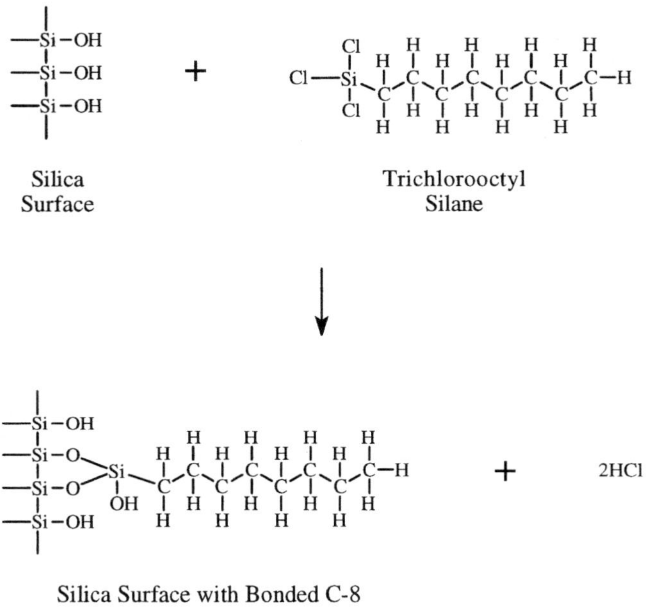

Figure 2.5. Synthesis of a trifunctional bonded reversed-phase sorbent (C-8).

Apparently, trifunctional reagents will react with adjacent silanol groups and derivatize them, which in turn creates a more hydrophobic surface on the silica gel. The greater stability of the sorbent at low pH is an important consideration for the isolation of acidic compounds. For example, organic acids require that the pH be lowered to two pH units below the pK_a for adequate sorption; thus, the pH must be lowered to ~2.0 for most carboxylic acids with a pK_a of about 4. In this case, the trifunctional-derivatized silicas are recommended because trifunctional SPE bonded phases will be more stable at acid pH.

The trifunctional derivatization process does not totally react all free silanol groups present on the silica surface. This fact was realized in liquid chromatography and the concept of endcapping was developed by derivatizing the acidic hydroxyl groups at the surface of the silica gel to prevent unwanted interactions between the analytes and the silanols of the silica gel. Endcapping consists of reacting the derivatized silica gel with trimethylchlorosilane reagent (Fig. 2.6). The endcapping process makes the overall surface of the silica gel somewhat more hydrophobic, thereby creating a more uniformly nonpolar stationary phase.

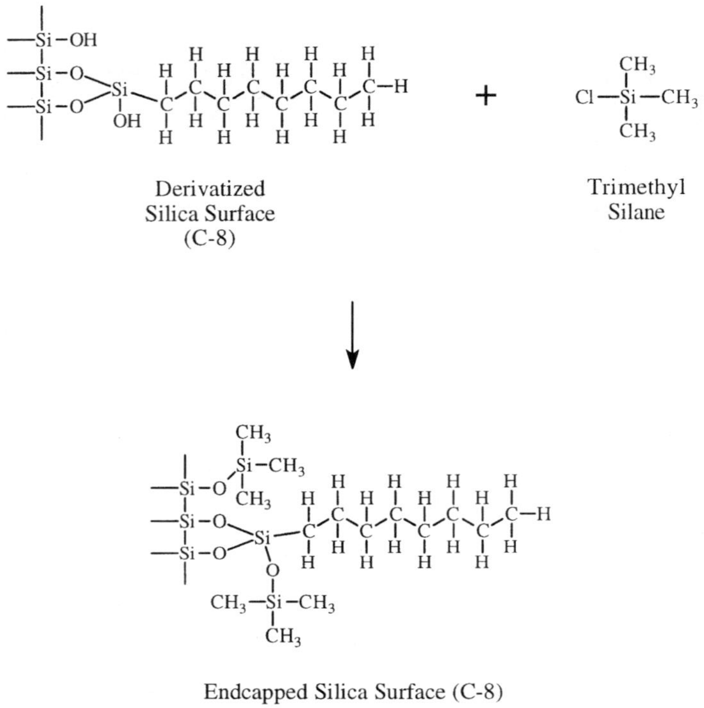

Derivatized
Silica Surface
(C-8)

Trimethyl
Silane

Endcapped Silica Surface (C-8)

Figure 2.6. Endcapping of free silanols with trimethylsilane for a C-8 sorbent.

Some SPE packing materials do not receive endcapping; thus, it is important to check for endcapping when choosing an SPE sorbent for a specific analyte. For example, in the case when the R group is C-18, the solute would experience reversed-phase sorption to the C-18 and in some cases sorption to the hydroxyl groups. When the interaction between the two groups is intentional, this is called a mixed-mode application of SPE. More examples of mixed-mode SPE will be shown later in this chapter.

Examples of the three common types of bonded phases for SPE are shown in Figure 2.7. They include the normal-phase sorbents that are bonded with the cyano and amine functional groups, the reversed-phase sorbents (Fig. 2.7), and the ion-exchange sorbents, which are the cation (SCX, strong cation exchanger) and anion types (SAX, strong anion exchanger). The sulfonic-acid functional group is used in the SCX, and the quaternary amine functional group is used in the SAX. In all cases, these examples show the monofunctional derivatization to the silica sorbent.

Methyl, C1

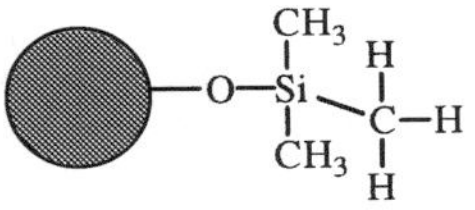

Butyl, C4

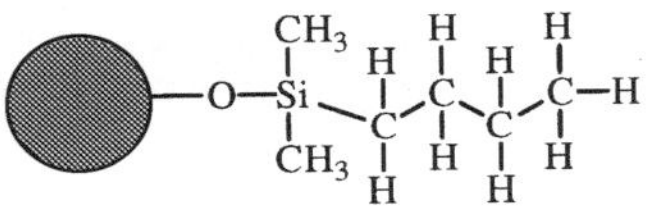

Hexyl, C6

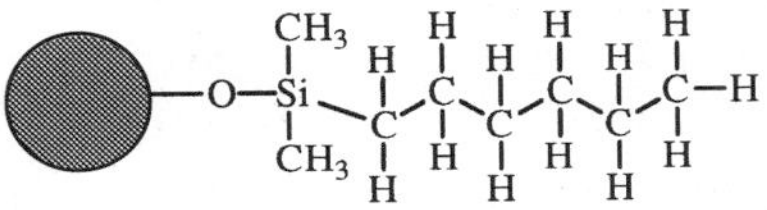

Octyl, C8

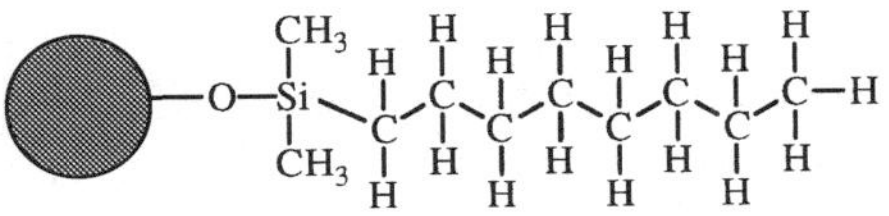

ODS, C18

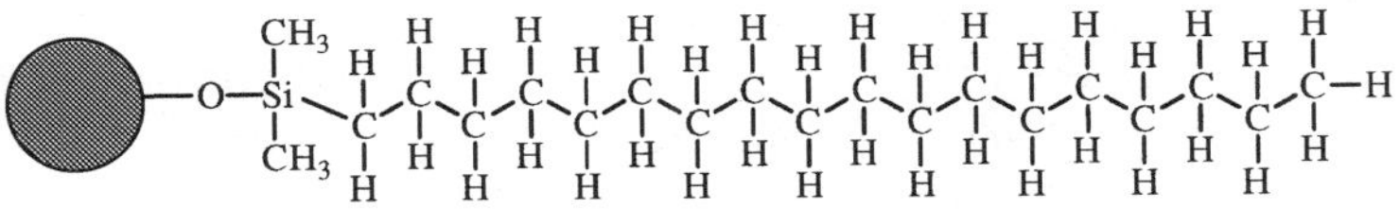

Phenyl

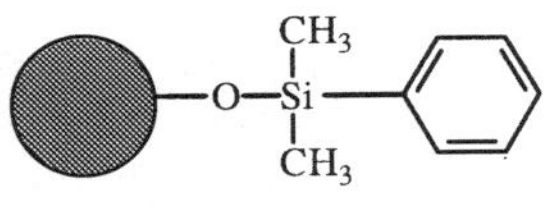

Figure 2.7. (*contd.*)

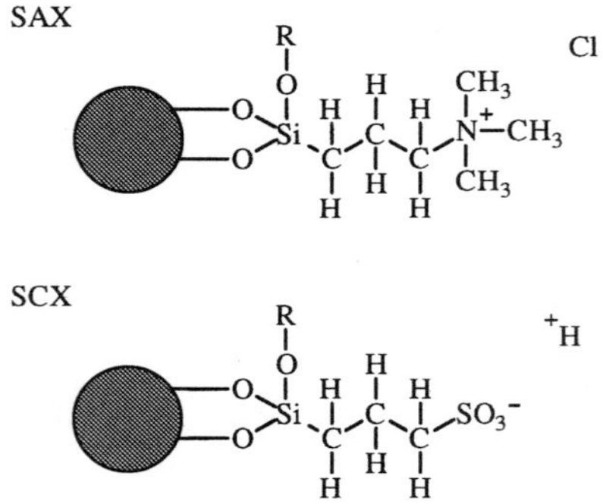

Figure 2.7. Examples of reversed-phase, normal-phase, and ion-exchange sorbents for SPE.

It is quite important to know the pore size of the silica gel (which controls surface area), the amount of liquid phase or functional groups bonded to the silica, the type of derivatization reagent (mono- or trifunctional), and the extent of endcapping. This information was realized by some of the manufacturers of SPE products; thus, they supply customers with quality control and quality assurance (QA) data of their SPE products. Figure 2.8 shows an example of such a QA report from IST (International Sorbent Technology).

This example of a Quality Assurance Report by IST gives the basic information on the C-18 SPE product. The sorbent columns contain 1 g of endcapped C-18 sorbent. A trifunctional derivatization reagent has been used and the material meets all the specifications of the manufacturer. For example, the average pore size (65 Å), specific surface area (550 m²/g), average particle size (60 μm), and particle size distribution are given. The carbon loading is shown as percent C (19%), as well as surface pH, column flow resistance, ultraviolet (UV) absorbance of methanol, capacity factor, and selectivity checks. The manufacturer does not describe what the capacity factor and selectivity check mean, but it probably refers to the sorption of the HPLC test probes: uracil, acetophenone, nitrobenzene, benzene, and toluene. Because this sorbent is not an ion-exchange resin, no exchange capacity is given.

Finally, the advantages of the bonded silica sorbents are that they are stable to organic solvents, stable to acid solution (trifunctional sorbents to pH 2.0), and they are rigid packing materials with good flow characteristics. They wet easily with both polar and nonpolar solvents. They are reasonably inexpensive and come packed in syringe barrels and cartridges containing from 50 mg to as much as 10 g of sorbent. Finally, a wide range of functional groups are available in R groups (Table 1.1 and Fig. 2.7), which makes these SPE

IST
INTERNATIONAL
SORBENT TECHNOLOGY

Quality Assurance Report

Product Description: ISOLUTE™ C18(EC)   1g/6mL Column Reservoir

Part Number: 221-0100-C   Lot Number: 5213506CC

Box of 30 high performance ISOLUTE SPE columns.

The columns are packed with 1g of END-CAPPED sorbent.

This sorbent is manufactured using trifunctional silane under carefully controlled conditions.

The ISOLUTE SPE product has been subjected to the following Q.C. tests.

BASE SILICA TESTING

Average Pore Size (Å)	65
Specific Surface Area (m^2/g)	550

PARTICLE SIZE ANALYSIS

Ion Exchange Capacity meq/g	N/A

Average Particle Diameter (μ)	60
% > 75μ	2.8
% <10μ	<0.03
% 30-75μ	96.8

Carbon Flow Resistance	PASS
Surface pH	7

BONDED SILICA HPLC SELECTIVITY

HPLC Test Probes

URACIL
ACETOPHENONE
NITROBENZENE
BENZENE
TOLUENE

Capacity Factor (k')	PASS
Selectivity Check	PASS
End-capping Test	PASS
FTIR Sorbent Check (CN Phase only)	N/A

(SAX, SCX, PRS, CBA, HCX & HAX only)

This ISOLUTE™ product has passed IST's rigorous Q.C. tests.

IST welcomes feedback on all quality issues.

Figure 2.8. An example of the QA report on the C-18 bonded phase of IST. [Published with permission of International Sorbent Technology (IST). Catalog copyright of International Sorbent Technology Ltd. ISOLUTE and IST are registered trademarks of International Sorbent Technology Ltd.]

sorbents popular for sample preparation. There are, however, polymeric sorbents that are useful for SPE with an increased nonpolar surface area when compared to the reversed-phase bonded silica sorbents. For example, there are polar compounds that have low or near zero capacity on the bonded silica sorbents but which may often be sorbed effectively onto polymeric sorbents.

2.2.3. Polymeric Sorbents

Although the literature chiefly examines bonded-phase SPE sorbents, there are applications of organic polymers as important phases for SPE. The two most common sorbents are styrene–divinylbenzene (SDB) and activated carbon. These two sorbents do not contain silica but are entirely organic polymers (Fig. 2.9). Typically they have large surface areas (600–1200 m²/g), greater capacity than the bonded phases because of higher carbon percentage, and a more hydrophobic surface. They have their major application in the area of reversed-phase SPE.

The polymeric sorbents have a long history of use in the extraction of both drugs and environmental compounds and have been in use since the early 1970s. Recently, many manufacturers have purified these polymeric sorbents when compared to the early forms, in order to improve the quality of the sorbent and to reduce the contamination leaching from the SPE sorbent. In the 1970s, the XAD (polymeric sorbents) resins gained a bad reputation because they gave many contaminant peaks in gas chromatography that were not present in the sample, but were contaminants of the sorbent itself. The early XAD polymers were synthesized as resins to be used in industrial processes, not in analytical methods. Unfortunately, this bad reputation slowed the use of these materials for many environmental applications involving gas chromatography/mass spectrometry (GC/MS). However, the use of these materials for analytical methods was revived in the mid-1980s and the XAD-2 resin (styrene–divinylbenzene) structure was used for analytical-grade HPLC columns. The free packing material also was available so that these materials were incorporated and used in SPE columns. Because the sorbents were synthesized with reagent-grade materials, the cleanliness of the sorbents is considerably greater today than in the past.

The polymeric sorbents are available from several vendors (Hamilton, 3M, and IST). In all cases, these sorbents are the styrene–divinylbenzene structure but vary in surface area. Generally speaking the greater the surface area, the greater the capacity of the sorbent for trace organic compounds. Furthermore, the aromatic rings of the matrix network permit electron-donor interactions between the sorbent and π bonds of the solute, which may further increase analyte–sorbent interactions, which increases the energy of sorption. Thus, the

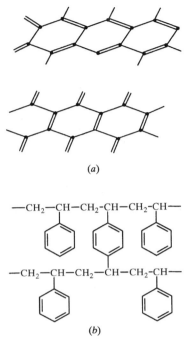

(a)

(b)

Figure 2.9. Structure of (a) activated carbon and (b) styrene-divinylbenzene (SDB) sorbents. Published with permission of Supelco.

polymeric sorbents are more retentive than the C-18 reversed-phase sorbents. Furthermore, the polymeric sorbents are often "doped" with a hydrophilic group, such as a sulfonic acid, to enhance water movement into the sorbent and to improve mass transfer, which generally makes the sorbents more effective. Thus, some of the SPE manufacturers will lightly sulfonate the styrene–divinylbenzene matrix, which creates a low content of sulfonic acid groups, or cation-exchange sites. These sites impart some polar character to the matrix of the sorbent and allow more effective mass transfer and better sorption and elution characteristics to the polymeric sorbents. For example, IST does sulfonate lightly its polymeric sorbents (personal communication with IST), and Waters adds an *N*-vinylpyrrolidone group to the styrene matrix for a similar purpose (personal communication with Waters).

The polymeric sorbents have considerably more capacity for polar compounds. For example, Hennion and Pichon (1994) have reported capacities of 20 to 40 times greater for the polymeric phases than the most hydrophobic

C-18 phases when isolating polar aromatic compounds from water. This increased capacity is quite important when low detection limits are required. Another advantage of the polymeric sorbents over the silica-based sorbents is their tolerance for both high and low pH. These polymeric sorbents are stable at pHs from 2.0 to 12.

The graphitized carbon sorbents also have had a comeback in the area of trace enrichment of contaminants from water samples. A series of studies by Di Corcia and colleagues (Di Corcia et al., 1994) have shown the usefulness of carbon sorbents for the trace enrichment of many classes of pesticides (acidic, neutral, and basic). These sorbents are sometimes the only sorbents capable of trace enrichment of many extremely polar organic solutes. Apparently, they not only have the hydrophobic effect and high surface areas for sorption, but also the potential for specific interactions with the analyte. It is thought that graphitized carbon contains carbon–oxygen complexes (positively charged sites) that act as anion-exchange sites in the presence of acidified water samples and allow for the sorption of extremely polar analytes. Thus, the carbon polymer acts as a mixed-mode sorbent (reversed-phase and anion exchange) with high capacity. A negative aspect of the carbon sorbents is the irreversible sorption of analytes to specific sorption sites on the graphitized carbon. Furthermore, the analytes are somewhat retained even in organic solvents. For these reasons, elution of the cartridge in the reverse direction has been advocated. For example, Di Corcia and co-workers (1994) recommend a reverse-elution step so that there are good recoveries of both polar and nonpolar solutes. Back elution is reversing the flow of solvent to the direction that the sample was pumped. However, the back elution step is not always convenient in many applications of SPE.

Examples of the use of these sorbents and their applications will be given in Chapter 7. Finally, a good review of examples of multiresidue extractions of compounds of environmental interest using polymeric and graphitized-carbon sorbents is given by Font and co-workers (1993).

2.3.   MECHANISMS OF SORPTION

2.3.1.   Reversed Phase

This mechanism involves the partitioning of organic solutes from a polar mobile phase, such as water, into a nonpolar phase, such as the C-18 sorbent (Fig. 2.10). The mechanism of isolation is a nonpolar interaction, called van der Waals, dispersion forces, or partitioning. The partitioning mechanism is a low-energy process (5 vs. 80 kcal/mol for ion exchange) and is analogous to a molecule being removed from water in a liquid–liquid extraction. The

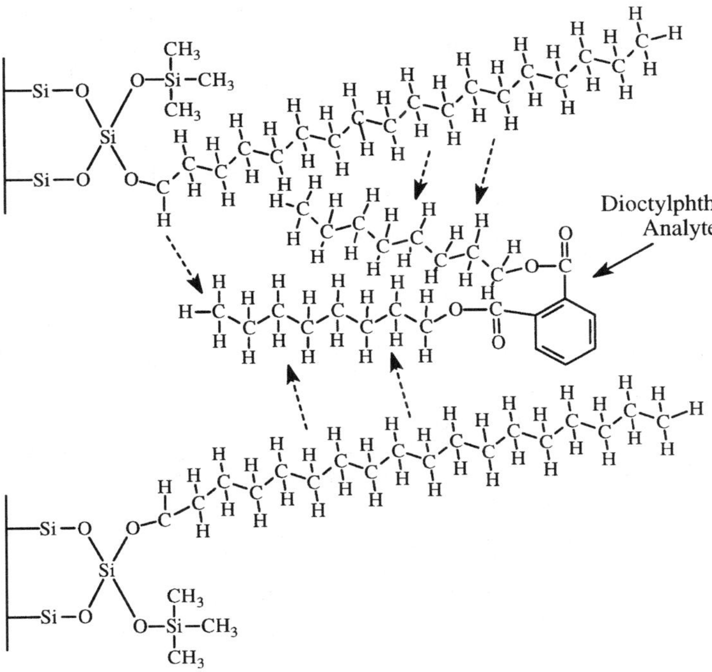

Figure 2.10. Reversed-phase mechanism of sorption of dioctylphthalate in SPE. [Reproduced in modified form from Zief and Kiser (1987) and published with permission.]

difference being that the organic phase is chemically bonded to the silica. The mechanism is called reversed phase because it was the opposite of early work where the stationary phase was polar and the mobile phase was non-polar. The early work was named "normal-phase" chromatography (see Chapter 1).

Common reversed-phase sorbents are C-8 and C-18 hydrocarbons (Fig. 2.7), and others phases include C-2, C-4, cyclohexyl, and phenyl groups. The analytes that are most hydrophobic have the greatest tendency to sorb by this mechanism and will have the greatest capacity or binding constant on a reversed-phase sorbent.

2.3.2. Normal Phase

Normal-phase SPE refers to the sorption of an analyte by a polar surface. It is called "normal" because this was the standard type of separation (prior to

1960) that was done by classical liquid chromatography. The mechanism of isolation is a polar interaction, such as hydrogen bonding, dipole–dipole interactions, π–π interactions, and induced dipole–dipole interactions. The mechanism involves the sorption of the functional groups of the solute to the polar sites of the packing material versus the solubility of the solute in the mobile phase of the column. The sorption by normal phase is a low to moderately strong interaction (Fig. 2.11). Figure 2.11 shows the sorption of a nitrobenzene molecule to the surface of a silica-bonded cyanopropyl sorbent. The mechanism of sorption is through hydrogen bonding to the cyano group through the amino groups of the nitroaminobenzene analyte.

The types of nonbonded phases used for normal-phase SPE are silica, alumina, and magnesium silicate (Florisil). The most popular phase is silica. Several bonded phases may also be used for normal-phase SPE, including aminopropyl, cyanopropyl, and propyldiol (Table 1.1, Fig. 2.7). Water is not used in the mobile phase in normal-phase SPE because it will sorb to the active sites of the sorbent and reduce the interaction between analyte and sorbent. Typically, normal-phase SPE is used as a clean-up procedure for organic extracts of water, soil, food, or other materials. Normal-phase SPE is also used for the isolation of analytes from organic liquids, such as oils.

The elution of the solute from a normal-phase sorbent is typically a function of the eluotropic strength and polarity of the eluting solvent. Table 2.1 shows the solvent eluotropic strength and polarity for a range of organic solvents typically used in normal-phase SPE. Silica will adsorb moderately polar

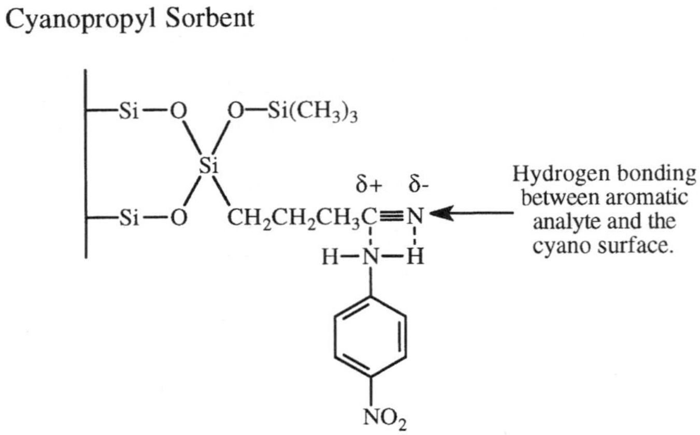

Cyanopropyl Sorbent

Figure 2.11. Normal-phase mechanism in SPE for the sorption of nitrobenzene. [Reproduced from Zief and Kiser (1987) and published with permission.]

Table 2.1. Solvent eluotropic strength, $E^{\phi}*$, and polarity, p'

Solvent	$E^{\phi}*$	p'
Acetic acid, glacial	> 0.73	6.2
Water	> 0.73	10.2
Methanol	0.73	6.6
2-Propanol	0.63	4.3
Pyridine	0.55	5.3
Acetonitrile	0.50	6.2
Ethyl acetate	0.45	4.3
Acetone	0.43	5.4
Methylene chloride	0.32	3.4
Chloroform	0.31	4.4
Toluene	0.22	2.4
Cyclohexane	0.03	0.0
n-Hexane	0.00	0.06

After Zief and Kiser (1987), published with permission of J. T. Baker, Inc.

compounds dissolved in organic solvents that have E^{ϕ} values less than 0.38 because the solvent will not compete with the analyte for the polar interactions with the solid phase (Table 2.1). Examples include compounds such as aldehydes, alcohols, and organic halides that are sorbed strongly by silica and that are eluted with solvents of greater eluotropic strength, such as methanol, or other solvents with eluotropic strengths greater than 0.6.

In general, basic compounds are retained more strongly on mildly acidic surfaces, such as silica or acidic alumina. Acidic compounds are retained on basic surfaces, such as basic alumina. Because both silica and alumina are hydroscopic, they adsorb water to their surface. This water greatly reduces the retention of organic solutes because it deactivates the hydrogen-bonding sites. Thus, it is important to keep the SPE sorbents dry and free from water. They may be stored in a dessicator prior to use. Very polar compounds, such as carbohydrates or amino compounds, are tightly bound to nonbonded normal-phase sorbents, such as silica and alumina. However, the use of cyanopropyl or aminopropyl phases often permit the recovery of these compounds when silica does not work.

In summary, retention of analytes by normal-phase SPE is facilitated by the dissolution of the sample in nonpolar solvents that do not compete for polar sorption sites on the solid sorbent. Elution of the sample from the sorbent is facilitated with polar solvents that can disrupt the hydrogen bonding between functional groups of the analyte and the sorbent surface. Examples of the use of normal-phase SPE include isolation of pesticides from food, soil, and other solid materials that are high in fat or organic matter and contain complex matrices.

2.3.3.   Ion Exchange

This mechanism involves the ion exchange of a charged organic solute from either a polar or nonpolar solvent onto the oppositely charged ion-exchange sorbent. This reaction follows classical ion-exchange theory, which is explained in more detail in Chapter 6. The mechanism of isolation is a high-energy, ionic interaction; thus, polar solutes may be effectively removed from polar solvents, including water as well as less polar organic solvents.

Sorbents are termed strong cation or anion exchangers if they have a permanent fixed charge, either positive or negative, respectively (Fig. 2.7). In the case of the strong cation-exchange site, a sulfonic-acid functional group is present with a proton as its counter ion. When another cation enters the vicinity of the cation-exchange site, there is a competition or exchanging of ions that depends on the selectivity of the cation for the site and the number or mass of cations that are competing for the site. Thus, the analyte to be concentrated must compete for sorption sites with the other cations in the sample. The mechanism is similar for anions, except that the strong anion site is a quaternary nitrogen atom with chloride as the most common counter ion (Fig. 2.12). The example in Figure 2.12 shows the anion exchange of 2,4-D, an anionic herbicide onto a strong anion-exchange resin. The resin is in its chloride form and exchanges one chloride ion for the single negative charge on a 2,4-D molecule.

The ion-exchange resin is weak if pH adjustment will remove the charge on the exchanger (Fig. 2.13). Weakly acidic or basic organic molecules may be

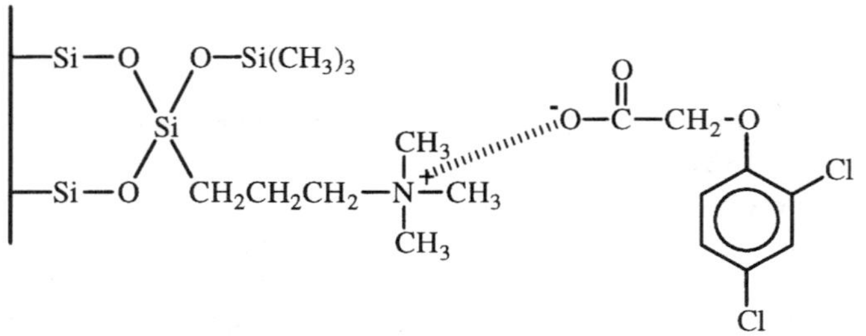

Figure 2.12. Mechanism of ion-exchange SPE for 2,4-D. [Reproduced in modified form from Zief and Kiser (1987) and published with permission.]

Cation

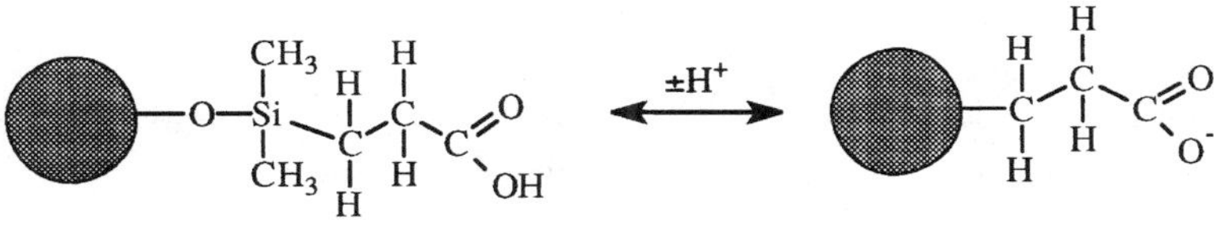

Anion

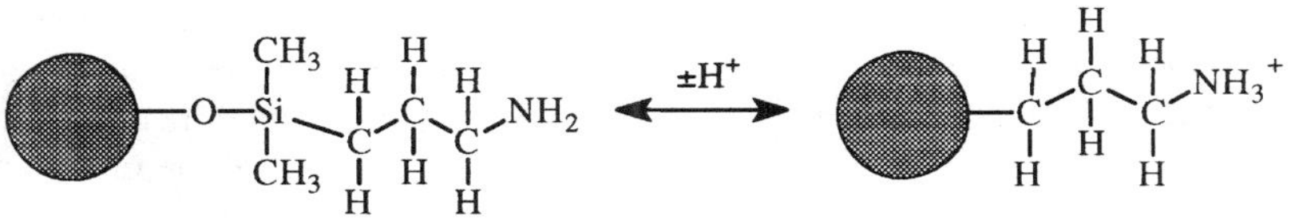

Figure 2.13. Examples of weak ion-exchange sorbents.

sorbed by ion exchange if the pH of the sample is adjusted near the pK_a of the analyte. The pK_a of an organic molecule is a critical factor in the use of ion exchange as is the competition of other ions with the solute, especially inorganic cations and anions. Basic compounds are 50% protonated as cations at a pH equal to their pK_a, and organic acids are 50% ionic as anions at a pH equal to their pK_a. Thus, at pH values near the pK_a organic ions are effectively removed by ion exchange from either polar or nonpolar solvents.

Factors that affect retention in ion exchange include the charge on the analyte that is being exchanged, as well as the charge on the competing ions. The greater the charge the more tightly bound is the ion. Thus, a singly charged anion is less tightly bound by anion exchange than a doubly charged anion. Second, the larger the anion the more tightly bound for the same charge. For example, sulfonic acids would be more tightly retained by anion exchange than carboxylic acids because they contain more oxygen groups and are larger in size (see discussion in Chapter 6). Generally speaking, retention of anions in ion chromatography is a good indication of retention in ion-exchange SPE. Ionic strength of the sample will affect the retention of the analyte in ion exchange by contributing inorganic ions to the sample that will compete for ion-exchange sites with the target organic analyte in the sample. Finally, note that wide-pore ion exchangers have been developed with pore sizes of 275 Å and are suitable for exchange of compounds with molecular weights greater than 2000 daltons.

2.3.4. Mixed Mode

The deliberate use of two different functional groups on the same sorbent is called mixed-mode SPE. Generally, the isolation involves both reversed phase and cation exchange. An example of the mixed mode is shown in Figure 2.14. In this example, the compound (a herbicide metabolite) is bound both by reversed-phase bonding on the hydrocarbon and aromatic side of the molecule, and at the same time, by the amino functional group, which is protonated by the strong cation-exchange site of the sorbent. The mechanism of isolation is both of low and high energy (reversed phase is low energy and the ion exchange is high energy); thus, solutes and interferences may be retained by one mechanism, but only the analyte is retained by the second mechanism and the interferences may be eluted from the SPE sorbent. One of the major applications of mixed-mode SPE is in the isolation of drugs and metabolites from urine and blood. In these examples, the SPE sorbent uses a combination of reversed phase (C-8 or C-18) and strong cation exchange. Chapter 8 will give examples of how these separations are carried out.

Occasionally, one inadvertently may use a mixed-mode mechanism in an SPE method. A good example is the sorption of triazine herbicides onto C-18 bonded phases that are monofunctional and are not endcapped. In this case, the basic compound has the potential for hydrogen bonding to the silanol sites of the silica gel, as well as reversed-phase sorption into the C-18 bonded phase.

Figure 2.15 shows how this may take place. The same triazine metabolite that is sorbed by reversed phase, and ion exchange in Figure 2.14 is now sorbed by reversed phase and by hydrogen bonding through a silanol group at the surface of the silica. This mechanism is a low-energy bond for both

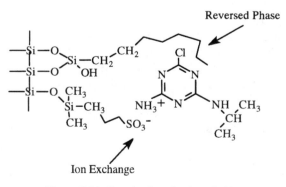

Figure 2.14. Sorption by mixed-mode SPE.

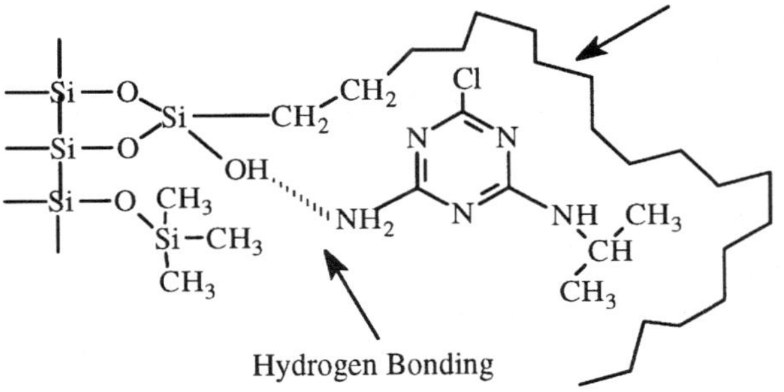

Figure 2.15. Example of silica mixed-mode sorption on partially endcapped C-18.

mechanisms in the case where the solvent is aqueous. In the case of herbicide metabolites in water, the retention of the triazine metabolite is significantly greater when the monofunctional sorbent is used because of this added mechanism of sorption. Because this triazine metabolite is water soluble (>1000 mg/L) its capacity is lower on C-18 bonded phases that are endcapped. Thus, it is possible to have unanticipated mixed-mode interactions, which may affect both sorption and recovery of analytes from reversed-phase sorbents.

2.4. ELUTION OF SORBENTS IN SOLID-PHASE EXTRACTION

2.4.1. Reversed Phase

The elution of analytes from reversed-phase sorbents is a rather simple process and consists of choosing a nonpolar solvent to disrupt the van der Waals forces that retain the analyte. Because the sorption process is a partitioning process, it is usually only necessary to allow the eluting solvent to have intimate contact with the bonded phase (e.g., C-18) in order to elute the analytes from the sorbent. Because the bonded phases consist of a silica matrix, they have an increased polarity compared to the original hydrophobicity of the C-18 alkane. Thus, the elution solvent must be capable of mutual solubility with the silica surface, as well as with the C-18 or other bonded phase.

Three solvents that are quite compatible with the reversed-phase sorbents are methanol, acetonitrile, and ethyl acetate. These solvents are capable of hydrogen bonding to the free silanols of the silica surface, they have a certain capacity to dissolve residual water that may be trapped in the bonded phase, and they are easily capable of breaking van der Waals interactions between the solute and the C-18 bonded phase. Thus, they are excellent solvents for elution of the reversed-phase sorbents. However, there may be hydrophobic solutes that will not effectively elute with these solvents. In this case, methylene chloride may be added (1:1) with the ethyl acetate for an effective elution solvent.

While there are other solvents that may also be used effectively with reversed-phase sorbents, these solvents are capable of dissolving the range of compounds that are generally isolated by SPE. If more hydrophobic solvents are used in bonded-phase SPE, then the sorbent must be carefully dried by vacuum to remove all traces of water in the silica matrix in order that the eluting solvent can interact with all areas of the sorbent and not be stopped by residual water trapped in the pores. If this is not done, then the hydrophobic solvent (let us say, methylene chloride) will not effectively wet the surfaces of the C-18 bonded phase and poor recoveries will result. If methanol, acetonitrile, or ethyl acetate is used, then complete drying of the sorbent is not required because the solvent will either be miscible with water (methanol and acetonitrile) or displace the water from the silica (ethyl acetate).

Furthermore, there are other advantages of the combined use of these two solvents in reversed-phase SPE. First, ethyl acetate will elute a wide range of solute polarities from relatively polar herbicides (triazines) to nonpolar insecticides (cyclodienes). Water needs only to be pushed from the sorbent without extensive drying because the ethyl acetate will displace water from the silica. However, water is not significantly soluble in the ethyl acetate and separates as a second phase. Second, ethyl acetate does not solubilize the ionic organic compounds, such as organic acids, because the counter ion of the organic acid (which is usually an inorganic cation) must be eluted also and is not soluble in ethyl acetate.

Thus, there is a separation that occurs on the C-18 sorbent that is quite useful between nonionic and ionic solutes that may be used to one's advantage in liquid chromatographic analysis of ionic species (Aga et al., 1994). Furthermore, ethyl acetate is an excellent solvent for gas chromatography, which is a commonly used analytical method coupled with SPE. Fourth, methanol may be used as a follow up solvent to elute the remaining polar and ionic solutes from the C-18 sorbent. These types of compounds are amenable to HPLC analysis, and methanol is a good solvent for HPLC. Because of the elution efficiency of these solvents, the term generic elution solvent is introduced for this solvent combination. Figure 2.16 shows the generic concept of ethyl acetate and methanol elution and it will be elaborated on in the following chapter. Finally, the following chapters that deal with methods development and specific applications will discuss the details of drying and elution of sorbents, again, with more discussion of the specific points of elution techniques.

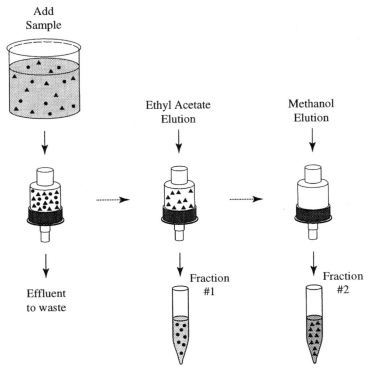

Figure 2.16. Generic method of elution of reversed-phase SPE sorbents using ethyl acetate and methanol. [After Aga et al. (1994), published with permission.]

2.4.2. Normal Phase

The elution of a solute from a normal-phase sorbent is typically a function of the eluotropic strength of the eluting solvent. Table 2.1 shows the solvent eluotropic strength and polarity for a range of organic solvents typically used in normal-phase SPE. Silica will adsorb moderately polar compounds dissolved in organic solvents with E^{ϕ} values less than 0.38 (Table 2.1). Examples include compounds such as aldehydes, alcohols, and organic halides, which could be dissolved in hexane for strong sorption by silica. To achieve elution solvents of greater eluotropic strength, such as methanol with an eluotropic strength greater than 0.6, must be used. Chapter 5 will give more extensive examples of how to elute normal-phase sorbents.

2.4.3.   Ion Exchange and Mixed Mode

Ion-exchange elution is a complicated problem and requires an entire chapter to explain and elucidate the concepts behind this process. Likewise, mixed mode will be addressed in Chapter 8 with a detailed explanation of how it is used.

SUGGESTED READING

Berthod, A. and Roussel, A. 1988. The role of the stationary phase in micellar liquid chromatography, *J. Chromatog.*, **449**: 349–360.

Busezewski, B. 1990. The influence of properties of packing materials upon the recovery of biological substances isolated from urine by solid-phase extraction, *J. Pharmaceut. & Biomed. Anal.*, **8**: 645–649.

Column Liquid Chromatography, Fundamental reviews, *Analytical Chemistry*, even years.

Falkenhagen, J. and Dietrich, P. G. 1994. New aspects concerning the stability of silica-based HPLC separation phases, *Am. Lab.*, September, 48–54.

Hennion, M. C. and Pichon, V. 1994. Solid-phase extraction of polar organic pollutants from water, *Environ. Sci. Tech.*, **28**: 576A–583A.

Kirkland, J. J., Glajch, J. L., and Farlee, R. D. 1988. Synthesis and characterization of highly stable bonded phases for high-performance liquid chromatography column packings, *Anal. Chem.*, **61**: 2–11.

Kohler, J., Chase, D. B., Farlee, R. D., Vega, A. J., and Kirkland, J. J. 1986. Comprehensive characterization of some silica-based stationary phases for high-performance liquid chromatography, *J. Chromatog.*, **352**: 275–305.

Liska, I., Krupcik, J., and Leclercq, P. A. 1989. The use of solid sorbents for direct accumulation of organic compounds from water matrices—A review of solid-phase extraction techniques, *J. High Resol. Chromatog.*, **12**: 577–590.

Lochmuller, C. H., Colborn, A. S., Hunnicutt, M. L., and Harris, J. M. 1983. Organization and distribution of molecules chemically bound to silica, *Anal. Chem.*, **55**: 1344–1348.

Majors, R. E. 1977. Recent advances in high performance liquid chromatography packings and columns, *J. Chromatog. Sci.*, **15**: 334–351.

Majors, R. E. 1980. Practical operation of bonded-phase columns in high-performance liquid chromatography. C. Horvath (Ed.) In *High-Performance Liquid Chromatography*, Vol. 1, Academic Press, Orlando, pp. 75–111.

Marko, V., Soltes, L., and Novak, I., 1990, Selective solid-phase extraction of basic drugs by C18-silica. Discussion of possible interactions, *J. Pharmaceut. Biomed. Anal.*, **8**: 297–301.

Marko, V., Soltes, L., and Radova, K. 1990. Polar interactions in solid-phase extraction of basic drugs by octadecylsilanized silica, *J. Chromatog. Sci.*, **28**: 403–406.

Ruane, R. J. and Wilson, I. D. 1987. The use of C18 bonded silica in the solid-phase extraction of basic drugs—possible role for ionic interactions with residual silanols, *J. Pharmaceut. Biomed. Anal.*, **5**: 723–727.

Simpson, N. and Van Horne, K. C. 1993. *Sorbent Extraction Technology Handbook*, 2nd ed. Varian Sample Preparation Products, Harbor City, CA.

Simpson, N. 1997. *Solid Phase Extraction: Principles, Strategies, and Applications*, Marcel Dekker, New York.

Unger, K. K., 1990. *Packings and Stationary Phases in Chromatographic Techniques*, Marcel Dekker, New York.

Weber, S. G. and Tramposch, W. G. 1983. Cation exchange characteristics of silica-base reversed-phase liquid chromatographic stationary phases, *Anal. Chem.*, **55**: 1771–1775.

Zief, M. and Kiser, R. 1987. *Solid-Phase Extraction for Sample Preparation*, J. T. Baker Inc., Phillipsburg, NJ.

REFERENCES

Abel, E. W., Pollard, F. H., Uden, P. C., and Nickless, G. 1966. A new gas-liquid chromatographic phase, *J. Chromatog.*, **22**: 23–28.

Aga, D. S., Thurman, E. M., Pomes, M. L. 1994. Determination of alachlor and its sulfonic acid metabolite in water by solid-phase extraction and enzyme linked immunosorbent assay, *Anal. Chem.*, **66**: 1495–1499.

Aiken, G. R., Thurman, E. M., Malcolm, R. L., and Walton, H. F. 1979. Comparison of XAD macroporous resins for the concentration of fulvic acid from aqueous solution, *Anal. Chem.*, **51**: 1799–1803.

Di Corcia, A., Samperi, R., and Marcomini, A. 1994. Monitoring aromatic surfactants and their biodegradation intermediates in raw and treated sewages by solid-phase extraction and liquid chromatography, *Environ. Sci. Tech.*, **28**: 850–858.

Font, G., Manes, J., Molto, J. C., and Pico, Y. 1993. Solid-phase extractions in multiresidue pesticide analysis of water, *J. Chromatog.*, **642**: 135–161.

Gilpin, R. K. and Burke, M. F. 1973. Role of trimethylsilanes in tailoring chromatographic adsorbents, *Anal. Chem.*, **45**: 1383–1389.

Halasz, I. and Sebastian, I. 1969. New stationary phase for chromatography, *Angew. Chem. Interact. Ed.*, **8**: 453–456.

Hennion, M. C. and Pichon, V. 1994. Solid-phase extraction of polar organic pollutants from water, *Environ. Sci. Tech.*, **28**: 576A–583A.

Horvath, C. and Melander, W. 1977. Liquid chromatography with hydrocarbonaceous bonded phases; Theory and practice of reversed phase chromatography, *J. Chromatog. Sci.*, **15**: 393–404.

Nawrocki, J. 1991. Silica surface controversies, strong adsorption sites, their blockage and removal, parts I and II, *Chromatographia*, **31**: 177–205.

Scott, R. P. W. and Simpson, C. F. 1992. A review of developments in bonded-phase synthesis carried out at Birkbeck College, London, *J. Chromatog. Sci.*, **30**: 59–64.

Weiss, A. and Weiss, A 1954. Proceedings of I.U.P.A.C. Colloquium, Munster, September, 1954, p. 41.

Wells, M. J. M. and Michael, J. L. 1987. Reversed-phase solid-phase extraction for aqueous environmental sample preparation in herbicide residue analysis, *J. Chromatog. Sci.*, **25**: 345–350.

Zief, M. and Kiser, R. 1987. *Solid Phase Extraction for Sample Preparation*, J. T. Baker Inc., Phillipsburg, NJ.

METHODS DEVELOPMENT

An understanding of the mechanisms of solid-phase extraction (SPE) are crucial for effective methods development. The four mechanisms outlined in Chapter 2 are sufficient for the majority of SPE and are an effective set of tools for methods development. The molecule's structure and the sample matrix are the main features used to choose a mechanism of isolation and separation. This chapter will discuss a six-step approach to methods development, how to execute the SPE recovery experiment, troubleshooting and optimizing conditions for the SPE recovery experiment, and how to critically evaluate previously published methods.

3.1. SIX-STEP APPROACH TO METHODS DEVELOPMENT

3.1.1. What is Solute Structure? The Clue to Effective Solid-Phase Extraction

The solute structure is the clue to effective isolation methods by SPE. What are the functional groups on the molecule? Is the solute ionic or nonionic? What is the aqueous solubility of the molecule? In what organic solvents is the molecule soluble? One must consider the pH of the sample, ionic strength, and the effect these matrix factors have on the solubility and structure of the analyte. When these questions are answered, then a mode (mechanism) of isolation may be planned, and an eluting solvent may be chosen that will sub-sequently unlock the mechanism of sorption.

Figure 3.1 shows the structure of the solute, atrazine, and lists four questions to ask initially when beginning methods development. Sometimes a simple picture of the solute, solvent, and sorbent phase are useful in analyzing the problem. From this point, five more steps are included to develop a viable method; thus, total methods development consists of six steps:

1. What is solute structure? The clue to effective SPE.
2. Identify the goal.
3. Obtain physical constants.

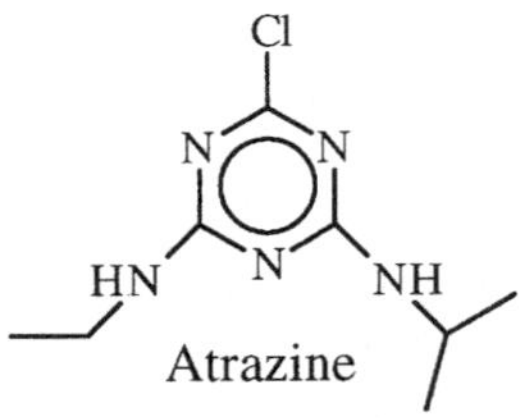

Figure 3.1. Solute structure (e.g., atrazine) is the key to sorbent selection.

4. Choose the mode (mechanism) of SPE.
5. Elute the SPE cartridge.
6. Perform the sorption experiment and determine breakthrough.

3.1.2. Identify the Goal

What are the compounds to be extracted? What is the matrix? What is the analytical procedure and instrument to be used, and in what solvent should the sample be stored for final analysis? These four questions should be answered as the first step in methods development. An example problem of how one might answer these questions is shown on the Methods Development Form (Fig. 3.2).

An herbicide, atrazine, which is commonly used on corn crops in the midwestern United States, was chosen as an example. During the spring, rainfall washes this compound from soil, and concentrations of herbicide in surface water will reach the Environmental Protection Agency's annual maximum contaminant level of 3 µg/L. It is necessary to detect this compound at concentrations as low as 0.05 µg/L. How would one isolate and purify this compound from water for analysis by gas chromatography/mass spectrometry (GC/MS) using SPE? Some questions and ideas that come to mind follow:

1. What is the aqueous solubility of atrazine?
2. Is the reversed-phase (partition) mechanism possible for atrazine?
3. Is a hydrogen-bonding mechanism possible for the amino nitrogens?
4. Could the pK_a of atrazine be a problem because of ionization of the analyte?

3.1.3. Obtain Physical Constants

Begin by drawing the chemical structure of the compound (atrazine), using, for example, the *Merck Index*. Enter this structure onto the Methods

Compound _____   Structure:
Aqueous Solubility _____
Organic Solubility (1) _____
Organic Solubility (2) _____
Organic Solubility (3) _____
Matrix _____   Important Functional Groups:
Sample Concentration_____
Mass of Solute for Detection_____ pK_a_____
Amount of Sample Needed _____   pH of Sample_____
Solid Phase Chosen_____
Mechanism of Retention_____
Percent Retained on SPE_____ %
Wash Solvents_____
Elution Solvent _____   Retention Graph
Percent Desorbed from SPE_____ %

                                          Method Recovery_____%

Signature_____
Date_____

Methods Development Worksheet for *Atrazine*

Compound_____*Atrazine*_____   Structure:
Aqueous Solubility _____*33 mg/L*_____
Organic Solubility (1) _*Methanol*_____
Organic Solubility (2) __*Ethyl acetate*__
Organic Solubility (3) _*Chloroform*____
Matrix _____*Water*_____   Important Functional Groups:
                                              2nd Amine, triazine ring, halogen

Sample Concentration _____*0.05 μg/L*____
Mass of Solute for Detection _*100 pg*___ pK_a ____*1.7*_____
Amount of Sample Needed ___*100 mL*__   pH of Sample___*5–8*_____
Solid Phase Chosen_____*C-18*_____
Mechanism of Retention *Reversed phase*
Percent Retained on SPE___*100*_____ %
Wash Solvents _____*None*_____
Elution Solvent___*Ethyl acetate*_____   Retention Graph
Percent Desorbed from SPE_____*95*___ %   Method Recovery___*95*_____%

Signature_____   Date _____

Figure 3.2. Methods Development Worksheet.

Development Worksheet for SPE (Fig. 3.2). The structure for atrazine shows that it contains a triazine ring with two secondary amines. The aqueous solubility is 33 mg/L, and the compound is quite soluble in methanol, ethyl acetate, and chloroform. List these solvents on the worksheet in Figure 3.2 as possible elution solvents. The pK_a of atrazine is not given in the *Merck Index*, but a quick literature search shows that the pK_a is less than 2.0. Furthermore, the compound is stable in water and the pH range of samples will be from 5 to 8. This information may be recorded on the worksheet together with the pK_a and pH of the sample.

Next, determine the mass of sample that is required for detection limits of 0.05 µg/L based on GC/MS analysis of atrazine. Detection limits for a typical GC/MS are calculated as follows. The GC/MS has a detection limit of 100 pg when 2 µL of solvent are injected. The microvials for the GC/MS require 100 µL of solvent in order to function properly on the autosampler of the GC/MS; therefore, the minimum mass of analyte required for a detection limit of 0.05 µg/L is given as:

$$\frac{2\,\mu L}{100\,\mu L} = \frac{1}{50} \tag{3.1}$$

Therefore, 50 times the detection limit of 100 pg is needed or

$$5000\ pg = 5\ ng\ \text{for minimum detection} \tag{3.2}$$

$$5\ ng/V = 0.05\ \mu g/L \tag{3.3}$$

Solving for the volume V results in a sample volume of 100 mL (3.4)

Thus, for a detection limit of atrazine of 0.05 µg/L by GC/MS, a minimum volume of 100 mL of sample is required, which must ultimately be concentrated to 100 µL for an injection of 2 µL into the GC/MS. There is now sufficient information to choose a mode or mechanism of separation and to design the SPE experiment.

3.1.4. Choose the Mechanism of Solid-Phase Extraction

The four mechanisms that could be considered for the isolation of atrazine (Fig. 3.1) include reversed phase, normal phase, ion exchange, and mixed mode. Normal phase is for the isolation of atrazine from an organic solvent, while the other three mechanisms may be used to isolate atrazine from aqueous samples. Because atrazine will protonate at low pH (pK_a of less than 2), all three mechanisms of interaction could be chosen for isolation.

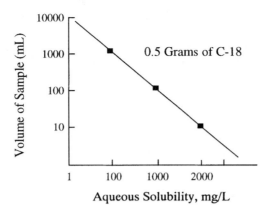

Figure 3.3. Anagram of volume of sample and aqueous solubility for a C-18 bonded phase.

However, in order for the cation-exchange mechanism to work on atrazine, the pH of the sample must be lower than the pK_a of the analyte. This would require lowering the pH of the sample to ~1.0, and a substantial amount of acid is required for this process. Furthermore, acidification to this low of a pH will protonate the organic acids naturally present in the water sample and increase the possibility of interferences when the sample is eluted from the SPE sorbent. Furthermore, the bonded-phase ion-exchange SPE cartridges are not capable of operating at this low of pH because of dissolution of the silica matrix, although it would be possible to use a polymeric sorbent.

Reversed phase is the next mechanism to consider. Because the water solubility of atrazine is only 33 mg/L, the reversed-phase mechanism should work well for a 100-mL sample. In fact, Figure 3.3 shows an anagram that uses aqueous solubility to estimate the volume of sample that can be applied to a C-18 column for approximately 95% retention. The anagram is an empirical relationship based on recovery experiments of triazines on C-18 (Mills and Thurman, 1994). Furthermore, Thurman and co-workers (1978) found that aqueous solubility could be used on polymeric sorbents to predict k' (the capacity factor of the solute for the sorbent) for a variety of compounds, including organic acids, phenols, aniline, and various aromatic and aliphatic hydrocarbons. There is a more detailed discussion of estimating capacity in the next chapter on chromatographic theory of reversed-phase SPE. The anagram, however, indicates that organic compounds with solubilities of greater than 1000 mg/L have breakthrough volumes of less than 100 mL. Thus, the C-18 column has sufficient capacity for samples of one liter or more for the example molecule of atrazine.

3.1.5 Elute the Solid-Phase Extraction Sorbent

An eluting solvent should be selected next from the worksheet solubility choices (Fig. 3.2). Because atrazine has sufficient solubility in three organic solvents, any could be chosen for elution. However, the elution solvent must be compatible with the final analysis method. In the example of atrazine, gas chromatography is being used for analysis; thus, either ethyl acetate or chloroform would make a good elution solvent from the standpoint of compatibility with the instrument. If liquid chromatography were being used, then methanol or acetonitrile would be a good solvent for elution.

Another consideration in efficient elution with an organic solvent is removal of water from the C-18 sorbent. In order for the elution solvent to effectively wet the bed of the sorbent, it is necessary that both water in the void volume, trapped water, and water sorbed to the silica be removed.

Figure 3.4 shows these three types of water and indicates that they may be important in effective elution of reversed-phase sorbents. Water is retained in the void spaces of the sorbent, as well as adsorbed to the silica surface through hydrogen bonding and dipole–dipole interaction. These two types of water

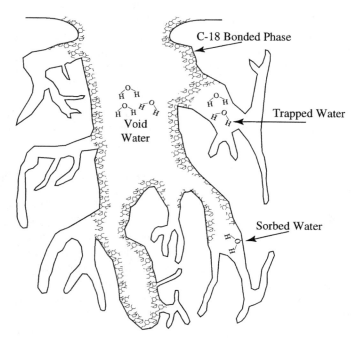

Figure 3.4. Three examples of the retention of water in C-18 silica bonded-phase sorbent.

need to be displaced depending on the type of desorption. Air or vacuum may be used to remove the water from the sorbent. This air will remove the majority of the water in the void volume. However, the water in the smallest pores and water that is sorbed to the silica surface will be tightly trapped or bound to the silica surface. This tightly bound water will greatly affect the recovery of analytes from the bonded phase, especially if the elution solvent is not miscible with water. For example, chloroform will have difficulty eluting atrazine completely from the resin if the residual bound water is not removed because of the immiscible nature of chloroform and water.

The bound water may be removed by drying the sorbent under vacuum (25 mm Hg) for at least 15 min. Another approach is to add methanol or acetonitrile to the chloroform (~ 20%) to allow for the dissolution of water into the eluent, which permits the infiltration of the chloroform–methanol eluent into the bonded C-18 phase. Because of retention of bound water in the C-18 sorbent, a review of the literature commonly shows that a polar solvent is mixed with a nonpolar solvent as an effective eluent in SPE. Later in such a method, the water will be removed by a drying agent such as anhydrous sodium sulfate.

Another approach is to go directly to an eluent that is miscible with water, such as methanol. This approach is most effective when the method of detection is liquid chromatography. The methanol also effectively wets the smallest pores and removes bound water, replacing it with methanol. The methanol effectively infiltrates the C-18 bonded phase and elutes the atrazine efficiently. A third choice is a solvent, such as ethyl acetate, that effectively replaces water bound at the surface of the sorbent and infiltrates the C-18 bonded phase. However, the ethyl acetate has a low solubility for water, so that the water is displaced from the sorbent as a second phase and is pushed out ahead by the ethyl acetate during the elution process. Thus, in the example of atrazine analysis by GC/MS, either the ethyl acetate or the chloroform–methanol are two examples of eluting solvents.

The effectiveness of the elution solvent may now be tested. The simplest method is to make up a standard solution of atrazine in the elution solvent, such as ethyl acetate, and pass the solvent and standard through a column that has been preconditioned in the normal manner (methanol and the methanol then displaced with ethyl acetate). Ethyl acetate was chosen as the eluting solvent because it is a generic elution solvent for reversed-phase sorbents (see discussion at the end of Chapter 2 on elution). A second aliquot of each volume is passed through the sorbent following the standard, and the two fractions are combined. The recovery of the eluent is now measured on the GC/MS and is compared to the original standard solution of atrazine before its passage through the column. If recoveries are 85% or greater, then the eluting solvent should be a good starting solvent for the SPE experiment.

In general, a good elution solvent will break the mechanism of interaction that holds a solute to the bonded phase, and at the same time the elution solvent must be compatible with the instrument that is used in the analysis. A general rule of thumb is that polar solutes will be best analyzed by HPLC and require polar solvents for elution, such as methanol and acetonitrile. Nonpolar solutes will often be analyzed by gas chromatography and are best eluted with nonpolar solvents such as ethyl acetate, chloroform, and hexane. If a solvent as nonpolar as chloroform or hexane is used, a bridging solvent is required to remove the water and the polar organic solvent from the sorbent prior to elution. An example is to displace water with ethyl acetate (the bridge solvent) and to displace the ethyl acetate with hexane (the elution solvent). Another approach that is commonly used is to dry the sorbent thoroughly under vacuum to remove all water. In this case, the more hydrophobic solvents may be directly passed through the sorbent.

Another concept to consider in reversed-phase elution is selective elution. Selective elution consists of using sequential elution solvents to selectively remove several classes of solutes, or using wash solvents to remove impurities that will interfere with the analysis. An example of selective elution is the separation and isolation of herbicides and their metabolites by a reversed-phase C-18 mechanism. Figure 3.5 shows the separation of alachlor, a herbicide, and its sulfonic-acid metabolite. In this method, both compounds are sorbed to the C-18 resin by a reversed-phase mechanism. Even the ionic sulfonic acid is bound to the C-18 bonded phase. The metabolite, whose structure is shown in Figure 3.5, is a surface-active compound and is bound by reversed phase with its ionic functional group solvated by the aqueous phase. The parent compound is eluted with ethyl acetate while the ionic metabolite stays bound to the C-18 resin. Apparently the solubility of the ionic metabolite in ethyl acetate is too low for dissolution. When methanol is applied to the column, the sulfonic acid metabolite elutes from the column. Thus, a fractionation is obtained by selective elution (Aga et al., 1994).

The actual volume of eluting solvent required is dependent on the void volume of the SPE sorbent and on the retention of the solute in the eluting solvent. A typical SPE cartridge that contains 500 mg of sorbent that is 60 μm in diameter will have a void volume of approximately 120 μL per 100 mg of sorbent (Hennion and Pichon, 1994), which is approximately 600 μL or 0.6 mL. Given that a k' (the column capacity factor) of many of the solutes could be 2 to 3 in the elution solvent, the prediction of the elution volume is given by the equation below:

$$V_r = V_0 (1 + k') \tag{3.5}$$

where V_r is the volume for the maximum concentration of the eluting peak, V_0

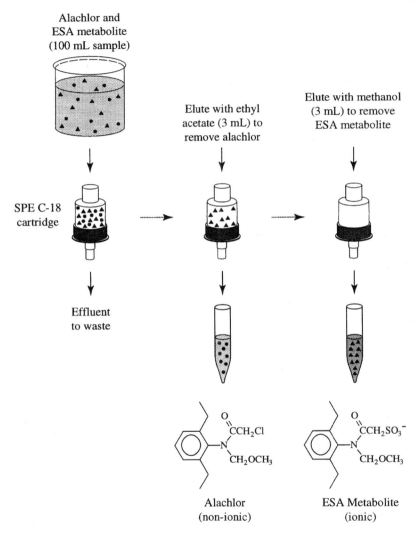

Figure 3.5. Example of selective elution of alachlor and its sulfonic-acid metabolite with ethyl acetate and methanol.

is the void volume, and k' is the column capacity factor for the elution solvent (the number of void volumes of a column that is required for the peak concentration to elute from the column). The k' is further defined in the next chapter in the section on theory of reversed-phase SPE. Substituting into Eq. (3.5) gives the following result:

$$V_r = 0.6\,(1 + 3) \tag{3.6}$$

$$V_r = 2.4 \text{ mL} \tag{3.7}$$

Thus, the maximum concentration of an analyte that has a k' of 3 on the C-18 sorbent would elute from 500 mg of sorbent in approximately 2.4 mL. A k' of 3 is a high value for most solutes so that this case should represent an extreme case of retention. Because the compound is at its maximum concentration at 2.4 mL, if this concentration were doubled, then recovery of the analyte should be nearly 100%. Thus, 5 mL would be a good first estimate for the volume of elution solvent that is required for elution. If the k' were less than 1, then an elution volume of only 2.5 mL would be required. Typically, experience shows that syringe barrels and cartridges containing 500 mg of C-18 sorbent require approximately 3 to 5 mL of elution solvent. From Eq. (3.5), one recognizes that the amount of void volume per 100 mg of sorbent is an important number and has a direct influence on the amount of solvent required to elute the SPE sorbent. Thus, it is important to pack SPE columns as efficiently as possible and to minimize the void volume in order that the least amount of solvent may be used to elute the sorbent.

3.1.6. Perform the Sorption Experiment and Determine Breakthrough

Once a mechanism for isolation and an elution solvent are selected, it is necessary to perform a simple breakthrough experiment (in Chapter 4 there will be a more detailed explanation of breakthrough experiments). The breakthrough experiment is simply a measure of the volume of sample that may be passed through the sorbent before the analyte is no longer retained.

The quickest method to determine if breakthrough is occurring is to pump the sample through a preconditioned column and measure the mass of compound that is sorbed. The sorbent is eluted with the chosen elution solvent, and the mass of analyte recovered is divided by the original mass of analyte that has been applied to the column. This figure times 100 is the percent recovery. This value is entered into the worksheet (Fig. 3.2).

An alternative method of determining if breakthrough is occurring is to measure the mass of the compound that passes through the column unretained.

To measure the mass of analyte that passes through the column and remains in the aqueous phase, it is necessary to have an alternate method of extraction ready to extract the aqueous phase. For example, the analysis of atrazine could be carried out on a water sample by liquid–liquid extraction. Thus, a 100-mL water sample is passed through a C-18 cartridge and the water is collected, a liquid–liquid extraction is performed with chloroform (organic solvent 3 in the worksheet), the chloroform evaporated to 100 μL, and the sample analyzed by GC/MS. The mass of atrazine not retained by the sorbent divided by the mass applied to the sorbent times 100 gives the percent breakthrough of the sorbent. This value is *not* the percent recovery but is the percent that passes through the SPE sorbent.

As an example, Figure 3.6 shows the breakthrough pattern for atrazine on the C-18 cartridge. Notice that several liters of water must pass through the sorbent before the analyte appears in the water phase (at 2 L 2.5% breakthrough has occurred and at 4 L 15% breakthrough has occurred, and at 6 L 80% breakthrough has occurred). Several liters more must pass the cartridge before the influent and effluent concentrations of the cartridge are equal. This graph is called a breakthrough curve. In Chapter 4 in the section on chromatographic theory of reversed-phase SPE, there is a more detailed description of the process of breakthrough and another example of how to calculate the capacity and volume of sample that a sorbent may concentrate, before loss of analyte occurs.

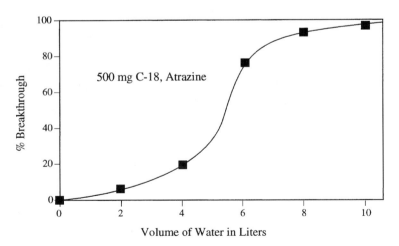

Figure 3.6. Breakthrough experiment for atrazine on C-18 sorbent.

Flow

Waste

Figure 3.7. Example of piggy-backing SPE cartridges in order to determine breakthrough capacity.

If you know that you have reasonably good retention of an analyte on the solid phase, an alternative shortcut method is to collect the sample after it has passed through a C-18 sorbent and reprocess the sample through another C-18 column. Or simply to place the cartridge piggy-back (Fig. 3.7) and measure the mass of analyte that appears on the second column relative to the mass of compound that was present in the original sample. To do this experiment it is necessary to have an elution solvent capable of removing the analyte from the C-18 bonded phase, which should have been established in the first stage of methods development. This completes the Methods Development Worksheet for SPE shown in Figure 3.2. The next section will review the process again and will discuss the execution of the SPE experiment in summary.

3.2. EXECUTING THE SOLID-PHASE EXTRACTION EXPERIMENT IN FIVE STEPS

The procedures for the SPE method have been chosen by following the previous steps of methods development. In order to test the efficiency of those choices, the developed method must be tested on real samples, and if it is successful, may also require further optimization. The following five steps are a practical and logical guide of how to test the method and sequentially eliminate potential problems with effective troubleshooting.

3.2.1. Step 1

The type of SPE sorbent has been chosen and an elution solvent selected. The first test is to determine the efficiency of the eluting solvent for that solid phase. Spike the compound into the eluting solvent of choice and pass it through a conditioned column. Compare recoveries to the same standard spiked into solvent and analyzed directly (i.e., standard has undergone no sample preparation losses due to SPE).

> *Result.* If 90 to 100% of the compound passes through the resin unretained in the eluting solvent, then the solvent is a good elution solvent.
>
> *Problem.* Incomplete elution, see Problem 1 in Section 3.3.1.

3.2.2. Step 2

Now carry out the planned SPE procedure at one concentration in distilled water or in the pure organic solvent needed for normal phase or ion exchange. Elute the compound from the cartridge with the elution solvent in step 1. Compare recovery to the same mass of standard spiked directly into the elution solvent and applied directly to the instrument (i.e., a sample that has undergone no sample preparation losses).

> *Result.* Excellent adsorption and elution. Recoveries are 90% or higher. The mechanism of retention is the correct one for this compound.
>
> *Problem.* Analyte not completely retained, see problems 2 and 3 in the Section 3.3.2.

3.2.3. Step 3

Next spike a standard into a real matrix (e.g., river water, urine, etc.) and repeat the procedure of step 2. Compare your recoveries with that of the spiked distilled-water sample. Don't forget to check for any background concentration of your compound in your sample matrix, that is, check the blank.

> *Result.* Excellent sorption and elution. Clean spectrum with no interfering substances. Interferences are unretained or not eluted.
>
> *Problem.* Poor recovery and analysis, see Problem 4 in Section 3.3.4.

3.2.4. Step 4

Make up a standard curve over the instrument's linear range in the matrix that will be used and process the standard curve through the SPE sorbent with the method that has been developed.

> *Result.* Standard curve plots as a straight line with correlation coefficient greater than 0.95.
> *Problem.* Poor correlation coefficient is obtained. See Problem 5 in Section 3.3.5.

3.2.5. Step 5

Try real samples of unknown concentration and compare with a previous method for accuracy and precision. Spike a sample and compare recovery to check for matrix interferences.

> *Result.* New method checks with previous method for accuracy and precision ($\pm 10\%$). Recoveries of spike are correct $95 \pm 10\%$. Methods development is complete.

3.3. TROUBLESHOOTING AND OPTIMIZING A METHOD

3.3.1. Problem 1: Incomplete Elution

The compound is not eluted from the sorbent with the elution solvent in step 1 of execution of the method. This problem suggests that the compound is sorbing to the bonded phase and that the elution solvent is not capable of breaking the bonding mechanism of retention.

Remedies

1. Increase the volume of eluting solvent or use back elution of the sorbent.
2. Change the elution solvent to one that will more effectively break the bonding mechanisms or change the polarity of the original solvent by mixing solvents to either increase or decrease solvent polarity.
3. Make pH adjustments to the eluting solvent to change the hydrophobicity or ionic state of the compound. Hydrogen-bonding interactions may also be occurring between the compound and active sites on the silica-based sorbent, requiring a basic solvent for elution.

4. Decrease the strength of the interaction between the compound and the solid phase. For example, if a C-18 sorbent is being used for a reversed-phase mechanism, decrease the energy of the interaction by going to a C-8 or a C-2 bonded phase or even a cyanopropyl bonded phase.

5. As a last effort completely change the mechanism of retention on the solid phase (e.g., from reversed phase to ion exchange). Select a new sorbent and begin the methods development process.

3.3.2. Problem 2: Breakthrough of Analyte

The analyte is not sufficiently retained on the sorbent, resulting in low recoveries (step 2 in the execution of the method).

Remedies

1. To determine the volume of sample at which breakthrough began, construct a breakthrough curve. Pass a known volume of sample and analyte through a cartridge, collecting the effluent in small aliquots (e.g., 10 mL) and passing each aliquot through a separate cartridge (Fig. 3.7) or analyzing the aliquot by a different method. This procedure will tell when the capacity of the sorbent for the analyte has been reached. Increase the amount of sorbent or decrease the sample volume in order to obtain needed recovery. Also check the example of reversed-phase sorption in the chromatographic theory section of the next chapter for other methods of determining breakthrough or capacity.

3.3.3. Problem 3: Significant Breakthrough of Analyte

The sorbent has little capacity for the analyte, resulting in very low recovery (step 2 in the execution of the method).

Remedies

1. Change solid sorbent to one that has a greater affinity for the compound from the sample matrix (i.e., water, urine, etc.). Try the same mechanism of sorption but with a stronger interaction. For example, a compound is too polar for the C-18 bonded phase. Try a styrene–divinylbenzene polymeric sorbent or activated-carbon sorbent.

2. Reduce the flow rate of the sample passing through the sorbent. Flow rates of 5 mL/min are slow enough for equilibrium for most

instances. In the case of ion exchange, flow rates may need to be decreased to less than 0.5 mL/min.

3. Change the form of the analyte. If the molecule is ionic, change the pH of the sample so that the molecule is nonionic; this step will increase the sorption of the solute for a reversed-phase sorbent.

4. Try salting out for reversed phase. Add 5 to 10% sodium chloride to the matrix, if aqueous. Hydration of the salt increases the polarity of the solvent and drives the analyte onto the reversed-phase sorbent.

5. Change the mechanism of sorption. For example, if reversed phase is being used, and the molecule can be made ionic, change the mechanism of isolation to ion exchange.

3.3.4. Problem 4: Interfering Substances

The chromatogram contains unwanted peaks and interfering substances that cause irreproducible results. Possible sources of interferences are from the original sample matrix, leaching of plasticizers from the SPE cartridge, impure solvents, or incorrect solvent used for the instrument of analysis.

Remedies for Sample Matrix

1. Design a wash step that will selectively elute the interferences from the solid phase and retain the analyte. An example is the removal of urine metabolites from a cation-exchange sorbent with methanol while retaining a drug that is sorbed by cation exchange. Many metabolites in urine are organic anions and are easily removed from the sample by a methanol wash.

2. Change the elution solvent so that the interferences remain on the solid phase and the analyte is eluted. An example is the sorption of a herbicide from water and its elution from a reversed-phase sorbent, C-18, using ethyl acetate rather than methanol. Ethyl acetate does not remove the majority of the natural organic substances (humic substances) from the sorbent, while methanol does. Thus, the chromatogram is considerably cleaner with the ethyl-acetate eluent.

3. Alternatively, clean up the eluent from the SPE sorbent, which contains the analyte and interferences, with another sorbent that uses a different mechanism of sorption for the interference, but does not retain the analyte. An example is a methanol extract of soil that contains an insecticide and an extract of the natural soil organic matter. The analyte is non ionic and the soil organic interferences are

anionic. Thus, a clean-up step with an anion-exchange resin will retain the natural organic matter and allow the pesticide to pass through for subsequent analysis. The pesticide is dissolved in methanol for this step.

Remedies for Solvents or Hardware

1. SPE hardware bleed. Wash the sorbent with the final eluting solvent during sorbent conditioning to remove interferences.
2. Check the pH of the sample to ensure that the degradation of the solid phase is not occurring due to extremes of pH (e.g., acid or base hydrolysis).
3. Change the solvent being used in order to eliminate bleed from the SPE hardware, to eliminate impure solvents, or to stop the bleed from the instrument used for analysis.

3.4. CRITICALLY EVALUATE PREVIOUSLY PUBLISHED METHODS

There are many published methods for SPE in the literature. There are four questions to keep in mind when reading and adapting previously used methods for SPE.

1. *Was the published method developed in distilled water or in a real sample matrix?* Distilled water has unusual properties with respect to mechanisms of ion exchange and to some degree for reversed phase. It is critical that the method is robust and that it is capable of working across a wide range of pH and ionic strengths.
2. *What was the volume of sample that was passed through the sorbent?* Were breakthrough experiments conducted to establish the capacity of the cartridge for the compounds isolated? If this procedure was not done, a simple cartridge breakthrough experiment might be carried out to check the capacity of the analyte for the sorbent.
3. *How was the sorbent conditioned?* What are the solvents used for conditioning? Is the sorbent properly ready for the mechanism of sorption and have any interferences been removed from the sorbent?
4. *Were matrix interferences present?* Was the analytical instrument blind to interferences (e.g., GC/MS under selected ion mode) or was sample clean-up effective? Is a different method of analysis going to be used from that published? If so, will the solvent used for elution still be effective in the new instrumental method?

1. Choose Mechanism of SPE for Aqueous Sample

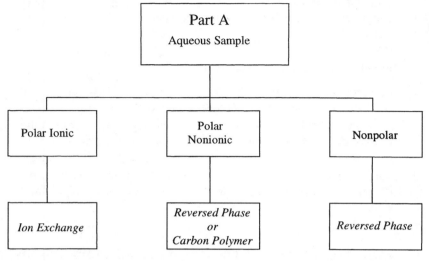

1. Choose Mechanism of SPE for Nonaqueous Sample

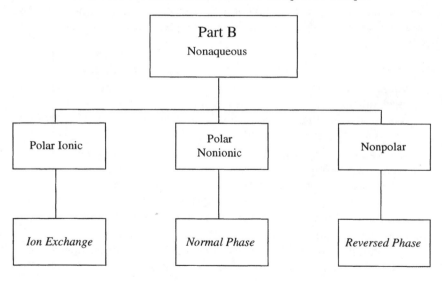

2. Run SPE Experiment.

3. Troubleshoot and Optimize.

Figure 3.8. Flowchart for SPE methods development for aqueous and nonaqueous samples.

3.5. SIMPLE SUMMARY OF METHODS DEVELOPMENT

The fundamental step in SPE is recognizing that the hydrophobicity versus the polarity of an analyte controls the mechanism of sorption that may be selected. Thus, step one is to select a mechanism of sorption, and Figure 3.8 is a guide to this process. Part A shows the process for aqueous samples and part B for non-aqueous samples. With aqueous samples, the decision for the sorbent is based on the polarity and ionic character of the analyte. If the analyte is polar and ionic, then ion exchange is the preferred sorbent. If the analyte is polar but nonionic, then reversed phase by either C-18 or a polymeric phase is the best sorbent. If the analyte is nonpolar, then a reversed-phase sorbent such as C-8 or C-18 is chosen.

With nonaqueous samples (part B of Fig. 3.8), the decisions for sorbents are somewhat reversed. For example, an analyte that is polar and ionic is best recovered with ion exchange. This example is similar to that of an aqueous sample. If the analyte is polar but nonionic, then the sorbent choice could be reversed phase or normal phase. The choice depends on the organic solvent, either polar or nonpolar, respectively. Finally, if the analyte is nonpolar, the sorbent choice is reversed phase. The second step in methods development is to execute the SPE experiment. Lastly, one has to optimize and troubleshoot the SPE method.

SUGGESTED READING

McDonald, P. D. and Bouvier, E. S. P. 1995. *Solid Phase Extraction Applications Guide and Bibliography, A Resource for Sample Preparation Methods Development*, 6th ed., Waters, Milford, MA.

Simpson, N. 1997. *Solid Phase Extraction: Principles, Strategies, and Applications*. Marcel Dekker, New York.

Simpson, N. and Van Horne, K.C., Eds. 1993. *Varian Sorbent Extraction Technology Handbook*, Varian Sample Preparation Products, Harbor City, CA.

Zief, M and Kiser, R. 1988. *Sorbent Extraction for Sample Preparation*, J. T. Baker, Phillipsburg, NJ.

Zief, M. and Kiser, R., 1990. Sample preparation, *Am. Lab.*, January, 70–82.

REFERENCES

Aga, D. S., Thurman, E. M., and Pomes, M. L. 1994. Determination of alachlor and its sulfonic acid metabolite in water by solid-phase extraction and enzyme linked immunosorbent assay, *Anal. Chem.*, **66**: 1495–1499.

Hennion, M. C. and Pichon, V. 1994. Solid-phase extraction of polar organic pollutants from water, *Environ. Sci. Tech.*, **28**: 576A–583A.

Mills, M. S. and Thurman, E. M. 1994. Reduction of nonpoint-source contamination of surface and ground water by starch-encapsulated atrazine, *Environ. Sci. Tech.*, **28**: 73–79.

Thurman, E. M., Malcolm, R. L., and Aiken, G. R. 1978. Prediction of capacity factors for aqueous organic solutes adsorbed on a porous acrylic resin, *Anal. Chem.*, **50**: 775–779.

REVERSED-PHASE SOLID-PHASE EXTRACTION

4.1. INTRODUCTION

Reversed-phase solid-phase extraction (SPE) involves the partitioning of organic solutes from a polar mobile phase, such as water, into a nonpolar solid phase, such as the C-18 sorbent (Fig. 4.1). Partitioning involves the interaction of the solute within the chains of the stationary phase, which may be a C-18 hydrocarbon, C-8 hydrocarbon, or the polymeric sorbents (such as styrene–divinylbenzene). The word hydrophobic mechanism is commonly

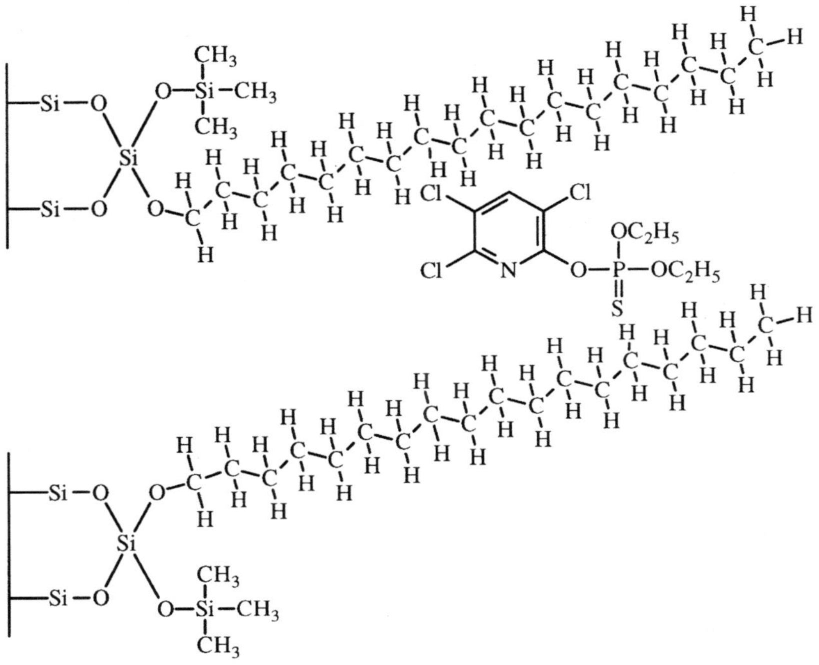

Figure 4.1. Reversed-phase mechanism of sorption of chloropyrifos, commonly called partitioning.

used for this process, although it may not be strictly correct because hydrophobicity only considers the insolubility of the analyte in water and does not consider the interaction of the analyte with the solid phase. Thus, partitioning is a
more correct term (Dorsey and Cooper, 1994). Nonetheless, the term
hydrophobic is commonly used in the literature. For more information,
Dorsey and Cooper (1994) review the recent literature on partitioning theory
and the reversed-phase mechanism. There is also an issue of the *Journal of
Chromatography* (1993, Vol. 656) that is dedicated to this topic.

The partitioning involves van der Waals or dispersion forces. The partitioning is regulated by the difference in the chemical potential of the solute
between the two phases (Dorsey and Cooper, 1994). This partitioning mechanism is a low-energy process (5 vs. 80 kcal/mol for ion exchange) and is analogous to a molecule being removed from water by an organic solvent in a
liquid–liquid extraction. The difference being that the organic phase is chemically bonded to the silica and is now a solid. Because the partitioning process
involves the differences in solubility of the analyte between the two phases,
the polarity of the analyte plays an important role in estimating how effective
the reversed-phase sorbent will be. This process will be examined later in this
chapter on predicting the capacity that an analyte has on a reversed-phase SPE
sorbent.

4.2. STRUCTURE OF REVERSED-PHASE SORBENTS

The most common reversed-phase bonded sorbent is the C-18 hydrocarbon
(Table 4.1), based on many thousands of SPE applications (see References in

Table 4.1. Common Sorbents Available for SPE

Sorbent	Structure	Typical Loading (%C)
	Bonded Phases C-18	
Octadecyl (C-18) (Varian)	—$(CH_2)_{17}CH_3$, endcapped, trifunctional, 40 μm, 60-Å pores, irregular silica	18
Octadecyl (C-18/OH) (Varian)	—$(CH_2)_{17}CH_3$, no endcapping, 40 μm, 60-Å pores, irregular silica	13.5
Octadecyl (C-18) (J. T. Baker)	—$(CH_2)_{17}CH_3$, endcapped, trifunctional, 40 μm, 60-Å pores, irregular silica	17.2

(contd.)

Table 4.1. (*continued*)

Sorbent	Structure	Typical Loading (%C)
Octadecyl (C-18)-Light (J. T. Baker)	—$(CH_2)_{17}CH_3$, not endcapped, 40 μm, 60-Å pores, irregular silica	12
Octadecyl (C-18)-Polar Plus (J. T. Baker)	—$(CH_2)_{17}CH_3$, not endcapped, 40 μm, 60-Å pores, irregular silica	16.1
Octadecyl (C-18) (Waters)	—$(CH_2)_{17}CH_3$, endcapped, monofunctional, 55–105 μm, 125-Å pores, irregular silica	12
Octadecyl (*t*C-18) (Waters)	—$(CH_2)_{17}CH_3$, endcapped, trifunctional, 37–55 μm, 125-Å pores, irregular silica	17
Octadecyl (C-18)EC (IST)	—$(CH_2)_{17}CH_3$, endcapped, trifunctional, 70 μm, 55-Å pores, irregular silica	17.9
Octadecyl (C-18) (IST)	—$(CH_2)_{17}CH_3$, not endcapped, trifunctional, 70 μm, 55-Å pores, irregular silica	~16
Octadecyl (C-18)MF (IST)	—$(CH_2)_{17}CH_3$, endcapped, monofunctional, 70 μm, 55-Å pores, irregular silica	~16
	Bonded Phases C-8	
Octyl (C8) (Varian)	—$(CH_2)_7CH_3$, endcapped, 40 μm, 60-Å pores, irregular silica	12.3
Octyl (C8) (J. T. Baker)	—$(CH_2)_7CH_3$, endcapped, 40 μm, 60-Å pores, irregular silica	14
Octyl (C8) (Waters)	—$(CH_2)_7CH_3$, endcapped, monofunctional, 37–55 μm, 125-Å pores, irregular silica	9
Octyl (C-8)EC (IST)	—$(CH_2)_7CH_3$, endcapped, trifunctional, 70 μm, 55-Å pores, irregular silica	~12
Octyl (C-8) (IST)	—$(CH_2)_7CH_3$, not endcapped, trifunctional, 70 μm, 55-Å pores, irregular silica	~12

(*contd.*)

Table 4.1. (*continued*)

Sorbent	Structure	Typical Loading (%C)
Bonded Phases C-2		
Ethyl (C-2) (Varian)	—CH$_2$—CH$_3$, endcapped, 40 μm, 60-Å pores, irregular silica	5.7
Ethyl (C2) (J. T. Baker)	—CH$_2$—CH$_3$, endcapped, 40 μm, 60-Å pores, irregular silica	4.8
Ethyl (*t*C2) (Waters)	—CH$_2$—CH$_3$, endcapped, 37-55 μm, trifunctional 125-Å pores, irregular silica	2.7
Ethyl (C2)EC (IST)	—CH$_2$—CH$_3$, endcapped, trifunctional, 70 μm, 55-Å pores, irregular silica	~5
Ethyl (C2) (IST)	—CH$_2$—CH$_3$, not endcapped, trifunctional, 70 μm, 55-Å pores, irregular silica	~5
Other Bonded Phases		
Methyl (C-1) (Varian)	—CH$_3$, endcapped, 40 μm, 60-Å pores, irregular silica	4.3
Cyclohexyl (Varian)	Cyclohexyl ring, endccapped, 40 μm, 60-Å pores, irregular silica	9.7
Cyclohexyl (J. T. Baker)	—CH$_2$CH$_2$-Cyclohexyl ring, endcapped, 40 μm, 60-Å pores, irregular silica	12
Cyclohexyl EC (IST)	—CH$_2$CH$_2$-Cyclohexyl ring, endcapped, trifunctional, 70 μm, 55-Å pores, irregular silica	~12
Phenyl (Varian)	Phenyl ring, endcapped, 40 μm, 60-Å pores, irregular silica	10.5
Phenyl (J. T. Baker)	—CH$_2$CH$_2$CH$_2$-Phenyl ring, endcapped, 40 μm, 60-Å pores, irregular silica	10.6
Phenyl EC (IST)	Phenyl ring, endcapped, trifunctional, 70 μm, 55-Å pores, irregular silica	~10
Phenyl (IST)	Phenyl ring, not endcapped, trifunctional, 70 μm, 55-Å pores, irregular silica	~10

(*contd.*)

Table 4.1. (*continued*)

Sorbent	Structure	Typical Loading (%C)
CN-E (Varian)	—$CH_2CH_2CH_2CN$, endcapped, 40 μm, 60-Å pores, irregular silica	8.5
CN-EC (IST)	—$CH_2CH_2CH_2CN$, endcapped, trifunctional, 70 μm, 55-Å pores, irregular silica	~8
C-4 Wide Pore (J. T. Baker)	—$CH_2CH_2CH_2CH_3$, 275-Å pores, 40 μm, irregular silica, endcapped	5.9
C-3 Wide Pore Hydrophobic Interaction (J. T. Baker)	—$CH_2CH_2CH_3$, not endcapped, 40 μm, irregular silica	11.7
	Polymeric Sorbents	
Graphitized Carbon (Supelco)	Aromatic C throughout	
Copolymer (IST)	Styrene–divinylbenzene	
	Disk Sorbents	
C-18 Disk (3M)	8–10 μm, Teflon fibril, trifunctional, 60 Å	16–18
C-18 Disk (Toxi-Lab)	30 μm, 70-Å pores diameter, 7 μm particles	
C-18 AR Disk (Toxi-Lab)	30 μm, 70-Å pores diameter	
C-18 Speedisk (J. T. Baker)	—$(CH_2)_{17}CH_3$, endcapped, trifunctional, 40 μm, 60-Å pores, irregular silica	17.2
C-18 Speedisk XF (J. T. Baker)	—$(CH_2)_{17}CH_3$, endcapped, trifunctional, 40 μm, 60-Å pores, irregular silica	17.2
C-8 Disk (Toxi-Lab)	30 μm, 70-Å pore diameter	
C-8 Disk (3M)	8–10 μm, Teflon web	~12
C-8 Speedisk (J. T. Baker)	—$(CH_2)_7CH_3$, endcapped, trifunctional, 40 μm, 60-Å pores, irregular silica	14
C-2 (3M)	8–10 μm, Teflon web	~5
SDB-XC Disk (3M)	Styrene–divinylbenzene copolymer, 80-Å	
SDB-RPS Disk (3M)	Styrene–divinylbenzene copolymer, lightly sulfonated, 80-Å	
Speedisk DVB (J. T. Baker)	Styrene–divinylbenzene copolymer	

Chapter 1). The widespread use of the C-18 for reversed-phase sorption is a result of the popular use of C-18 columns in high-pressure liquid chromatography (HPLC). For this reason, it gained immediate popularity for SPE applications. Other reversed phases include, C-8, C-4 (wide pore), C-2, C-1, cyclohexyl, and phenyl groups. The most commonly used reversed-phase polymeric sorbent in SPE is the styrene–divinylbenzene, or SDB. Graphitized carbon is also used.

Table 4.1 and the Appendix list the major vendors of bonded-phase sorbents for reversed phase as well as the more detailed information on the chemistry of the bonded phase. The list of C-18 phases to choose from is quite long, and this list does not include all of the possible vendors.

Among the C-18 phases from the different vendors, the carbon loading (percent of the R group bonded to the surface of the silica) will vary from as low as 12% to as high as 18%. Also the pore diameter may vary from the average value of 60 Å to as much as 125 Å. The smaller the pore size the larger the available surface area of the sorbent. Furthermore, larger molecules (> 2000 molecular weight) will be excluded from the 60-Å pore but may sorb more effectively in the 125-Å pore. For the largest molecules, the use of the wide-pore (275-Å) sorbents using C-3 or C-4 bonded phase are recommended. These wide-pore sorbents are available with both light (5.9%) and heavier carbon loadings (11.7%) and are also available in both endcapped and non-endcapped varieties.

The C-18 sorbents are available in both trifunctional and monofunctional forms. If the pH is adjusted to acid values, it is critical to use the trifunctional silica in order to prevent hydrolysis of the hydrocarbon group from the surface of the sorbent. However, if polar molecules are to be sorbed and they contain basic sites, the monofunctional sorbents and the nonendcapped sorbents will have greater capacity because of the free silanol sites that are available for secondary interactions. For these reasons, the C-18 sorbents vary from manufacturer to manufacturer. Thus, the source of the C-18 sorbent may be selected to enhance the isolation that is required.

When the sorbent is endcapped, which is typically done with trimethylsilane, the trimethylsilane will increase the carbon loading by bonding to any available silanol sites that have not been derivatized by the C-8 or C-18. Because of the smaller size of the endcapping reagent, it can derivatize the more sterically hindered sites on the silica. This process makes the sorbent more hydrophobic and increases the carbon loading as much as several percent.

Similar to the C-18 bonded phases, the C-8 sorbents are available in a variety of forms. Generally, they have considerably reduced carbon loadings, varying from 9 to 14% (Table 4.1). The C-2 bonded phase has the least carbon loadings, from 2.7 to 5.7%. The utility of the lower carbon loadings is that the most hydrophobic analytes can be sorbed and effectively eluted without the use of large volumes of elution solvents. Generally, there have not been many published applications of the shorter-chain bonded-phase sorbents, and this topic is an area for further study and applications.

Some of the other bonded phases that may have specific applications include the cyclohexyl and the phenyl phases. The cyclohexyl does show an ability to sorb phenolic compounds from aqueous solutions with a greater affinity than expected from their lower carbon loadings, when compared to C-18. It is hypothesized that the configuration of the cyclohexyl plays a role in this increased sorption. Likewise, the phenyl sorbent shows a greater capacity for aromatic compounds because of increased interaction between the aromatic rings of the analytes and the aromatic rings of the bonded phase, because of the sharing of π electrons between the aromatic rings.

The polymeric phases commonly are used in reversed-phase applications. They are especially useful for the isolation of the most polar solutes by reversed phase because of their increased surface area and carbon loading over C-18. They also have the possibility of π–π interactions between the aromatic structure of the sorbent and available π electrons in the analyte. Phenols and other water-soluble compounds often have considerably more capacity when using the polymeric sorbents. The extensive literature on XAD (polymeric sorbent) resins, especially the XAD-2 resin, serve as a guide for applications of polymeric sorbents.

The disk technology also is available in a number of reversed-phase sorbents including C-18, C-8, and SDB. Several vendors now produce these disks, generally in a 8- to 10-µm particle size. The advantage of the disk technology is that because the disks incorporate sorbent of small particle size, the disks have a large surface area that results in good mass transfer and fast flow rates, with as much as 100 mL/min through a 10-µm sorbent (500 mg of C-18). The various disks have different flow characteristics because of the design of the disks and their particle size (see Chapter 11). Some disks contain Teflon fibrils that decrease the flow rate while others use glass fibers as a matrix and have more rapid flow rates. Both types are quite effective for large-volume samples. The disks may be used in the same applications as other C-18 packings, but it is important to obtain detailed information on the chemistry of the sorbents used in the disks for the most effective methods development.

4.3. REVERSED PHASE AS A PARTITIONING MECHANISM

The reversed-phase isolation nearly always involves the partitioning of the organic compound from water or an aqueous solvent. The affinity of the solute for the stationary phase is related to the log of the octanol–water partition coefficient of the molecule or in a simpler sense the water solubility of the compound. The log of the octanol–water partition coefficient is the partition equilibrium constant when an analyte is partitioned between water and the eight-carbon alcohol, octanol. It is a common modeling tool used in environmental chemistry to predict analyte behavior based on its hydrophobicity. In essence, the octanol phase competes for the solute with the water phase. The more polar the solute, the more readily it is dissolved in water. Thus, in simple terms, the log of the octanol–water partition coefficient is a measure of the relative affinity of a compound for octanol versus water. This relates to SPE in that the loading of the organic phase onto the silica, which is measured as the percent carbon, is analogous to the volume of organic solvent used in an extraction of an organic solute from water. Thus, a larger carbon loading will have more capacity for an analyte with the same R group (i.e., C-18).

In order to predict or estimate the affinity that a new analyte may have for a reversed-phase sorbent, the capacity factor may be useful. The capacity factor of an analyte is a function of the distribution coefficient of the analyte between the sorbent and water (mobile phase). The distribution coefficient is similar in concept to the log octanol–water partition coefficient except that the solid phase replaces the octanol phase. The distribution coefficient of the solute for the sorbent, K_d, for the reversed phase is given by the following equation:

$$K_d = \frac{C_{B(\text{bonded phase})}}{C_{S(\text{solution phase})}} \qquad (4.1)$$

where the concentration on the bonded phase is C_B and the concentration in solution is C_S. The distribution coefficient is calculated from a series of batch experiments at constant temperature and is plotted as a so-called isotherm. A batch experiment is a series of equilibrium sorption experiments where the concentration of an analyte is measured both before the experiment begins and at the end of the experiment when the solution has reached equilibrium (Fig. 4.2). The mass of sorbent and the volume of water in each flask is held constant, and the concentration is increased over a large range. The mass of analyte that has sorbed onto a known amount of sorbent is calculated and the resulting data are plotted in a form that is called an isotherm (Fig. 4.2)

The slope of the line of the isotherm is the distribution coefficient, K_d. The K_d has units of milliliters/gram. Furthermore, the isotherm demonstrates that the capacity of the sorbent for a compound is a function of the concentration that is introduced into the column. For example, if a low concentration is introduced, let us say 1 µg/L, the capacity (mass sorbed) will be considerably less than if the concentration is 1000 µg/L. There is a point where the isotherm becomes nonlinear and the capacity of the sorbent has been reached. This point is sometimes used by manufacturers to express the capacity of the

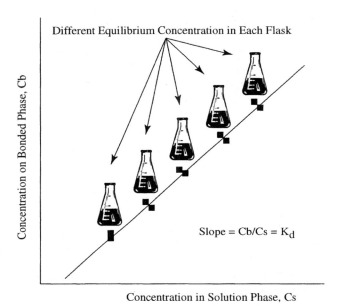

Figure 4.2. An example of an equilibrium batch experiment used to calculate an isotherm for determination of K_d.

sorbent in milligrams/gram. The compound that has been used commonly is caffeine. The use of capacity in this sense is not rigorously correct because the capacity in this sense is merely one point on the isotherm and could very well not be the concentration for which the sorbent is being used. Moreover, the word *capacity* in the general sense may be used for either the mass of analyte (a small volume of high concentration of analyte) or the volume of sample (a large volume of a low concentration of analyte) that may be sorbed onto a SPE sorbent. The K_d theoretically predicts either of these values for the analyte and the SPE sorbent, although the isotherm is rarely used in practice. Rather, the breakthrough experiment commonly is used to determine capacity or volume of sample that may be applied to an SPE sorbent.

In summarizing reversed-phase SPE, both the K_d and the log of the octanol–water partition coefficient of the compound are related to the aqueous solubility of the analyte, although the relationship is not always straightforward because there are other factors that affect solubility of an analyte but that do not affect sorption, such as crystal lattice energy for solids. In spite of this factor, one could theoretically estimate the K_d from the solubility of the analyte and relate this solubility to the capacity of the solute for the reversed-phase sorbent. The ability to use solubility to predict capacity is also addressed in more detail later in this chapter. A general rule of thumb is that solutes with aqueous solubilities less than 1000 mg/L will work well with reversed-phase bonded silica sorbents, such as C-18. If the aqueous solubility is greater than 1000 mg/L, then the graphitized carbon and the styrene–divinylbenzene sorbents are recommended because they have considerably more capacity than the C-18.

4.4. CHROMATOGRAPHIC PLATE THEORY AND REVERSED-PHASE SOLID-PHASE EXTRACTION

4.4.1. Introduction

The following discussion of chromatographic plate theory uses the terminology defined below:

influent = solvent entering the column
effluent = solvent exiting the column
eluent = eluant = solvent used for elution
eluate = solution resulting from elution
C_0 = concentration of influent
C = concentration of effluent
Saturation of column = capacity of column when C_0 equals C
$P = N$ = total number of theoretical plates in a column
v = eluate volume at any given point
v_0 = void or interstitial volume of a column; that volume occupied by liquid between the particles.

If one considers that the contents of an SPE cartridge could be opened and placed in a beaker with analytes present and shaken, then sorption would occur between the analytes in solution and the solid-phase sorbent (Fig. 4.3). All of the sorbent has an equal opportunity to interact with the solute in solution at the same time, and the whole mixture reaches the same equilibrium at once. This process may be described by considering the sorption of a single analyte, which is shown by Eq. (4.2) as a single step process.

$$S_{aq} = S_S \qquad (4.2)$$

where S_{aq} is the solute in the aqueous phase, and S_S is the solute in the sorbed phase. The equilibrium equation that describes the process is

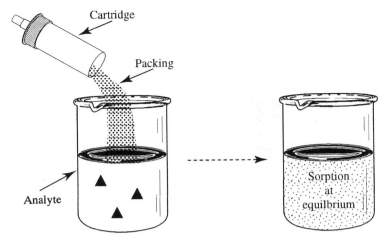

Figure 4.3. One-plate sorption model, also called a batch experiment, where a sorbent phase sorbs analytes from solution onto the solid phase.

shown below:

$$K_{eq} = [S_S]/[S_{aq}] \tag{4.3}$$

This single-step process could be thought of as a "single plate" or single sorption step (Fig. 4.3). The word *plate* comes from the early theory used in chromatography to describe sorption and refers to distillation columns that contain plates.

If we imagined that the same sorbent in the beaker in Figure 4.3 were placed into a column and the solute passed through the column then a *series* of sorption steps would occur. Rather than the general equilibration between all of the sorbent and the analyte at once, a series of equilibrations would occur as mobile phase gradually moved the analytes down through the column, and analyte is continually coming into contact with new "layers" or "plates" of sorbent deeper in the column (Fig. 4.4).

As solution containing the solute to be sorbed is added to the column, solute will equilibrate section by section, or plate by plate, as shown in Figure 4.4. Once solute has sorbed in the first plate, the water, which is now depleted of analyte, continues to move down the column into the next plate until equilibrium is reached in the first plate. When this occurs, the solution with analyte flows into the next plate, and so forth, until the entire column is at equilibrium, but in a multistage batchlike process. Although the distribution coefficient (K_d) is the same in each plate, the process is established in increments. Because the

Sample

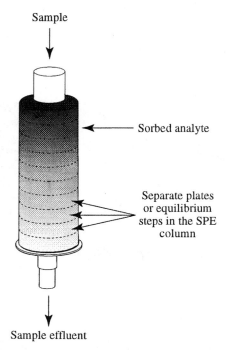

Sorbed analyte

Separate plates
or equilibrium
steps in the SPE
column

Sample effluent

Figure 4.4. Sorption of a solute in a theoretical column by steps or plates. Each step or plate is an equilibration that is occurring between the solute and the sorbent.

K_d for solutes may be very large on the C-18 resin (i.e., a large affinity of the solute for the solid phase), then a large capacity (large volume of low concentration of analyte) may exist for solutes in an SPE column, even on a small column with a few plates. Typically, an SPE column will contain only 20 plates or less, while an HPLC column will contain 10,000 plates! For this reason, SPE is generally not used for the separation of one analyte from another along the length of the column due to different K_ds', rather SPE is used for the simple sorption and isolation of analytes in an on/off mode.

Elution of the SPE column also is similar to sorption in that the analytes move in a series of platelike equilibrations down the column, depending on the k' (column equilibrium constant) of the solute in the elution solvent. Therefore, the amount of solvent needed to elute the analyte also is affected by the number of plates. That is, the more theoretical plates that an SPE column contains, the smaller the volume that is required to elute the analyte. This is because the column with more plates has less dispersion of solute during the elution step and less elution solvent is needed. Obviously, as the mass of packing material increases, more solvent would be required for elution.

4.4.2. Theoretical Aspects of Plate Theory

There are at least two different approaches used to explain the concept of theoretical plates. The first is the discontinuous approach, as in the distillation column with discrete plates and equilibration occurring step by step in the horizontal sections of the distillation column. Second is the continuous process, which is a more precise explanation. Both will be discussed.

The plate theory as applied to chromatography is derived from the mathematical theory of distillation columns. Figure 4.5 shows a distillation column.

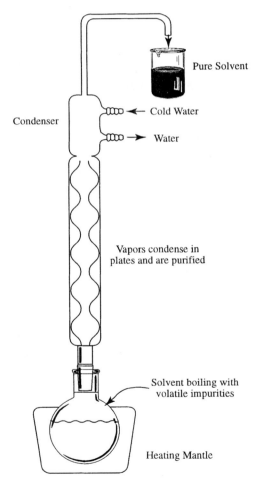

Figure 4.5. Concept of plates as derived from a distillation column.

As the vapors rise and recondense, plate by plate, the analyte being separated from the others in the distillation column becomes more pure, until the final vapors that leave the distillation column are pure. Thus, the more plates there are, the better is the purification and separation.

In chromatographic theory, a plate is a transverse section of a cross-sectional area equal to the cross-sectional area of the column. The thickness of the plates is called the HETP, or the height equivalent of a theoretical plate. It is derived from the number of plates divided by the length of the column:

$$\text{HETP} = \text{length of column/number of theoretical plates} \qquad (4.4)$$

The plate concept was first adapted to chromatography by Martin and Synge (1941), who derived the theory and equations that will be shown below. The equations assume the following:

1. The solution phase reaches equilibrium with the exchange of each plate before going to the next plate.
2. The distribution between the solid phase and solvent is the same throughout the column, that is, constant K_d spatially within the column.
3. Concentration of solute that is being separated is low so that there is no isotherm effect on the K_d, that is, the K_d is constant within the concentration range tested.

4.4.2.1. Discontinuous Approach

The solute partitions in the first plate between sorbent and liquid phase such that

$$K_d = M_r/M_s \qquad (4.5)$$

where M_r is the mass in the sorbent and M_s is the mass in solution. Furthermore, one assumes that the sum of the mass on the sorbent and in solution is equal to 1:

$$M_r + M_s = 1 \qquad (4.6a)$$

or

$$M_r = 1 - M_s \qquad (4.6b)$$

Next, Eq. (4.5) is rearranged to give

$$M_r = K_d M_s \tag{4.7}$$

Substituting Eq. (4.6b) into Eq. (4.7) gives

$$1 - M_s = K_d M_s \tag{4.8a}$$

or

$$K_d(M_s) + M_s = 1 \tag{4.8b}$$

$$M_s (K_d + 1) = 1 \tag{4.9}$$

Therefore,

$$M_s = 1/(K_d + 1) \tag{4.10}$$

Thus, the mass of analyte in solution is now expressed in units of K_d. This is the first important derivation. Next, for the quantity M_r in units of K_d, the following substitutions are done. First, Eq. 4.5 is rearranged to give

$$M_s = M_r/K_d \tag{4.11}$$

and substituting back into Eq. (4.6a) gives

$$M_r + M_r/K_d = 1 \tag{4.12}$$

Simplifying to K_d as the common denominator gives

$$K_d M_r/K_d + M_r/K_d = 1 \tag{4.13}$$

which may be rewritten as

$$(K_d M_r + M_r)/K_d = 1 \tag{4.14}$$

which is equal to

$$M_r(K_d + 1)/K_d = 1 \tag{4.15}$$

Therefore,

$$M_r = (K_d)/(K_d + 1) \tag{4.16}$$

This is the second important derivation and expresses the mass of analyte on the sorbent in units of K_d. Next, we consider Figure 4.6 and look at the mass of material that is equilibrating between the sorbent (M_r) and the solution (M_s), in terms of the K_d [Eqs. (4.10) and (4.16)]. Plate 1 in Figure 4.6 begins with the total mass in the solution phase before equilibrium being equal to 1 and 0. Then, at equilibrium the mass of solute in solution is $1/K_d + 1$, which is the derivation in Eq. (4.10). The mass sorbed to the resin is $K_d/K_d + 1$, which is the derivation in Eq. (4.16). As the solution phase containing $1/K_d + 1$ of analyte moves into the second plate, the reequilibration begins again for the solid phase and results in the equations shown in Figure 4.6 for Plate 2. These equations are derived in the same manner as before using the algebra shown in Eq. (4.10) through (4.16) and may be continued for each step or plate shown in Figures 4.6 and 4.7. Basically, this process of equilibration from plate to plate may be extended down the chromatographic column in algebraic terms using only K_d. The result is that the expansion of the K_d terms may be modeled with a binomial expansion.

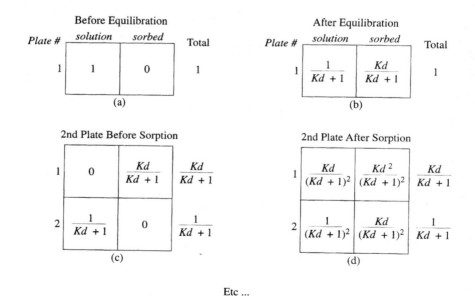

Figure 4.6. Example of the discontinuous approach of theoretical plates in a chromatographic column, which results in the binomial expansion after Khyme (1974). [Published with permission of Prentice-Hall, Inc.]

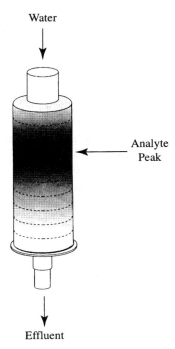

Water

Analyte
Peak

Effluent

Figure 4.7. Dark coloration showing the peak of an analyte in a chromatographic column that is eluted only with water.

4.4.2.2. Continuous Approach

The individual plates could also be envisioned as a series of histograms (Fig. 4.7), where a minuscule amount of analyte is added to the top of the chromatographic column and the analyte is then eluted from the column with only water as the mobile phase, and each increment goes through the equilibration described above with expansion of the distribution equation using the binomial expansion.

The chromatographic peak that results from this histogram has been shown by the work of Glueckauf (1955) to be described by the Gaussian distribution (Fig. 4.7). The major conclusion of the plate theory approach is that the peak maximum in the distribution is equal to the K_d. In liquid chromatographic theory, K_d is described as k', which is simply the column distribution coefficient. It is from this elegant theoretical work that the practical approach of measuring peak height and peak width of a chromatographic peak is used to calculate the number of theoretical plates or the efficiency of a chromatographic column (these calculations using peak height and peak width are shown later in this chapter).

Thus, with all the considerations of plate theory, there are several parameters that affect the performance of an SPE column because of plate number. They are as follows:

1. *Flow Rate.* Flow rates that are too rapid do not allow "equilibrium" to be reached, which reduces plate number and the efficiency of the column. Equilibrium is placed in quotation marks because the time that an analyte is present in an SPE column may not, in fact, be enough time for true equilibrium to be reached.

2. *Particle Size.* Smaller particles increases plate number per unit length or unit mass of an SPE column because they have more surface area. Therefore, less dispersion between the particles takes place.

3. *Column Length.* Column length increases the number of plates. However, in SPE, the volume (or mass of sorbent) per unit length is important also because it increases the capacity (mass) of analyte that may be sorbed.

4. *Temperature.* Increasing the temperature of the sorbent increases the number of theoretical plates. This practice is generally not practical in sorption of SPE but may be useful in elution of SPE sorbents because of less dispersion during elution.

4.5. CHROMATOGRAPHIC PLATE THEORY: ITS APPLICATION TO REVERSED-PHASE SOLID-PHASE EXTRACTION

4.5.1. Introduction

There are several simple chromatographic calculations that are fundamental to HPLC and to SPE. Solid-phase extraction can be considered to be a simple form of liquid chromatography with a low number of theoretical plates. The sorbent is the stationary phase, and the mobile phase is water or the organic solvent in which the analyte is dissolved. Figure 4.8 shows the separation of two chromatographic peaks, peak 1 and 2. The efficiency of a liquid chromatographic column to separate the two peaks may be expressed by Eq. (4.17):

$$N = 5.54 \, (t_{r1}/w_1^{1/2})^2 \tag{4.17}$$

where N = plate number

$\quad t_{r1}$ = retention time of a retained compound (peak 1) in minutes

$\quad w_1^{1/2}$ = peak width measure at one-half peak height (in units of time)

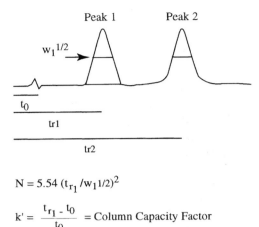

$$N = 5.54 \, (t_{r_1}/w_1 1/2)^2$$

$$k' = \frac{t_{r_1} - t_0}{t_0} = \text{Column Capacity Factor}$$

Figure 4.8. Chromatogram used for calculation of column efficiency N, also called the number of theoretical plates.

Equation (4.17) is derived from the Gaussian distribution discussion of the preceding section. The value of N (plate number) is quite important in liquid chromatography and in SPE because it has a direct effect on the performance of the column. Typically an HPLC column will contain many more plates than the SPE column. Thus, the efficiency of the SPE column is quite low, but there are sufficient plates for sorption to take place and for the analyte to be removed from the sample matrix in an effective manner.

The particle size of the packing material is one of the major reasons that the SPE column contains less theoretical plates and is less efficient than the HPLC column. The SPE column will have particles that vary from 40 to 80 μm in diameter, while HPLC columns typically have particles that range in diameter from 3 to 5 μm. Typically, the number of plates will increase by the square root of 2 (1.4 times) for every halving of particle size.

The second factor that affects the SPE column is the packing density of the material. Typically, an SPE column is dry packed, while the HPLC column is slurry packed under pressure and kept wet. This results in a close packing arrangement with greatly improved efficiency for the HPLC column. Next, the particle size is closely controlled in an HPLC column with just a slight difference among particles. Less care is generally taken in SPE and a wider range of particle sizes is commonly present, which leads to less efficient packing and some channeling of sample flow through the column.

A second important equation in reversed-phase liquid chromatography, which is related to the chromatogram in Figure 4.8, is the capacity factor, called k'. It is related to the elution time of a retained analyte relative to a non-retained analyte. It is also expressed in a comparison of time units; thus, it is a dimensionless quantity. The equation is

$$k' = (t_{r1} - t_0)/t_0 \qquad (4.18)$$

where t_0 = retention time of a nonretained compound, typically an inorganic analyte such as chloride, and t_{r1} = retention time of the first retained compound. The k' is essentially the column distribution coefficient of a compound for the stationary phase given the conditions of the mobile phase. These equations hold equally well in SPE as they do in HPLC.

Another important concept, which is used in SPE but is not used in HPLC, is the relationship of breakthrough of an analyte in what is called frontal chromatography (Fig. 3.6). Frontal chromatography is the process where a sample is continually applied to a column until some of the solute begins to appear in the effluent of the column. Breakthrough occurs when a solute is no longer retained by the sorbent because the capacity of the sorbent has been reached. This concept is especially important in SPE because sample is applied to an SPE column continuously and the sorbent must retain all the solute. If the solute is not sorbed but is gradually passing through the column, the shape of the breakthrough curve is quite important in determining recovery and usefulness of the sorbent.

Breakthrough volume can be measured by monitoring the ultraviolet (UV) signal of a water sample spiked with traces of a solute, S, which has an initial absorbance, A_0. The spiked sample is passed through an SPE column. If the compound is retained by the sorbent, the effluent will have an absorbance of zero. A frontal or breakthrough curve is recorded (Fig. 4.9) beginning at a volume, V_b, usually defined as 1% of A_0, up to a volume of V_m, defined as 99% of A_0, where the effluent has the same composition as the spiked water sample (Hennion and Pichon, 1994).

Under ideal conditions, this curve has a logarithmic shape, where the inflection point is the retention volume, V_r, of the analyte. This V_r may be related to a chromatographic separation retention volume, V_r, by the following reasoning. If the same SPE column were used in a chromatographic sense, that is, if the analyte were injected onto the column in a microliter volume with water as a mobile phase, the peak would be detected at a volume of V_r. This is shown in Figure 4.10.

The parameter, V_b (1% breakthrough), is a key variable for the preconcentration of the analyte and may be estimated by at least three methods. All three methods use different ways of estimating or measuring V_r. The similarities between SPE and liquid chromatography indicate that data generated by liquid

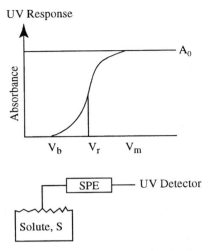

Figure 4.9. Frontal chromatography application in SPE. [After Hennion and Pichon (1994), published with permission.]

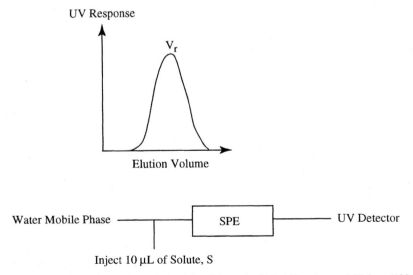

Figure 4.10. Elution chromatography and breakthrough. [After Hennion and Pichon (1994), published with permission.]

chromatography for measuring or estimating V_r are useful. The first method involves the use of an approximate V_b in order to estimate the V_r.

4.5.2. Determining and Predicting Breakthrough Volumes

Recording breakthrough curves is time consuming and reading V_b at 1% of A_0 is difficult and not always accurate. Moreover, the sample should be spiked at a trace level in order to not overload the sorbent, and the UV signal of the effluent should be monitored at low absorbances, which leads to problems with baseline drift and poor reproducibility. Moreover, some compounds have low UV absorbances.

A faster method, which is easily performed using an on-line SPE procedure, consists of preconcentrating samples of increasing volume, with each sample containing the same mass of analyte. Thus, as the sample volume increases, the analyte concentration decreases, provided that breakthrough does not occur. The amount concentrated on the SPE column remains constant and the peak areas in the on-line chromatograms are constant. When breakthrough occurs, the amount extracted decreases and the elution peak area decreases. Corresponding recoveries can be calculated by dividing peak areas obtained after breakthrough by those obtained before (Figs. 4.9 and 4.11). An advantage of this method is that the V_b values of several compounds may be estimated simultaneously by preconcentration and on-line HPLC under the real conditions of sample analysis.

Another method approximates the breakthrough volume using V_r (retention volume), which is related to chromatographic data and precolumn characteristics by

$$V_r = V_0 (1 + k'_w)
\qquad (4.19)$$

where V_0 is the void volume of the SPE column and k'_w is the capacity factor of the solute eluted by water; V_0 may be calculated from the porosity of the sorbent and the geometric volume of the SPE column or sorbent bed in the cartridge. Values of k'_w are estimated from chromatographic measurements using analytical columns packed with reversed-phase sorbents, such as a bonded C-18 that are eluted with a mobile phase composed of water–methanol mixtures. The advantage of this method is that experimental data are obtained rapidly by measuring the capacity factor k'_w of the analyte in methanol–water phases. Over a range of methanol concentrations, often between 15 and 90%, there is a linear relationship between the logarithm of k' and the percentage of methanol. This result has been observed for bonded alkyl silicas, nonpolar styrene–divinylbenzene copolymers, and porous graphitic carbon.

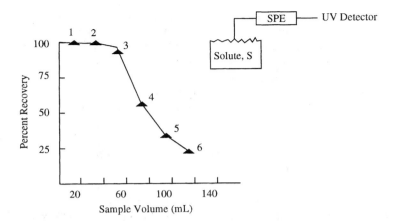

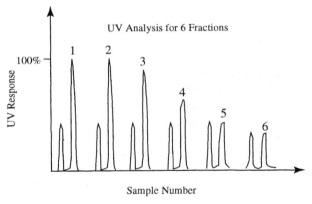

Figure 4.11. Experimental determination of breakthrough to determine V_b or 1% breakthrough. [After Hennion and Pichon (1994), published with permission.]

From rapid measurements with three or four mobile phases containing different methanol concentrations, k'_w can be estimated by graphically extrapolating to zero methanol content. In Figure 4.9, the experimental values of log k'_w should be similar to predicted values because the relationship is linear over the entire range. Werkhoven-Goewie and co-workers (1981) found good agreement between V_r values for some chlorophenols derived from experimental breakthrough curves and values calculated from extrapolated k'_w values. Differences were on the order of 10 to 20%. The extrapolated value is only an approximate value, but for trace enrichment studies this is acceptable.

Another consideration concerning the breakthrough volume is the addition of methanol to the sample which is sometimes recommended in reversed-phase SPE. A small percentage of organic solvent can be added to the raw sample without decreasing the recovery, only if the retention is high for an analyte and if the sample volume percolated through the sorbent is less than the V_r value calculated from the corresponding log k'_w for this organic solvent concentration. Solutes with low k'_w for the sorbent phase will be affected by the addition of methanol and will have low recoveries.

4.5.3. Polarity and Breakthrough

Because the relative difference in the terms polar and nonpolar is somewhat undefined, a general parameter to define hydrophobicity is useful. Such a parameter is the octanol–water partition coefficient, P_{oct}, which was defined earlier in this chapter. This parameter plays an important role in correlating phenomena of physicochemical, biological, and environmental interest. Many log P_{oct} values are available in the literature and calculation methods have been reported. Hennion and Pichon (1994) define polarity from the log P_{oct} scale, with values below 1 as polar, 1 to 3 as moderately polar, and above 3 as nonpolar.

Hennion and Pichon (1994) recommend C-18 for compounds in the range of log P_{oct} from 1 to 3. For compounds with log P_{oct} greater than 3, a sorbent with a lower carbon coating could be used (i.e., C-2, C-4, or C-8). Another choice is to add methanol to the mobile phase to decrease the k'_w of the nonpolar solute for the C-18 sorbent. Finally, for the most polar compounds, with a log P_{oct} of less than 1, the C-18 may not have adequate capacity for trace enrichment. In this case, the polymeric sorbents and the graphitized carbon are useful (Hennion and Pichon, 1994; also see discussion in Chapter 2 on polymeric sorbents).

Braumann (1986) has gathered many log k'_w values that were obtained with different C-18 bonded phases using methanol–water mobile phases. A linear relationship was found between the average log k'_w values and log P_{oct} for closely related compounds and even for compounds having different polarity and chemical properties. For example, 60 compounds covering a wide range of structures from a polar compound, such as aniline, to a very hydrophobic compound, p,p'-DDT (log $P_{oct} = 6.2$) are related by

$$\log k'_w = 0.988\,(\pm 0.051)\log P_{oct} + 0.02\,(\pm 0.060) \qquad (4.20)$$

All the relationships published in Braumann (1986) and other studies have similar coefficients, showing that values of log k'_w and log P_{oct} are similar in value. Thus, with C-18 bonded silicas, k'_w may be approximated without any additional measurements.

4.5.4. Example Calculation of Breakthrough Volume Using k_w'

An example of how to calculate the V_r is shown below. Hennion and Pichon (1994) use the example of chlorophenols, which have log P_{oct} values from 5.0 for pentachlorophenol to 1.5 for phenol itself. The appropriate constants show that the V_0 is the product of the geometric volume of the SPE column and the porosity of the C-18 bonded silica. An average porosity value is between 0.65 and 0.7. The average density of the sorbent is 0.6 g/mL and a V_0 is calculated as 0.12 mL per 100 mg of sorbent. Hennion and Pichon (1994) found good agreement between these calculated values and chromatographically determined values for their SPE columns.

The log P_{oct} value is used to calculate the log k_w' from Eq. 4.20:

$$\log k_w' = 0.988\,(5) + 0.02 \tag{4.21}$$

$$\log k_w' = 4.96 \tag{4.22}$$

$$k_w' = 91{,}200 \tag{4.23}$$

Assume that the sorbent is 100 mg of C-18 silica with a porosity of 0.65, a density of 0.6 g/mL, and a V_0 of 0.12 mL per 100mg.

$$V_r = V_0\,(1 + k_w') \tag{4.24}$$

$$V_r = 0.12\,(91{,}201) \tag{4.25}$$

$$V_r = 10{,}944 \tag{4.26}$$

or approximately 11 L of water.

This now may be used to calculate the V_b (1% breakthrough) value from the number of theoretical plates in the SPE column by the following equation:

$$V_b = V_r - 2\sigma_v \tag{4.27}$$

$$\sigma_v = V_0(1 + k_w')/N \tag{4.28}$$

Given that an average SPE column contains about 20 plates (for an N value), it gives:

$$\sigma_v = 0.12(91{,}200)/4.5 \tag{4.29}$$

$$\sigma_v = 2.4\ \text{L} \tag{4.30}$$

$$V_b = 11 - 4 \tag{4.31}$$

$$V_b \sim 7\ \text{L} \tag{4.32}$$

One can rapidly obtain V_r from log P_{oct} and determine whether the C-18 is suitable for the extraction. These data reveal a significant difference between nonpolar analytes and moderately polar ones. The V_b values are 7 L for the pentachlorophenol, 0.35 L for 3,5-dichlorophenol, 16 mL for 2-chlorophenol, and < 5 mL for phenol. Thus, one quickly sees the effect of increasing polarity on the retention of the solute. One can increase the amount of sorbent to increase capacity, but this is only effective for the moderately polar compounds. A compound as polar as phenol is not affected much. The approach of Hennion and Pichon (1994) is an effective method for breakthrough determinations and capacity calculations when using on-line SPE.

These calculations also work effectively on polymeric sorbents. The work of Hennion and Pichon (1994) shows that polymeric sorbents may be 20 to 40 times greater in capacity for reasons discussed in Chapter 2. Thus, they are a good choice for isolation of polar compounds. Graphitized carbon may sorb with even greater efficiencies than the polymeric sorbents because of other interactions, due to impurities in the carbon (oxygen atoms that are positively charged). These positively charged sites may interact by ion exchange and increase the retention of analytes. For these reasons, the above calculations do not work with graphitized carbon, according to the work of Hennion and Pichon (1994). But the graphitized carbon does present a good last choice for very polar analytes that cannot be isolated by either C-18 or polymeric sorbents. There are graphitized-carbon sorbents that are made from the polymerization of resins that contain few impurities and ion-exchange sites. These sorbents show possible usefulness for SPE applications of the most polar compounds.

4.5.5. Practical Effects of Plate Number on Solid-Phase Extraction

The effect of N, or plate number, becomes important when considering both the V_b volume and the elution volume in SPE. Figure 4.12 shows how the breakthrough curve varies with plate number. The values are taken from a technical brochure by Waters Chromatography that compares several of their SPE cartridges with different particle sizes. The V_r for both columns was the same value of 200 mL. However, the V_b varied from 25 mL for the column with 10 plates to 125 mL with the column with 50 plates. Thus, although the distribution coefficient, k'_w, was the same for both columns, the efficiency of the column was better for the column with 50 plates. This result means that the column with more plates will recover more analyte than the less efficient column.

Plate number also affects the flow rate under which an SPE column may be operated. For example, Figure 4.13 shows the plot of flow rate versus the loss of efficiency for two columns that start out with a difference in plate number of

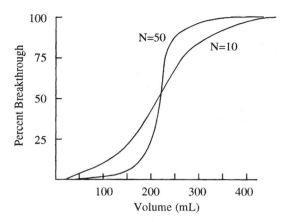

Figure 4.12. Variation in breakthrough with plate number or column efficiency. [Published with permission of Waters, after Bouvier (1994).]

about 2.5 times. Note that the flow rate is expressed in milliliters per minute, a better unit might be bed volumes per minute. Both columns contained approximately 360 mg of packing material and have bed volumes of approximately 0.5 mL each. Thus, these flow rates are fast compared to HPLC flow rates.

Thus, controlling flow rate during the SPE run is important. Using the vacuum manifold approach does not control flow rates well; thus, the positive pressure approach is favored for the most reproducible results. Many of the

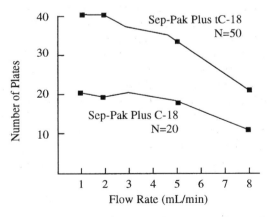

Figure 4.13. Loss of efficiency for two columns as a function of flow rate. [Published with permission of Waters, after Bouvier (1994).]

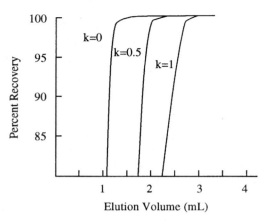

Figure 4.14. The effect of k' on elution volume in SPE. [Published with permission of Waters, after Bouvier (1994).]

automated workstations are capable of controlled flow rates and, for this reason, they improve reliability of an SPE method.

Likewise, the elution step of the cartridge will be affected by the plate number, N, of the column. The greater the number of plates in the column, the more efficiently the analytes will be eluted from the column. More importantly though is the k' of the solute in the elution solvent. Figure 4.14 shows the shape of the elution curve for different low k' values that might be expected for a solute on a C-18 column with 20 theoretical plates. The efficiency of elution being affected by the slight amount of retention of the analyte by the C-18 column. In this case, a shift to a reversed-phase column with less carbon loading and a shorter chain length may help. Thus, a C-8 column might more efficiently be eluted for the same analyte with a smaller volume of eluent.

A final consideration is the ionic state of the analyte. If the molecule is non-ionic, then the compound will behave according to the partition theory explained above. However, if the molecule is ionic, then two different k's must be considered, the k'_{ionic} and $k'_{nonionic}$. Generally speaking, when the molecule is ionic, its capacity factor for the C-18 phase is greatly reduced. Furthermore, the pH at which the ionic or nonionic character of the compound becomes important is a function of the pK_a of the compound. The equation that relates these two species to one another is the Henderson–Hasselbach equation:

$$pH = pK_a + \log A/HA \tag{4.33}$$

where A is the conjugate base of the acid, HA. This equation shows that when the pH = the pK_a the ratio of A to HA is one and the log term becomes zero in Eq. (4.33). When the pH is two units above the pK_a, then the molecule is chiefly in the *A* state (99%). When the pH is two units below the pK_a the molecule is chiefly in the *HA* state (99%). For an organic acid, *A* is the negatively charged species, and it has a much decreased retention on a reversed-phase sorbent. For an organic base, *A* is the uncharged basic species, which is much more retained by reversed phase than the positively charged, *HA* species. Equation (4.33) and the *k'* values for the two species are needed in the determination of capacity of the compound for the SPE column. It is a good idea to place the solute in the nonionic state for the best sorption performance on a reversed-phase column.

4.6. EXAMPLES OF METHODS DEVELOPMENT WITH REVERSED-PHASE SOLID-PHASE EXTRACTION

The selection and development of a reversed-phase method can be a straightforward process, and for this reason, many methods have been developed on reversed-phase SPE, especially the C-18 sorbent due to its reliable nature. The following examples describe the reversed-phase extraction of compounds of differing polarities from various matrices, as an indication of the broad extent to which reversed-phase SPE can be applied.

4.6.1. Trace Enrichment of Pesticides from Groundwater

The trace enrichment of pesticides from groundwater is a straightforward problem for reversed-phase SPE. The following example uses the EMPORE disk (C-18) to isolate organochlorine pesticides including aldrin, chlordane, endrin, heptachlor, lindane, methoxychlor, pentachlorophenol, and toxaphene from groundwater (Fig. 4.15).

SPE Conditions Used

Sample	Groundwater sample, 1 L.
Solutes	Organochlorine pesticides, Figure 4.15.
Sorbent	47-mm disk EMPORE C-18. Condition with 5 mL methanol, 5 mL of methylene chloride/ethyl acetate. Let stand on disk for 3 min. to soak into disk. Draw remaining solvent through the disk. Dry under vacuum for 1 min. Next add 5 mL methanol and let stand for 3 min. to condition disk before applying sample.
Eluent:	Elute with 5 mL ethyl acetate, then 5 mL of methylene chloride. Combine extracts and evaporate for GC/MS analysis.
Reference	Bakerbond Application Note EMP-005 Extraction of Organochlorine Pesticides from Water (see Suggested Reading, Chapter 1).

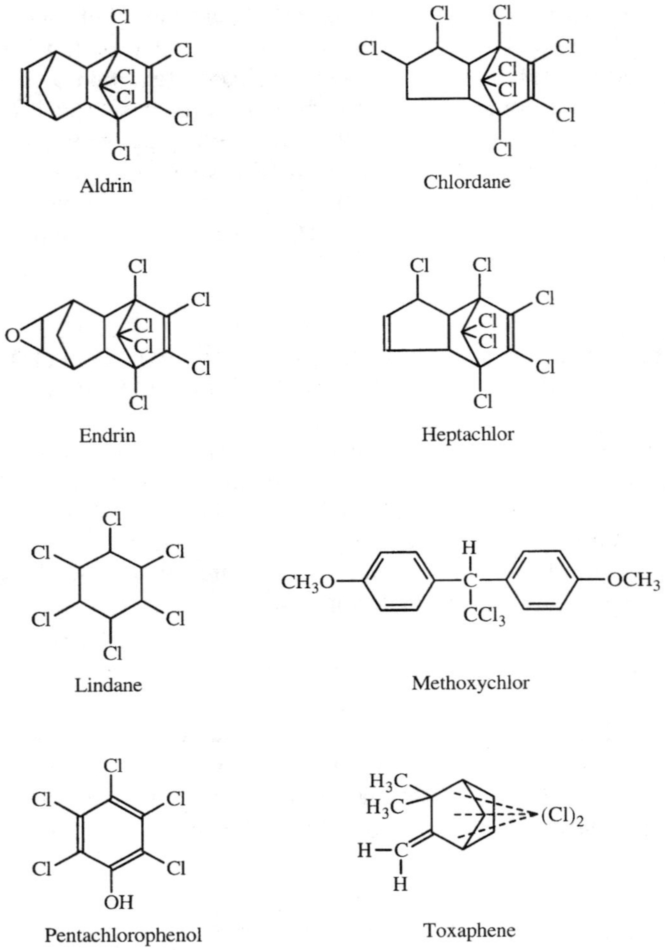

Aldrin

Chlordane

Endrin

Heptachlor

Lindane

Methoxychlor

Pentachlorophenol

Toxaphene

Figure 4.15.   Structures of organochlorine pesticides.

4.6.2.   Extraction of Protein from Dilute Aqueous Solution

Bovine serum albumin (BSA) may be extracted from aqueous solution by reversed-phase sorption onto a wide-pore butyl (C-4) column. The high-molecular-weight protein is able to penetrate into the wide pores of the sorbent where reversed-phase sorption occurs.

SPE Conditions Used

Sample	20-mL sample of BSA solution (1 mg/mL) in 0.025 M potassium phosphate buffer at pH 7.
Solutes	Bovine serum albumin (BSA).
Sorbent	Wide-pore butyl (C-4) sorbent, Bakerbond. Condition the column with 10 mL of methanol, followed by 5 mL 0.5 M potassium phosphate buffer (pH 7), followed by 6 mL of 0.025 M potassium phosphate buffer (pH 7).
Eluent	After addition of sample wash with 4 mL of 0.025 M phosphate buffer (pH 7) and elute with 2 × 0.5 mL isopropanol/water/tri-fluoroacetic acid (60 : 40 : 0.1).
Reference	Bakerbond Application Note Bi-001 Extraction and Concentration of BSA Protein from Dilute aqueous Solution (See Suggested Reading, Chapter 1).

4.6.3. Extraction and Purification of Synthetic Oligonucleotides

Synthetic 5′-DMT oligonucleotide fragments (sDNA) may be isolated from aqueous samples using a reversed-phase mechanism on a phenyl column. Because a 60-Å pore is used for the phenyl column, the fragments should be less than 50 nucleotides in length. After synthesis do not remove the final 5′-DMT. Wash support on a 2-mL medium sintered glass funnel with acetonitrile and air dry. For methyl phosphoramidites add 1 mL thiophenol/triethyl-amine/dioxane (1 : 2 : 2) and incubate at 20 °C for 60 min. Decant and wash 5 times with acetonitrile. For 2-cyanoethyl phosphoramidites, proceed directly to the next step. Remove the fragment from the support by treatment with concentrated aqueous NH_4OH for 60 min at 20 °C.

SPE Conditions Used

Sample	Synthetic oligonucleotides from aqueous solution.
Solutes	Synthetic oligonucleotides.
Sorbent	Phenyl, 3-mL column. Condition the column with 10 mL of methanol, followed by 10 mL methanol/100 mM triethylammonium acetate (TEAA), pH 7.4 (1:1), followed by 10 mL 20 mM TEAA.
Eluent	After addition of sample wash with 10 mL 20 mM TEAA, pH 7.4, followed by 10 mL methanol/100 mM TEAA, pH 7.4 (3:7). Elute with 4 mL methanol/100 mm TEAA, pH 7.4 (3:1).
Reference	Bakerbond Application Note Bi-004 Extraction and Purification

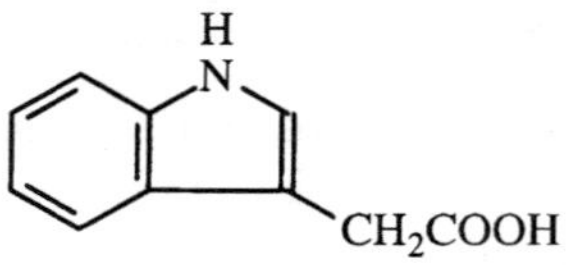

Indole-3-acetic acid

Figure 4.16. Structure of indole-3-acetic acid.

of Synthetic Oligonucleotides and the reference by Horn and Urdea (1988) (see Suggested Reading, Chapter 1).

4.6.4.   Extraction of Indole-3-Acetic Acid

Indole-3-acetic acid (Fig. 4.16) may be extracted from plant tissue after first homogenizing a 0.1-g sample and extracting the plant tissue with 75 mM potassium phosphate dibasic. After extraction, the pH is adjusted to 2.7 with 2.8 M phosphoric acid and the sample is dialyzed with 15 mL 0.1 M potassium phosphate dibasic. The pH adjustment is necessary to protonate the acetic acid functionality for good retention on the C-18 sorbent. Use a trifunctional sorbent that is resistant to acid hydrolysis.

SPE Conditions Used

Sample	Aqueous plant tissue extract.
Solutes	Indole-3-acetic acid.
Sorbent	C-18 sorbent, 6 mL column. Condition the column with 6 mL of methanol, followed by 6 mL of 0.1 M dibasic potassium phosphate (pH 2.7).
Eluent	After addition of sample wash with 2×1 mL methanol/water $(1:4)$ at pH 2.7. Dry column for 3 min. Elute with 2×1 mL methanol/water $(4:1)$.
Reference	Bakerbond Application Note Bi-006 Extraction of Indole-3-Acetic Acid from Plant Tissue and the reference by Liu and Tillberg (1983) (see Suggested Reading, Chapter 1).

4.6.5.   Extraction of Catecholamines from Urine

Catecholamines are extracted from urine using reversed-phase sorbents and are determined by HPLC.

SPE Conditions Used

Sample	Urine sample, pH 8.5 with 2 M ammonium hydroxide.
Solutes	Catecholamines.
Sorbent	C-18 sorbent, 1mL column (100 mg). Condition the column with 2 × 1 mL of methanol, followed by 2 × 1 mL of ammonium chloride/0.5% EDTA, pH 8.5.
Eluent	After addition of sample wash with 2 × 1 mL 0.2 M ammonium chloride, pH 8.5, followed by 1 mL ammonium chloride/methanol (80 : 20), pH 8.5. Air dry for 2 min and elute with 2 × 1 mL 0.08 M acetic acid.
Reference	Bakerbond Application Note Bi-007 Extraction of Catecholamines from Urine (see Suggested Reading, Chapter 1).

Many more examples of reversed-phase are given in Chapter 7.

SUGGESTED READING

Aga, D. S., Thurman, E. M., and Pomes, M. L. 1994. Determination of alachlor and its sulfonic acid metabolite in water by solid-phase extraction and enzyme linked immunosorbent assay, *Anal. Chem.*, **66**: 1495–1499.

Bouvier, E. S. P. 1994. Solid-phase extraction: A chromatographic perspective, *Waters Column*, **V** (1): 1–5.

Buszewski, B. 1991. Properties and biomedical applications of packings with high-density coverage of C18 chemically bonded phase for high-performance liquid chromatography and solid-phase extraction, *J. Chromatog.*, **538**: 293–301.

Crescenzi, C., Dicorcia, A., Passariello, G., Samperi, R., and Carou, M. I. T. 1996, Evaluation of 2 new examples of graphitized carbon-blacks for use in solid-phase extraction cartridges, *J. Chromatog.*, **733**: 41–55.

Gelencser, A., Kiss, G., Krivacsy, Z., Varga-Puchony, Z., and Hlavay, J. 1995. A simple method for the determination of capacity factor on solid-phase extraction cartridges, *J. Chromatog.*, **693**: 217–225.

Guenu, S. and Hennion, M. C. 1996. Evaluation of new polymeric sorbents with high specific surface-areas using an online solid-phase extraction liquid-chromatographic system for the trace-level determination of polar pesticides, *J. Chromatog.*, **737**: 15–24.

Gustafson, R. L., Albright, R. L., Jeisler, J., Lirio, J. A., and Reid, Jr., O. T. 1968. Adsorption of organic species by high surface area styrene-divinylbenzene copolymers, *I & EC Prod. Res. Dev.*, **7**: 107–115.

Junk, G. A., Avery, M. J., and Richard, J. J. 1988. Interferences in solid-phase extraction using C-18 bonded porous silica cartridges, *Anal. Chem.*, **60**: 1347–1350.

McDonald, P. D. and Bouvier, E. S. P. 1995. *Solid Phase Extraction Applications*

Guide and Bibliography, A Resource for Sample Preparation Methods Development, 6th ed., Waters, Milford, MA.

Pichon, V., Coumes, C. C. D., Chen, L., Guenu, S., and Hennion, M. C. 1996. Simple removal of humic and fulvic-acid interferences using polymeric sorbents for the simultaneous solid-phase extraction of polar acidic, neutral, and basic pesticides, *J. Chromatog.*, **737**: 25–33.

Simpson, R. M. 1972. The separation of organic chemicals from water, Third Symposium of the Institute of Advance Sanitation Research, International (Rohm and Haas).

Simpson, N. and Van Horne, K. C., Eds. 1993. *Varian Sorbent Extraction Technology Handbook*. Varian Sample Preparation Products, Harbor City, CA.

Thurman, E. M., Malcolm, R. L., and Aiken, G. R. 1978. Prediction of capacity factors for aqueous organic solutes adsorbed on a porous acrylic resin, *Anal. Chem.*, **50**: 775–779.

Zief, M and Kiser, R. 1988. *Sorbent Extraction for Sample Preparation*. J. T. Baker, Phillipsburg, NJ.

REFERENCES

Bouvier, E. S. P. 1994. Solid-phase extraction: A chromatographic perspective, *Waters Column*, **V** (1): 1–5.

Braumann, T. 1986. Determination of hydrophobic parameters by reversed-phase liquid chromatography, theory, experimental techniques, and application in studies on quantitative structure activity relationships, *J. Chromatog.*, **373**: 191–225.

Dorsey, J. G. and Cooper, W. T. 1994. Retention mechanisms of bonded-phase liquid chromatography, *Anal. Chem.*, **66**: 857A–867A.

Glueckaut, E. 1955. Ion Exchange and its Applications, Society of Chemical Industry, London, 34p.

Hennion, M. C. and Pichon, V. 1994. Solid-phase extraction of polar organic pollutants from water, *Environ. Sci. Tech.*, **28**: 576A–583A.

Horn, T. and Urdea, M. S. 1988. Solid supported hydrolysis of apurinic sites in synthetic oligonucleotides for rapid and efficient purification on reverse-phase cartridges, *Nucleic Acids Res.*, **16**: 11559–11571.

Khym, J. X. 1974. Analytical Ion-Exchange Procedures in Chemistry and Biology, Prentice Hall, Englewood Cliffs, NJ, 257p.

Liu, S. and Tillberg, E. 1983. 3-Phase extraction and partitioning with the aid of dialysis—A new method for the purification of indole-3-acetic and abscisic acids in plant materials, *Physiol. Plants*, **57**: 441–447.

Martin, A. J. P. and Synge, R. L. M. 1941. A new form of chromatogram employing two liquid phases, *J. Biochem.*, **35**: 1358–1368.

Werkhoven-Goewie, C. E., Brinkman, U.A., and Frei, R.W. 1981. Trace enrichment of polar compounds on chemically bonded and carbonaceous sorbents and application to chlorophenols, *Anal. Chem.*, **53**: 2072–2080.

CHAPTER

5

NORMAL-PHASE SOLID-PHASE EXTRACTION

5.1. INTRODUCTION

Normal-phase solid-phase extraction (SPE) refers to the mechanism of sorption where a polar surface sorbs an analyte from a nonpolar solvent based on polar interactions. It is called "normal" because this was the classical type of separation (prior to 1960) in liquid chromatography. The mechanism of isolation is a polar interaction, such as hydrogen bonding, dipole–dipole interactions, π–π interactions, and induced dipole–dipole interactions. The mechanism involves the sorption of the functional groups of the solute to the polar sites of the packing material versus competing against the solubility of the solute in the mobile phase of the column. Functional groups that are capable of hydrogen bonding and polar interactions include hydroxyls, amines, carbonyls, aromatic rings, sulfhydryls, double bonds, and groups containing heteroatoms such as oxygen, nitrogen, sulfur, and phosphorus.

The sorption energy is a low to moderately strong interaction; for example, Figure 5.1 shows the sorption of chlorophenol to the surface of an aminopropyl bonded sorbent. The mechanism of sorption is through hydrogen bonding from the hydroxyl group of the phenol to the amine group of the sorbent. A hydrogen bond is a sharing of a hydrogen atom between two electronegative atoms, typically with one pair of unshared electrons (Fig. 5.1). Commonly, hydroxyl groups and amines are donors of hydrogen to the hydrogen-bonding mechanism, and atoms that accept a hydrogen atom include nitrogen, sulfur, and oxygen. These polar interactions between analyte and sorbent are facilitated by nonpolar solvents because the solvent does not contain functional groups that hydrogen bond with the sorbent. If an oxygen-bearing solvent is used, it is capable of disrupting the hydrogen bonding that is occurring between the analyte and the surface of the sorbent, which prevents sorption. Polar interactions will also be disrupted by samples of high ionic strength or by the presence of other polar analytes or salts.

For these reasons, normal-phase SPE does not use polar solvents, such as water or alcohols, but works extremely well with nonpolar solvents such as hydrocarbons (hexane), chlorinated solvents (methylene chloride), or ether solvents (petroleum ether). Thus, normal-phase SPE is a popular method for applications that require the removal of organic analytes from nonpolar

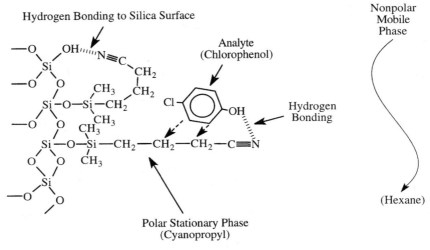

Figure 5.1. Hydrogen bonding normal-phase mechanism in SPE using a cyanopropyl sorbent.

organic solvents. Examples include the sorption of organochlorine pesticides from soil and sediment extracts, pesticides from fatty foods and tissues, fractionation of hydrocarbons from crude oil, and extraction of polychlorinated biphenyls (PCBs) from transformer oil. Other applications will be given in the last section of this chapter, as well.

5.2. STRUCTURE AND SORPTION OF NORMAL-PHASE SORBENTS

The types of nonbonded phases used for normal-phase SPE are typically silica, alumina, and magnesium silicate (called Florisil). The most popular phase is silica. Specific adsorption to silica is caused by hydrogen-bonding interactions with the hydroxyl groups of the silica and the solute, which is shown in Figure 5.2. The silica contains four types of bonds. There are siloxane bonds, which are the hydrophobic bonds between silicon and oxygen atoms. Then there are three types of silanols, "free silanols," vicinal silanols, and geminal silanols (Fig. 5.2). There has been controversy in the literature (Dorsey and Cooper, 1994) concerning the properties of these silanol groups and which play the most important role in attachment of the bonded phase and in secondary interactions. The result is that for normal-phase SPE many manufacturers will rehydroxylate the silica surface to try to obtain the most homogeneous surface possible. The silica surface has an acidic pH and is a

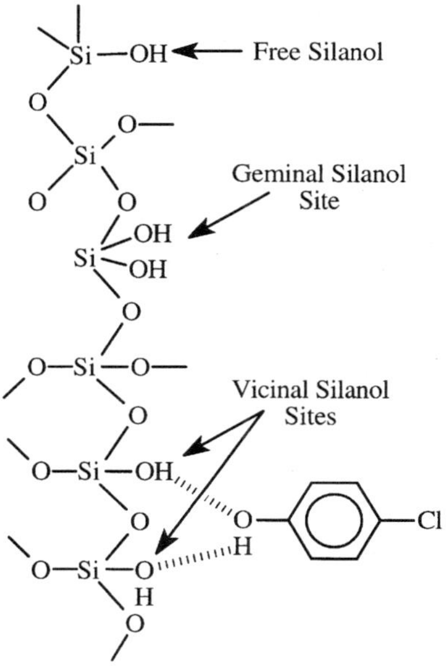

Figure 5.2. Specific sorption sites present in silica are vicinal, free silanols, and geminal hydroxyl groups, which can hydrogen bond to analytes.

weak ion exchanger, which is capable of donating a proton with a pK_a of approximately 4 to 5, and estimates are that the silica surface contains approximately 8 ± 1 μmol/m^2 of silanols.

Alumina may also be used as a sorbent in normal-phase SPE. It may be acidic, neutral, or basic depending on the pH of the wash solution used on the alumina. It too may bind solutes by hydrogen bonding, weak ion exchange, and other polar reactions. The type of alumina generally used will depend on the analytes, with basic solutes sorbed to acidic alumina and acidic analytes sorbed to basic alumina. The magnesium silicate, Florisil, may also be used in a similar fashion to silica.

Table 5.1 shows the various grades of these materials that are available for normal-phase SPE. The sorbents come in a variety of diameters from 25 to 200 μm, with sizes typically between 60 and 100 μm. Because of the long history of silica use, it has the most applications in the literature for normal-phase isolation and separation.

Although there have been many important uses for silica in normal-phase SPE, there is an important shift to newer silica-based bonded-phase sorbents for methods development. The reason for this shift is because of the occurrence of strong, often irreversible sorption sites on silica gel caused by free

Table 5.1. Common Normal-Phase Sorbents Available for SPE

Sorbent	Structure	Typical Loading
	Nonbonded (Adsorption) Phases	
Silica gel (J. T. Baker)	—SiOH, 60 Å, 40 µm, irregular	
Silica gel (Varian)	—SiOH, 60 Å, 40 µm, irregular silica	
Silica gel (Waters)	—SiOH, 125 Å, 55–105 µm, irregular silica, high activity	
Kieselguhr (diatomaceous earth) (J. T. Baker)	—SiOH, 40 µm, irregular	
Florisil (J. T. Baker)	$MgSiO_3$, 73–140 µm, irregular	
Florisil (Varian)	$MgSiO_3$, no information	
Florisil (Waters)	$MgSiO_3$, 60 Å, 50–200 µm, pH = 8.5	
Alumina (neutral) (J. T. Baker)	Al_2O_3, 50–200 µm, irregular	
Alumina (neutral) (Varian)	Al_2O_3, 25 µm, irregular, pH = 7.5	
Alumina (acid) (Varian)	Al_2O_3, 25 µm, irregular, pH = 4.5	
Alumina (base) (Varian)	Al_2O_3, 25 µm, irregular, pH = 10.0	
Alumina (neutral) (Waters)	Al_2O_3, 120 Å, pH = 7–8, 50–300 µm	
Alumina (acidic) (Waters)	Al_2O_3, 120 Å, pH = 4–5, 50–300 µm	
Alumina (basic) (Waters)	Al_2O_3, 120 Å, pH = 9–10, 50–300 µm	
	Bonded Phases	
Cyano (CN) (J. T. Baker)	—$(CH_2)_3CN$, 60 Å, endcapped, 40 µm, irregular silica	10.5%C, 2.4%N
Cyano (CN)-E (Varian)	—$(CH_2)_3CN$, 60 Å, endcapped, 40 µm, irregular silica	8.5%C
Cyano (CN)-U (Varian)	—$(CH_2)_3CN$, 60 Å, not endcapped, 40 µm irregular silica	8.0%C
Cyano (CN) (Waters)	—$(CH_2)_3CN$, 125 Å, endcapped, 55–105 µm, irregular silica, difunctional	6.5%C
Cyano CN(EC) (IST)	—$(CH_2)_3CN$, endcapped, 70 µm, trifunctional, 55-Å pores, irregular silica	
Cyano (CN) (IST)	—$(CH_2)_3CN$, not endcapped, 70 µm, trifunctional, 55-Å pores, irregular silica	
Propylamino (NH_2) (J. T. Baker)	—$(CH_2)_3NH_2$, 60 Å, endcapped, 40 µm, irregular silica	6.4%C, 2.2%N
Propylamino NH_2 (Varian)	—$(CH_2)_3NH_2$, 60 Å, not endcapped, 40 µm, irregular silica, pK_a = 9.8	5.5%C
Propylamino NH_2 (Waters)	—$(CH_2)_3NH_2$, 125 Å, not endcapped, 55–105 µm, irregular silica	3.5%C

(contd.)

Table 5.1. (*continued*)

Sorbent	Structure	Typical Loading
Propylamino NH₂ (IST)	—(CH₂)₃NH₂, not endcapped, 70 μm, trifunctional, 55-Å, pores, irregular silica	
Diol (COHCOH) (J. T. Baker)	—(CH₂)₃OCH₂CH(OH)CH₂(OH), 60 Å, endcapped, 40 μm, irregular silica	8.6%C
Diol (COHCOH) (Varian)	—(CH₂)₃OCH₂CH(OH)CH₂(OH), 60 Å, not endcapped, 40 μm, irregular silica	6.5%C
Diol (COHCOH) (Waters)	—(CH₂)₃OCH₂CH(OH)CH₂(OH), 300 Å, not endcapped, 37–55 μm, irregular silica	2%C
Diol (COHCOH) (IST)	—(CH₂)₃OCH₂CH(OH)CH₂(OH), not endcapped, 70 μm, trifunctional, 55-Å pores, irregular silica	
	Disk Sorbents	
Silica-Si (Toxi-Lab)	No information	
Primary and secondary amino (Toxi-Lab)	No information	
Aminopropyl (Toxi-Lab)	No information	
Cyanopropyl (Toxi-Lab)	No information	

silanol groups, such as the vicinal and geminal sites. It is sometimes more reproducible in normal-phase SPE to use the bonded sorbents, which have all of the same type of bonded sites and have been endcapped, in some cases, to protect the analytes from the surface hydroxyls of the silica. The result is that although the interactions are not as strong, they are more easily and subtly controlled by manipulation of hydrogen bonding and the solubility of analytes into the bonded phase, which is a second advantage of the bonded-phase sorbents in normal-phase SPE. Thus, methods using bonded-phase SPE may be easily reproduced in the laboratory without concern about the differences in silanols that occur in silica gel.

There are three general types of bonded normal-phase sorbents: amino-propyl, cyanopropyl, and diol (Fig. 5.3). All three are derivatized as a propyl hydrocarbon, which places them near the surface of the silica. Only the cyanopropyl sorbent is available in either endcapped or nonendcapped forms (Table 5.1). Thus, analytes have the opportunity to interact not only with the bonded phase but also with the underlying silica sorbent. Furthermore, because the hydrocarbon chain is short, only three carbons in length, the

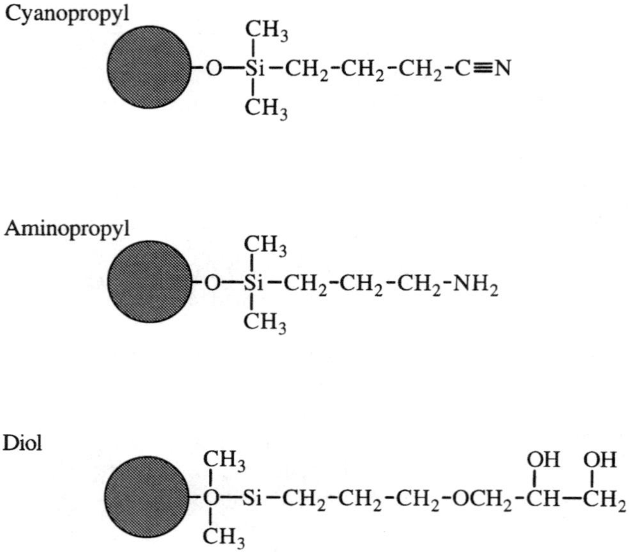

Cyanopropyl

$$O-Si-CH_2-CH_2-CH_2-C\equiv N$$

with CH_3 groups above and below the Si

Aminopropyl

$$O-Si-CH_2-CH_2-CH_2-NH_2$$

with CH_3 groups above and below the Si

Diol

$$O-Si-CH_2-CH_2-CH_2-OCH_2-CH-CH_2$$

with CH_3 groups above and below the Si, and OH OH on the terminal carbons

Figure 5.3. Three types of bonded normal-phase sorbents for SPE.

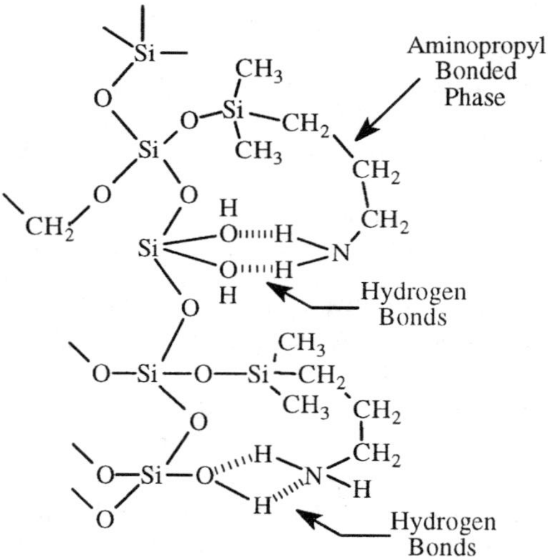

Figure 5.4. Hydrogen bonding of the active functional group of normal-phase bonded sorbents with the underlying silica surface.

bonded phase may interact with the underlying silica through hydrogen bonding (Fig. 5.4). The proximity of the functional group to the surface of the silica gives rise to this hydrogen-bonding interaction (Dorsey and Cooper, 1994).

The use of endcapping on the cyanopropyl bonded phase is an important feature and reduces the binding of analytes through secondary interactions with the silica gel. Thus, in methods development, the cyanopropyl sorbent will usually be a weaker hydrogen-bonding sorbent, and this may be used to one's advantage in the isolation and separation of strongly polar analytes that could sorb irreversibly with the silanol sites of the silica gel. Furthermore, there is one manufacturer, Toxi-Lab, that is now producing disk sorbents for normal-phase SPE. These sorbents may prove quite useful in applications of solid materials. Because many lipophilic substances must be extracted from solid phases with organic solvents, there is the difficulty associated with solvent handling and particulates. The disks can function to both filter the organic liquid and to sorb the analytes present. This procedure will simplify sample handling and could be quite useful for clean up of environmental samples, food, and natural products.

5.3. ELUTION OF NORMAL-PHASE SORBENTS

The elution of the solute from a normal-phase sorbent is typically a function of the eluotropic strength of the eluting solvent. Table 5.2 shows the solvent eluotropic strength and polarity for a range of organic solvents typically used in normal-phase SPE. Silica will adsorb moderately polar compounds dissolved in organic solvents with E° (eluotropic strength) values less than 0.38 (Table 5.2). Examples include compounds such as aldehydes, alcohols, and organic halides, which could be dissolved in hexane. These compounds would be strongly sorbed by silica. These compounds would be eluted with solvents of greater eluotropic strength, such as methanol, or other solvents with eluotropic strengths greater than 0.6. Very polar compounds such as carbohydrates or amino compounds are tightly bound to normal-phase sorbents, such as silica and alumina, and even high eluotropic strength solvents cannot break the interactions between these analytes and the sorbent. However, the use of cyanopropyl or aminopropyl phases often permit the recovery of these compounds when silica interacts too strongly.

In general, basic compounds are retained more strongly on mildly acidic surfaces, such as silica or acidic alumina. Acidic compounds are retained on basic surfaces, such as basic alumina. Because both silica and alumina are hydroscopic, they adsorb water to their surface. This water greatly reduces the retention of organic solutes because it sorbs to active sorption sites and

Table 5.2. Solvent Eluotropic Strength, $E^{\phi*}$, and Polarity, p'

Solvent	$E^{\phi*}$	p'
Acetic acid, glacial	> 0.73	6.2
Water	> 0.73	10.2
Methanol	0.73	6.6
2-Propanol	0.63	4.3
Pyridine	0.55	5.3
Acetonitrile	0.50	6.2
Ethyl acetate	0.45	4.3
Acetone	0.43	5.4
Methylene chloride	0.32	3.4
Chloroform	0.31	4.4
Toluene	0.22	2.4
Cyclohexane	0.03	0.0
n-Hexane	0.00	0.06

Alfer Zief and Kiser (1987), published with permission of J. T. Baker, Inc.

competes with the analyte. Thus, it is important to keep the SPE sorbents dry and free from water. Normal-phase sorbents commonly are stored in a dessicator prior to use.

In summary, retention of analytes by normal-phase SPE is facilitated by the dissolution of the sample in nonpolar solvents that do not compete for polar sorption sites on the solid phase. Elution of the sample from the SPE sorbent is facilitated with polar solvents that can disrupt the hydrogen bonding between functional groups on the analyte and the sorbent surface. Examples of the use of normal-phase SPE include isolation of pesticides from food, soil, and other solid materials that are high in fat or organic matter and contain complex matrices.

5.4. EXAMPLES OF METHODS DEVELOPMENT WITH NORMAL-PHASE SOLID-PHASE EXTRACTION

Normal-phase SPE methods development is a straightforward process, and for this reason, many methods have been developed on silica sorbents, especially with the broad and comprehensive literature available in thin layer chromatography (TLC). The development of the silica bonded phases has introduced a new and important aspect to methods development with normal-phase SPE.

5.4.1. Extraction of Aromatic Hydrocarbons from Crude Oil

The extraction of aromatic hydrocarbons (Fig. 5.5) from crude oil uses two sorbents in series, first a cyanopropyl column attached to a second silica column. In this procedure, aromatic hydrocarbons are sorbed on both the cyanopropyl sorbent and on the silica sorbent. The heteroatom hydrocarbons (containing nitrogen, oxygen, and sulfur) are trapped on the cyanopropyl sorbent and eluted as a separate fraction from the silica column. Because the major interaction of aromatic heterocyclic hydrocarbons is through hydrogen bonding to the surface of the sorbent, the cyanopropyl sorbent is easier to elute than a silica sorbent alone. For this reason, the cyanopropyl column is used before the silica column. The separation of hydrocarbons from crude oil is an example of normal-phase chromatography that has been performed for many years on silica gel prior to the introduction of SPE.

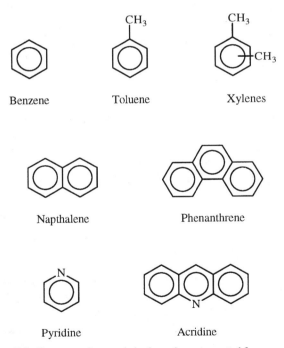

Figure 5.5. Structure of aromatic hydrocarbons separated from crude oil.

SPE Conditions Used

Sample	25 to 35 mg of crude oil.
Solutes	Aromatic hydrocarbons, and aromatics containing N, O, and S.
Sorbent	Silica (SiOH) 1 g, cyanopropyl (CN) 1 g, and filter for crude oil. Condition both columns with 6 mL of petroleum ether. Connect the cyanopropyl column to the top of the silica column with an adapter and attach filtration column above the cyanopropyl column.
Eluent	Dissolve the sample in 6 mL of petroleum ether and then gently force the oil through the filter and the two columns with pressure. Wash the columns with 2×0.5 mL of petroleum ether. The insoluble fraction of the oil is trapped on the filter and weighed. Wash the two columns with 2×1 mL of petroleum ether. Disconnect the two columns and elute both with 2×1 mL benzene/petroleum ether $(1:3)$, collecting both fractions and combining. Evaporate for the weight of aromatic hydrocarbons. Elute the cyanopropyl column with 2×1 mL benzene/methanol $(1:1)$ into a separate flask. Evaporate the eluate of the cyanopropyl column and weigh for the fraction of heterocyclic aromatic hydrocarbons.
Reference	Bakerbond Application Note EN-004 Extraction of Hydrocarbons from Crude Oil (See Suggested Reading, Chapter 1).

5.4.2. Extraction of Polychlorinated Biphenyls from Transformer Oil

The extraction of PCBs (Fig. 5.6) from transformer oil is an important analysis that is completed before the solid-waste disposal of transformers. The separation is accomplished on Florisil. It involves the sorption of the PCBs directly onto the Florisil sorbent and subsequent elution with hexane.

SPE Conditions Used

Sample	200 mg of transformer oil.
Solutes	PCB mixture.
Sorbent	Florisil sorbent, 1 g. Condition with 2 mL of hexane.
Eluent	Pore the transformer oil directly into the column. No wash steps are needed. Elute with 25 mL of hexane and evaporate for gas chromatography/mass spectrometry (GC/MS) analysis.
Reference	Bakerbond Application Note EN-014 Extraction of PCBs from Transformer Oil (see Suggested Reading, Chapter 1).

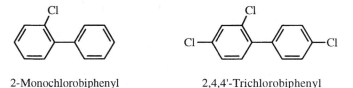

2-Monochlorobiphenyl   2,4,4'-Trichlorobiphenyl

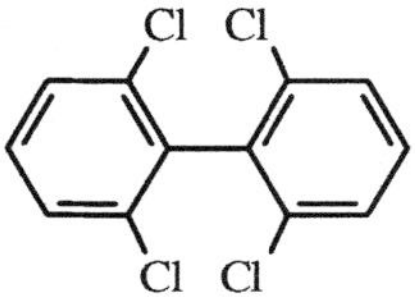

2,2',5,5'-Tetrachlorobiphenyl 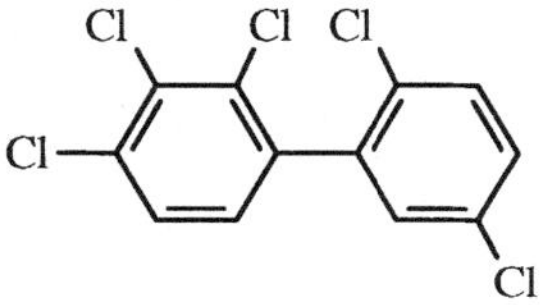   2,2',3,4,5'-Pentachlorobiphenyl

Figure 5.6. Structure of PCBs separated from transformer oil.

5.4.3. Extraction of Carotenes from Plant Tissue

The extraction of carotenes (Fig. 5.7) from plant tissue is carried out on kiesel-guhr, which is also called diatomaceous earth (a natural silica). The procedure consists of homogenizing the plant material in a petroleum ether and acetone mixture. This extraction dissolves the carotenes in the plant. Next the plant extract is filtered and applied to a column of kieselguhr. The column is washed with petroleum ether and also eluted with petroleum ether. In this procedure, the kieselguhr is used to purify the extract and to remove other plant pigments. The carotenes pass through the column and are analyzed by high-pressure liquid chromatography (HPLC).

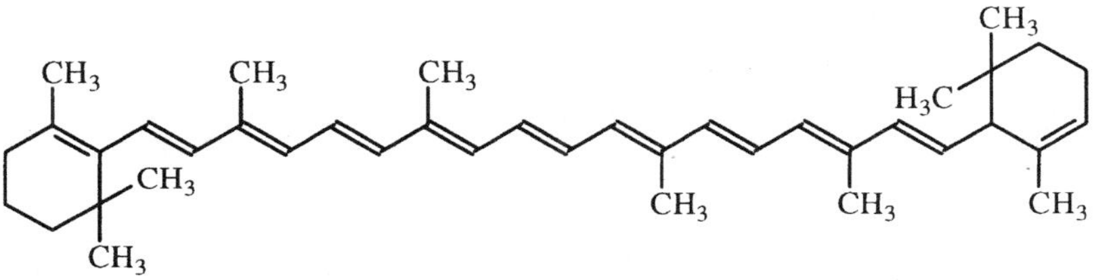

Figure 5.7. Structure of β-carotene extracted from plant material.

SPE Conditions Used

Sample	10 g of plant sample is extracted with 100 mL of petroleum ether/acetone (95 : 5) in blender. Then solids are centrifuged.
Solutes	Carotene.
Sorbent	Kieselguhr sorbent, 1 g. Condition column with 6 mL of petroleum ether, leave 2 mL of petroleum ether in column.
Eluent	Pore 3 mL of the plant extract on the column and collect any eluate. Wash and elute the column with 2 × 1 mL of petroleum ether and collect the wash/eluate fraction and combine with the first eluent. In this procedure the wash and eluate are the same fraction.
Reference	Bakerbond Application Note FF-003 Extraction of Carotenes from Plant Tissue (See Suggested Reading, Chapter 1).

5.4.4. Extraction of the *N*-Nitrosamine (*N*-Nitrosopyrrolidine) from Bacon

The extraction of *N*-nitrosopyrrolidine (Fig. 5.8) from bacon involves the use of a cyanopropyl column, and the mechanism is hydrogen bonding between the amino nitrogen of the *N*-nitrosopyrrolidine and the underlying silica surface, as well as an interaction between the propyl carbons and the analyte. The sample is applied as a solution of hexane and methylene chloride, which solubilizes the bacon and does not interfere with the separation of the analytes by hydrogen bonding. The eluent contains methanol, which breaks up the bonding and effectively solubilizes the *N*-nitrosopyrrolidine.

SPE Conditions Used

Sample	Dissolve 1.0 g of bacon grease in 5 mL of hexane/methylene chloride (9 : 1).
Solutes	*N*-nitrosopyrrolidine.
Sorbent	Cyanopropyl 0.5 g. Condition column with 3 mL hexane/methylene chloride (9 : 1). Wash column with 2 × 2 mL hexane/methylene chloride (9 : 1).

Figure 5.8.   Structure of *N*-nitrosopyrrolidine extracted from bacon.

Eluent   Elute column with 2×0.5 mL methanol/methylene chloride (95:5).

Reference   Bakerbond Application Note FF-009 Extraction of *N*-Nitrosopyrrolidine from Bacon (See Suggested Reading, Chapter 1).

5.4.5. Extraction of Aflatoxins from Corn, Peanuts, and Peanut Butter

The extraction of aflatoxins (Fig. 5.9) from corn, peanuts, and peanut butter involves the dissolution of the aflatoxins in an aqueous methanol extract, followed by extraction with chloroform and clean-up on a silica-gel column. The extraction is carried out with 50 g of food material and 200 mL of methanol/water (85:15) for 30 min. Filter 40 mL of the extract and collect the filtrate. Add 40 mL of 10% sodium chloride to salt out the aflatoxin and extract with 2×25 mL of chloroform. Evaporate to dryness and redissolve with 3 mL of methylene chloride. Sample is ready for addition to the silica-gel column that contains 0.5 g of packing material. The mechanism of sorption involves hydrogen bonding between the oxygen atoms of the aflatoxin and the silica gel. Sorption is from methylene chloride and elution is carried out with chloroform/acetone (9:1).

SPE Conditions Used

Sample   Dissolve 50 g of corn, peanuts, or peanut butter in an aqueous methanol extractant. Salt out the aflatoxins and extract into chloroform. Evaporate to dryness and reconstitute in 3 mL of methylene chloride.

Solutes   Aflatoxins.

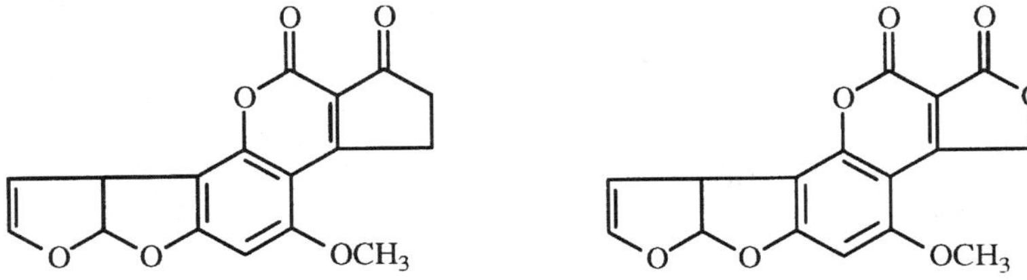

Aflatoxin B₁   Aflatoxin G

Figure 5.9. Structure of aflatoxins extracted from corn and peanuts.

Sorbent   Silica gel 0.5 g. Condition with 3 mL hexane, followed by 3 mL
             of methylene chloride. Sample is added from 3 mL of methy-
             lene chloride. Wash column with 3 mL hexane, 3 mL diethyl
             ether, and 3 mL of methylene chloride.
Eluent   Elute column with 6 mL of chloroform/acetone (9 : 1).
Reference   Bakerbond Application Note FF-013 Extraction of Aflatoxins
             from Corn, Peanuts, and Peanut Butter (See Suggested Reading,
             Chapter 1).

5.4.6. Extraction of Pyridonecarboxylic-Acid Antibacterials from Fish Tissue

The extraction of pyridonecarboxylic-acid antibacterials (PCAs) from fish tissue involves the isolation of oxolinic acid, nalidixic acid, and piromidic acid
(Fig. 5.10). These acids may be isolated from a hexane/ethyl acetate extract
using an aminopropyl column. The interaction involves a weak acid and weak
base interaction between the carboxylic acid groups of the antibacterials and
the amine group of the sorbent, as well as sorption of the aromatic ring and
hydrogen bonding to the silica surface. However, because of the polarity of the
extract of the fish tissue, there is only a weak hydrogen-bonding interaction

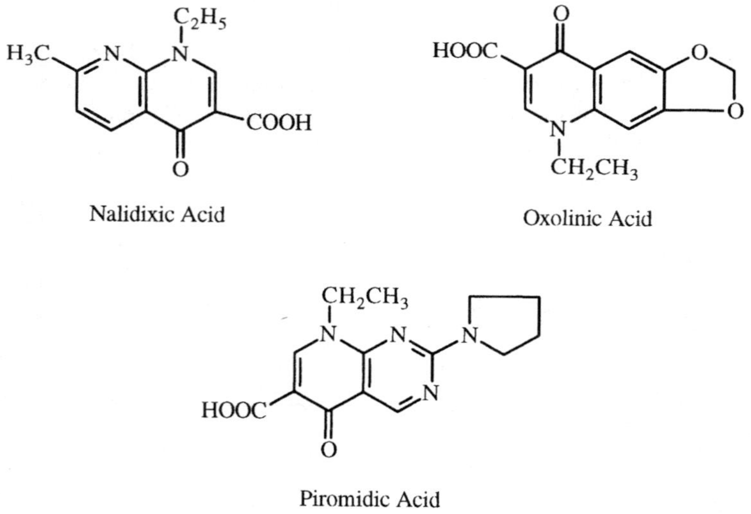

Nalidixic Acid

Oxolinic Acid

Piromidic Acid

Figure 5.10. Structure of oxolinic acid, nalidixic acid, and piromidic acid extracted from fish
tissue.

because of the disruption of this interaction by the solvent. There may be an anion-exchange mechanism involved in this procedure because of the effectiveness of the elution solvent, which uses oxalic acid as a "pusher" (competing ion in the ion-exchange sorbent).

SPE Conditions Used

Sample	Blend 5 g of sample with an extraction solvent of hexane/ethyl acetate (1 : 3) and 10 g of sodium sulfate. High speed blend and decant. Repeat and combine extracts.
Solutes	Pyridonecarboxylic-acid antibacterials.
Sorbent	Aminopropyl 0.5 g. Condition with 10 mL methanol, followed by 3 mL of extraction solvent. Add 2 to 3 mL of extraction solvent to the column, then add entire sample to the column. Wash with 5 mL of extraction solvent.
Eluent	Elute column with 10 mL of acetonitrile/methanol/0.01 M aqueous oxalic acid (pH = 3 with NaOH).
Reference	Bakerbond Application Note FF-017 Extraction of Pyridonecarboxylic Acid Antibacterials (PCAs) from Fish Tissue (See Suggested Reading, Chapter 1).

5.4.7. Extraction of Vitamin D from Serum

The extraction of vitamin D (Fig. 5.11) from serum involves sorption of the vitamin D onto silica from an organic-solvent extraction of serum using methylene chloride. Serum is liquid extracted first, followed by sorption of the

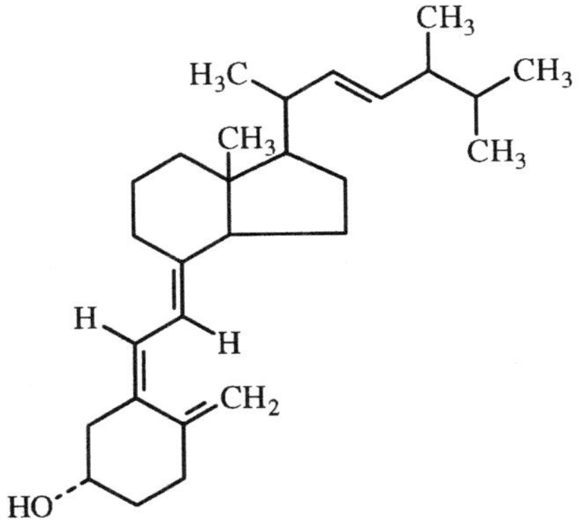

Figure 5.11. Structure of vitamin D$_2$ extracted from serum.

vitamin D from the methylene-chloride extract. The method involves the use of hydrogen bonding to sorb the vitamin D from a chlorinated nonpolar solvent.

SPE Conditions Used

Sample	Serum, 2 mL extracted with 7.5 mL of methylene chloride/methanol (33 : 67). Add 2.5 mL of methylene chloride and shake. Allow phases to separate and collect the lower methylene-chloride layer.
Solutes	Vitamin D.
Sorbent	Silica gel 0.5 g. Condition with 3 mL anhydrous ether/hexane (1 : 9).
Eluent	Add sample to column and wash with 10 mL of anhydrous ether/hexane (1 : 9). Elute with 7.5 mL anhydrous ether/hexane (33 : 67).
Reference	Bakerbond Application Note PH-026 Extraction of Vitamin D from Serum (See Suggested Reading, Chapter 1).

5.4.8.   Extraction of Antibiotics from Ointment

The extraction of chlortetracycline-HCl (Fig. 5.12) from ointment involves the use of a diol column and sorption from hexane. The mechanism of interaction involves the hydrogen bonding of the antibiotic to the hydroxyl groups of the diol.

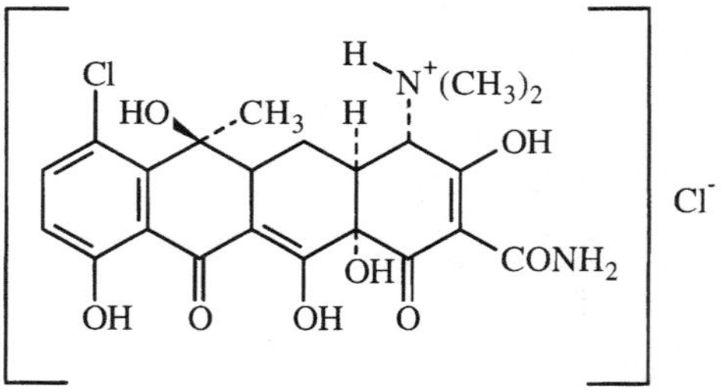

Figure 5.12. Structure of chlortetracycline-HCl extracted from ointment.

SPE Conditions Used

Sample	Ointment, 50 mg is extracted with 2 mL of hexane. The sample forms an insoluble suspension.
Solutes	Chlortetracycline-HCl.
Sorbent	Diol column, 0.5 g is conditioned with 3 mL of hexane. Then add the suspension to the column and wash with 2×1 mL of hexane. Air dry the column.
Eluent	Elute with 2×1 ml methanol/0.1 N HCl (1 : 1).
Reference	Bakerbond Application Note PH-037 Extraction of Chlortetracycline HCl from Ointment (See Suggested Reading, Chapter 1).

SUGGESTED READING

McDonald, P. D. and Bouvier, E. S. P. 1995. *Solid Phase Extraction Applications Guide and Bibliography, A Resource for Sample Preparation Methods Development*, 6th ed., Waters, Milford, MA.

Simpson, N. and Van Horne, K. C. 1993. *Varian Sorbent Extraction Technology Handbook*, Varian Sample Preparation Products, Harbor City, CA.

Zief, M and Kiser, R. 1988. *Sorbent Extraction for Sample Preparation*. J. T. Baker, Phillipsburg, NJ.

REFERENCES

Dorsey, J. G., and Cooper, W. T. 1994. Retention mechanisms of bonded-phase liquid chromatography, *Anal. Chem.*, **66**: 857A–867A.

Zief, M and Kiser, R. 1988. *Sorbent Extraction for Sample Preparation*. J. T. Baker, Phillipsburg, NJ.

CHAPTER

6

ION-EXCHANGE SOLID-PHASE EXTRACTION

6.1. INTRODUCTION

Ion exchange is a fundamental form of chromatography that was invented by Adams and Holmes in 1935. They realized that synthetic organic polymers, or resins, were capable of exchanging ions with the solution that contained them. The framework of the early resins consisted of a three-dimensional hydrocarbon network with ionizable groups that impart a hydrophilic nature to the resin. The ionizable groups are either acidic or basic.

Typically, ion-exchange resins consist of hydrophilic ion-exchange groups that are attached to a backbone of copolymerized styrene–divinylbenzene (SDB). The hydrophilic ion-exchange group is capable of exchanging either a cation or anion with free cation or anions in solution. If it exchanges cations, it is a cation-exchange resin; if it exchanges anions, it is an anion-exchange resin. Figure 6.1 shows the general structure of the styrene–divinylbenzene ion-exchange resins and the general process of exchange for both cations and anions. The two types of sites shown in Figure 6.1 are called strong ion-exchange sites because the pK_a of either functional group is such that the group is either a strong acid or a strong base. The strong cation- or anion-exchange site is always charged regardless of the sample pH. This means that the ion-exchange site is permanently fixed, and to displace the cation or anion that has been exchanged, another ion must take its place.

Weak cation- and anion-exchange resins also exist. These resins may have their ion-exchange site removed by adjustment of the pH of the solution. Figure 6.2 shows several types of weak cation- and anion-exchange sites. In the case of the carboxylic acid, it is a weak cation exchanger that prefers a hydrogen ion. Thus, it will loose its ion-exchange capability at pHs that are 2 pH units below the pK_a of the exchange site. In the case of the weak anion-exchange sorbent, the amine is protonated at pHs that are below the pK_a, but will loose its capacity for anion exchange at pHs that are 2 units above the pK_a. Thus, it is possible to displace ions from the weak exchangers by two means: ion exchange and by removal of the charge on the exchange resin by adjustment of pH.

Since the invention of ion exchange, it has found many applications and uses. Some of the applications that are possible with ion-exchange sorbents

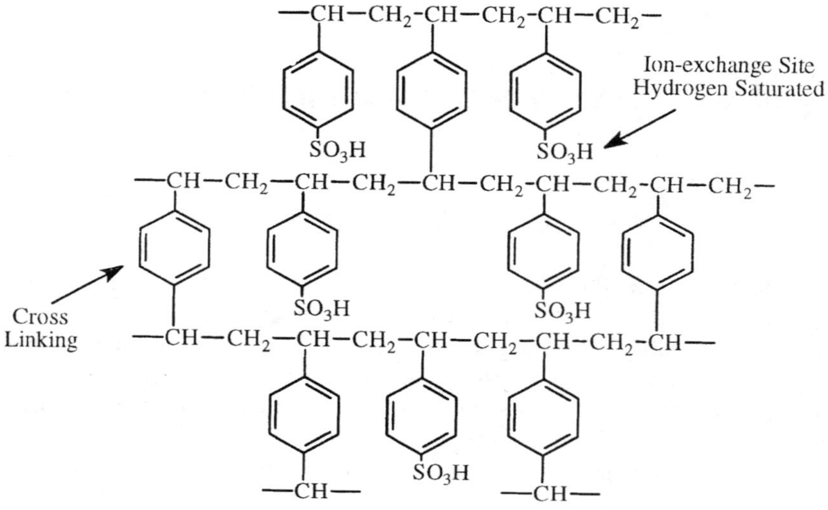

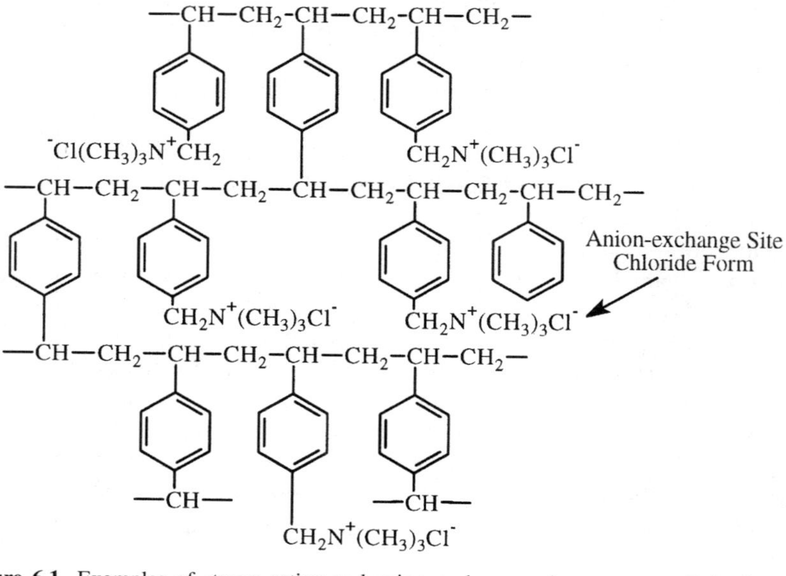

Figure 6.1. Examples of strong cation and anion exchange using a styrene–divinylbenzene resin.

Acrylicbenzene Matrix

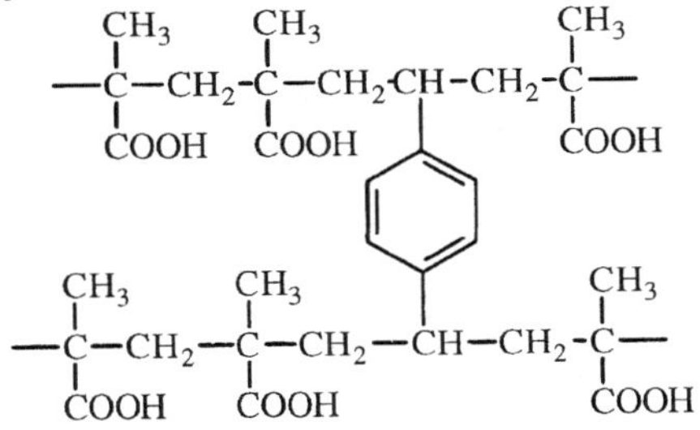

$$—\overset{\overset{\displaystyle CH_3}{|}}{C}—CH_2—\overset{\overset{\displaystyle CH_3}{|}}{\underset{\underset{\displaystyle COOH}{|}}{C}}—CH_2—CH—CH_2—\overset{\overset{\displaystyle CH_3}{|}}{\underset{\underset{\displaystyle COOH}{|}}{C}}—$$

$$—\overset{\overset{\displaystyle CH_3}{|}}{\underset{\underset{\displaystyle COOH}{|}}{C}}—CH_2—\overset{\overset{\displaystyle CH_3}{|}}{\underset{\underset{\displaystyle COOH}{|}}{C}}—CH_2—CH—CH_2—\overset{\overset{\displaystyle CH_3}{|}}{\underset{\underset{\displaystyle COOH}{|}}{C}}—$$

Silica Bonded-Phase Matrix

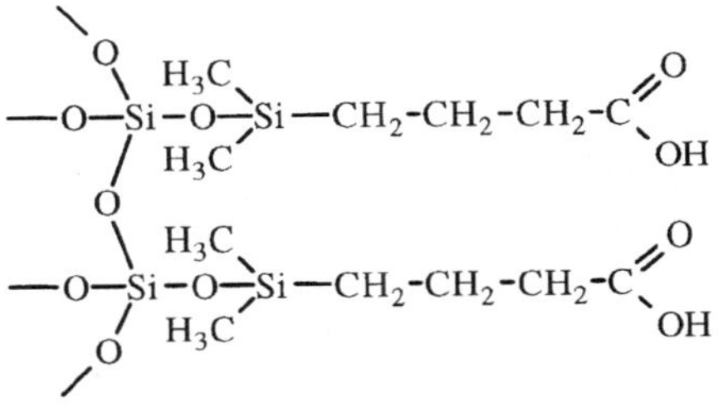

$$—O—\overset{\overset{\displaystyle O}{|}}{\underset{\underset{\displaystyle O}{|}}{Si}}—O—\overset{\overset{\displaystyle H_3C}{}}{\underset{\underset{\displaystyle H_3C}{}}{Si}}—CH_2\text{-}CH_2\text{-}CH_2\text{-}C\overset{\displaystyle O}{\underset{\displaystyle OH}{}}$$

$$—O—\overset{\overset{\displaystyle O}{|}}{\underset{\underset{\displaystyle O}{|}}{Si}}—O—\overset{\overset{\displaystyle H_3C}{}}{\underset{\underset{\displaystyle H_3C}{}}{Si}}—CH_2\text{-}CH_2\text{-}CH_2\text{-}C\overset{\displaystyle O}{\underset{\displaystyle OH}{}}$$

Styrene–Divinylbenzene Matrix

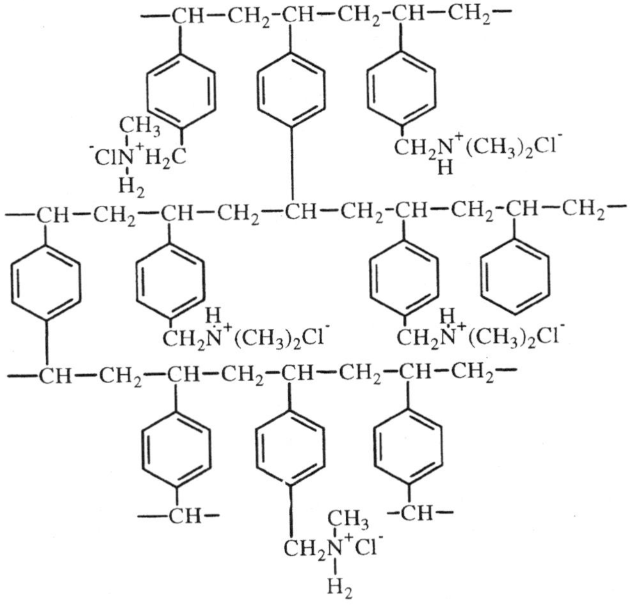

$$—CH—CH_2\text{-}CH—CH_2\text{-}CH—CH_2—$$

$$^-ClN^+H_2C\overset{\overset{\displaystyle CH_3}{|}}{\underset{\underset{\displaystyle H_2}{}}{}} \qquad \overset{}{\underset{\underset{\displaystyle H}{}}{CH_2N^+(CH_3)_2Cl^-}}$$

$$—CH—CH_2—CH—CH_2—CH—CH_2—CH—CH_2—CH—CH_2—$$

$$\underset{\underset{\displaystyle CH_2N^+(CH_3)_2Cl^-}{\overset{\displaystyle H}{}}}{} \qquad \underset{\underset{\displaystyle CH_2N^+(CH_3)_2Cl^-}{\overset{\displaystyle H}{}}}{}$$

$$—CH—CH_2—CH—CH_2—CH—CH_2—CH—CH_2—$$

$$—CH— \qquad \overset{\overset{\displaystyle CH_3}{|}}{\underset{\underset{\displaystyle H_2}{}}{CH_2N^+Cl^-}} \qquad —CH—$$

Figure 6.2. (*contd.*)

Silica Bonded-Phase Matrix

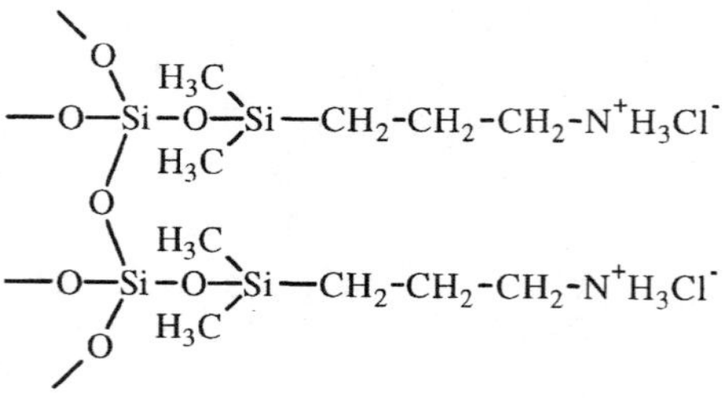

Figure 6.2. Examples of weak cation- and anion-exchange sorbents.

include conversion of salt solutions from one form to another, desalting a solution, trace enrichment, and the removal of ionic impurities or interferences. A variety of both inorganic and organic substances have been used as ion-exchange resins. Natural products, such as proteins, celluloses, common clays, and various mineral phases have been tried. Because of low capacities, these natural substances have had little commercial use.

This chapter deals with a number of facets of solid-phase extraction (SPE) ion exchange. They include the structure of ion-exchange sorbents for SPE, theory of ion exchange (such as ion exchange as a chemical reaction, selectivity, Donnan potential and the significance of these factors for SPE), kinetics of ion exchange and its effect on SPE, general principles of SPE methods development with ion exchange, and examples of the use of ion-exchange SPE.

6.2. STRUCTURE OF SOLID-PHASE EXTRACTION ION-EXCHANGE SORBENTS

6.2.1. Cartridges

The backbone of many of the ion-exchange sorbents used in SPE are based on bonded silica gel rather than on styrene–divinylbenzene. Silica is used because many different functional groups may be derivatized to the silica and a variety of ion-exchange sorbents are possible (Table 6.1).

For example, strong cation-exchange sorbents based on silica consist of either an aromatic sulfonic acid or a propane sulfonic acid that is bonded to the silica matrix. These sorbents typically are available in their hydrogen form and have different exchange capacities. The aromatic exchangers have approximately 1 meq/g, while the propane sulfonic acids have much less, about 0.3 meq/g. The strong cation-exchange sorbents are available in both endcapped

Table 6.1. Common Sorbents Available for Ion-Exchange SPE

Sorbent	Structure	Typical Loading
	Cation Exchange	
Weak cation exchange Carboxylic acid (J. T. Baker)	$-(CH_2)_2COOH$, 40 μm, 60 Å, endcapped	0.4 meq/g
Strong cation exchange Aromatic sulfonic acid (J. T. Baker)	$-(CH_2)_3$-phenyl-$SO_3^-H^+$, 40 μm, 60 Å, endcapped	1.0 meq/g
Weak cation exchange CBA (Varian)	$-CH_2CH_2COOH$, $pK_a = 4.8$, endcapped, 40 μm, 60 Å, 50-, 100-, 500-, 1000-mg sizes	0.35 meq/g, 7%C
Strong cation exchange Propane sulfonic acid PRS (Varian)	$-CH_2CH_2CH_2SO_3^-$ Na^+, $pK_a<1$, not endcapped, 40 μm, 60 Å	0.25 meq/g, 2%C
Strong cation exchange benzene sulfonic acid SCX (Varian)	$-CH_2CH_2CH_2$-phenyl-SO_3^- H^+, not endcapped, 40 μm, 60 Å, $pK_a<1$, available from 50-mg to 10-g columns	0.7 meq/g, 10%C
	Anion Exchange	
Weak anion exchange aminopropyl-(NH_2) (J. T. Baker)	$-(CH_2)_3NH_2$, 40 μm, 60 Å, endcapped	1.6 meq/g
Weak anion exchange primary and secondary amino (NH/NH_2) (J. T. Baker)	$-(CH_2)_3$-$NHCH_2NH_2$, 40 μm, 60 Å, endcapped	2.6 meq/g
Strong anion exchange quaternary amine (N^+) (J. T. Baker)	$-(CH_2)_3N^+(CH_3)_3Cl$, 40 μm, 60 Å, endcapped	0.7 meq/g
Weak anion exchange ethylenediamine-*N*-propyl primary and secondary amino PSA (Varian)	$-(CH2)_3$-$NHCH_2NH_2$, 7.5%C, not endcapped, 40 μm, 60 Å, $pK_a = 10.1, 10.9$	~1 meq/g
Weak anion exchange diethylaminopropyl DEA (Varian)	$-CH_2CH_2CH_2N(CH_2CH_2)_2$, 8.5%C, not endcapped, 40 μm, 60 Å, $pK_a = 10.7$	0.75 meq/g
Strong anion exchange SAX (Varian)	$-CH_2CH_2CH_2N^+(CH_2CH_2)_3Cl$, 6.5%C, not endcapped, 40 μm, 60 Å, available in 50-, 100-, 500-mg and 1-, 2-, 5-, 10-g columns	0.75 meq/g

(contd.)

Table 6.1. (*continued*)

Sorbent	Structure	Typical Loading
Size Exclusion and Ion Exchange		
Accell plus CM weak cation exchange (Waters)	Acrylic acid/acrylamide, copolymer, 5.5%C, on diol silica, not endcapped, 37–55 μm, 300 Å, —COO⁻ Na⁺	0.35 meq/g
Accell plus QMA weak anion exchange (Waters)	Acrylamide copolymer on diol, 6%C, silica, —C(O)NH(CH₂)₃N⁺(CH₃)₃Cl, 300 Å, not endcapped, 37–55 μm	0.22 meq/g
Wide-pore ion exchangers weak cation exchange CBX carboxylic acid (J. T. Baker)	—COO⁻ H⁺, 275 Å, 40 μm, 12.2%C, not endcapped	~ 0.5 me/g
PEI polyethyleneimine weak anion exchange (J. T. Baker)	—(CH₂CH₂NH)ₙ-, 275 Å, 6.3%C, 2.8%N, 40 μm, not endcapped	~ 0.3 meq/g
Disk Ion-Exchangers		
Strong anion exchange quaternary amine (N⁺) SAX *Speedisk* (J. T. Baker)	—(CH₂)₃N⁺(CH₃)₃Cl 40 μm, 60 Å, endcapped, 50 mm diameter	~ 0.35 meq/47-mm disk
Strong anion exchange quaternary amine (N⁺) Empore Disk (3 M)	SDB–(CH₂)₃N⁺(CH₃)₃Cl, styrene–divinylbenzene copolymer	~ 0.2 meq/47-mm disk
Strong cation exchange benzene sulfonic acid SCX (3 M)	SDB–phenyl-SO₃⁻ H⁺ styrene–divinylbenzene copolymer, $pK_a < 1$	0.35 meq/47-mm disk
Chelation disk styrene–divinylbenzene (3 M)	SDB-paired iminodiacetate ions	0.35 meq/47-mm disk
Strong cation exchange benzene sulfonic acid Novo-Clean disk (Alltech)	SDB–phenyl-SO₃⁻ H⁺, Ag⁺, Ba²⁺, styrene–divinylbenzene copolymer, $pK_a < 1$	0.35 meq/47-mm disk
Strong cation exchange benzene sulfonic acid SPEC (Ansys, Inc.)	Phenyl-SO₃⁻ H⁺, matrix chemistry not given, $pK_a < 1$	0.35 meq/47-mm disk
Strong anion exchange styrene–divinylbenzene SPEC (Ansys, Inc.)	Quaternary amine, matrix chemistry not given	~ 0.35 meq/47-mm disk

(*contd.*)

Table 6.1. (*continued*)

Sorbent	Structure	Typical Loading

Weak anion exchange
  Styrene–divinylbenzene
  SPEC
  (Ansys, Inc.)
	Aminopropyl,
	  matrix chemistry
	  not given
		~ 0.2 meq/47-mm disk

Mixed-mode sorbent
  SPEC
  (Ansys, Inc.)
	C-8 and phenyl-SO$_3^-$ H$^+$,
	  matrix chemistry
	  not given	

and nonendcapped silicas. They consist of 40-μm particles with a 60-Å pore diameter. Generally, these sorbents are derivatized with trifunctional silica reagents for resistance to hydrolysis from the acidic and basic ions that are used in ion exchange. This pore diameter is small and does not allow for the migration of larger molecules (molecular weights > 2000) to enter into the pores of the exchange resin. Thus, these sorbents are best used for small molecules and to exclude larger molecules, such as biological materials and natural humic substances, from their pores. Figure 6.3 shows an example of strong trifunctional bonded-phase cation-exchange sorbent.

If it is necessary to ion exchange the larger molecules, then the size-exclusion or wide-pore ion exchangers should be used (Table 6.1). They consist of weak cation and anion exchangers with pore diameters of 275 to 300 Å. These large pore diameters allow for proteinaceous substances and other large biological molecules to be included into the pores for ion exchange. Typically, these sorbents are not endcapped and consist of 40- to 60-μm particles. They

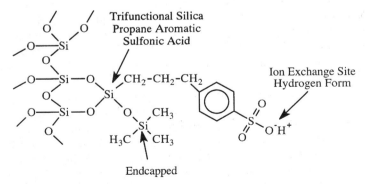

Figure 6.3. Strong cation-exchange sorbent with trifunctional bonded phase.

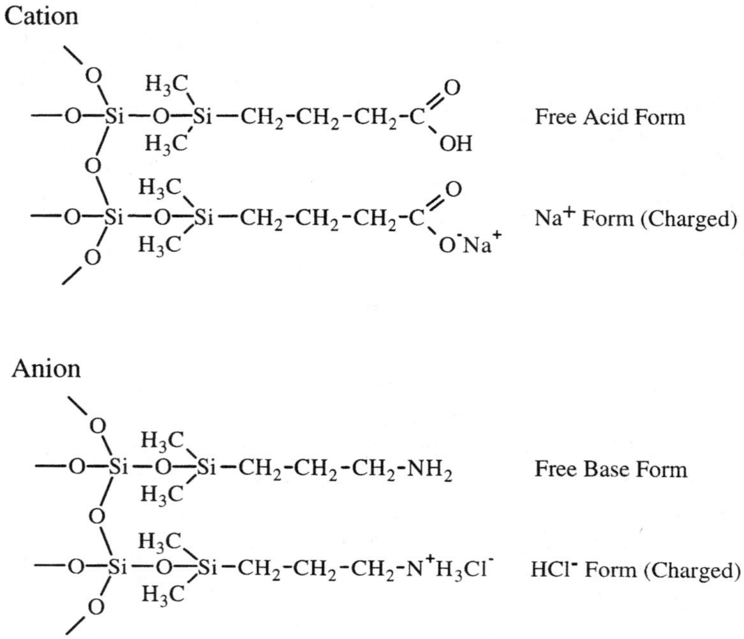

Cation

Free Acid Form

Na$^+$ Form (Charged)

Anion

Free Base Form

HCl$^-$ Form (Charged)

Figure 6.4. Example of the free and charged forms of both weak cation- and anion-exchange sorbents.

have exchange capacities of 0.2 to 0.5 meq/g and are available in sizes from 500-mg to 1-g columns.

Weak cation-exchangers consist of a propane carboxylic acid on a silica matrix and are packed in a hydrogen or in a sodium form (Fig. 6.4). The weak cation-exchange sorbents consist of 40-μm particles that have 60-Å pores and are typically endcapped. The exchange capacity is approximately 0.4 meq/g. The pK_a of the carboxylic acid group that is used is approximately 4.8, and it is in a good range for many types of samples including biological, environmental, and food and natural products. Thus, it is a useful sorbent for methods requiring cation exchange.

The strong anion-exchange sorbent consists of a quaternary amine, usually with propyl groups. It contains a permanent positive charge and is typically packed in its chloride form. The particle size is 40 μm and has a 60-Å pore, once again, size exclusion will occur for larger molecules and interferences that are high molecular weight will be excluded from the anion-exchange sorbent. The capacity of the strong anion-exchange sorbent is approximately 0.7 meq/g and is available in both endcapped and nonendcapped forms. The sorbents are available in columns from as little as 50 mg to as much as 10 g in the "Mega" columns. When additional ion-exchange capacity is required, these Mega columns are useful, both for cation and anion exchange. Thus, a single column could have from 5 to 10 meq of exchange capacity.

The weak anion-exchange sorbents consist of primary, secondary, and tertiary amines. The pK_a of these functional groups varies from 10.1 to 10.7, and these sites are basic. They are available in 40-μm particles with 60-Å pores and may be endcapped or nonendcapped. Because of their pK_a they typically act as anion exchangers at the pH of most aqueous samples (< pH 8). Because of the structure of the weak exchangers with several nitrogen atoms per active site, they have larger exchange capacities, from 1 to 1.5 meq/g. These sorbents are also available in sizes from 50 mg to 10 g. The weak-base sorbents usually come in a free-base form (Fig. 6.4), which consists of a functional group in a nonprotonated state or a non-ion-exchange state. In other words, the amine functional group must first be protonated and charged before it can perform as an ion-exchange sorbent.

6.2.2.   Disks

The silica-based ion-exchange resins have a disadvantage of being unstable at either low pH (< 2.0) or high pH (> 8.0). At these pH ranges, the silica gel will hydrolyze and decompose. The SDB ion-exchange sorbents do not have this problem and are stable across a pH range of 1.0 to 13. The ion–exchange disks have good kinetic properties because of small particle size. This fact is an important feature of the Empore disks because ion-exchange sorbents tend to have slow kinetics or exchange when compared to the C-18 sorbents that are typically used for SPE.

The ion-exchange disks are available from a number of manufacturers (Table 6.1 and Appendix Product Guide). The disks typically have from 0.2 to 0.4 meq/disk (47 mm) exchange capacity. The disks manufactured by 3 M (Empore disks) are made from a styrene–divinylbenzene copolymer; therefore, they do not have the problem of the silica-based disks with limited pH range. The Empore disks also come in a chelation disk for the removal of trace metals from solution, as well as in strong anion-exchange form (Fig. 6.1).

Disks also are available in nearly all of the silica-based forms and are marketed as SPEC products from Ansys, Inc. These disks include strong cation and anion exchange, weak anion exchange, and mixed-mode sorbents. The mixed-mode exchangers are explained in Chapters 2 and 8. Simply, they consist of both nonpolar interactions (C-8) as well as strong cation-exchange sites. The disks may also be purchased in other than hydrogen form; for example, Alltech makes strong cation-exchange disks that are either silver or barium saturated. These sorbents are used to desalt aqueous solutions that are high in chloride or sulfate by precipitation of either silver chloride or barium sulfate inside the disk itself at the ion-exchange site (Table 6.1).

6.3.   ION EXCHANGE AS A CHEMICAL REACTION

The formula of an ion-exchange sorbent may be viewed as a chemical equation when a salt solution is introduced into the sorbent. For example, the sulfonic-acid hydrogen form of the strong cation exchanger will exchange H^+ ion for K^+, if a KCl solution is introduced into the sorbent:

$$RSO_3^- H^+ + K^+ + Cl^- \Longleftrightarrow RSO_3^- K^+ + H^+ + Cl^- \tag{6.1}$$

Thus, the potassium ion is taken up by the sorbent, and HCl will appear in the effluent of the cartridge. The ion that is taken up is called the counter ion, that is, K^+. Likewise, for the anion exchanger, the anions will exchange, in this case:

$$RN(CH_3)_3^+ OH^- + K^+ + Cl^- \Longleftrightarrow RN(CH_3)_3^+ Cl^- + K^+ + OH^- \tag{6.2}$$

Chloride ion exchanges for the hydroxide ion, and KOH appears in the effluent of the column. Generally speaking, the exchange capacity of the silica-based ion-exchange sorbents is 0.5 meq/g, which is considerably less than the 1 to 1.5 meq/g values found in the styrene–divinylbenzene exchangers. The silica exchangers will expand somewhat when wetted, and this swelling is caused by the uptake of water into the ion-exchange sites.

6.4.   SELECTIVITY OF ION EXCHANGE

Ion-exchange reactions occur stoichiometrically; that is, for every ion equivalent removed from solution one ion equivalent of like charge must be released from the ion exchanger. The total number of equivalents that are capable of ion exchange per gram of sorbent is the exchange capacity. Typically these values are given in milliequivalents per gram dry weight of packing material. For example, typical SPE sorbents contain about 0.5 to 1 meq/g, which is about 1 meq/sorbent cartridge. This value is an important quantity to use in applying solutions to the sorbent, for it determines the capacity that the sorbent has and the volume of sample that may be treated.

When an exchange of ions is occurring, there is a distinct selectivity that may occur among the various counter ions. This selectivity is called the *selectivity sequence* and the typical sequence among monovalent cations in increasing order is shown below with Li^+ least retained and Ag^+ most retained by the strong cation-exchange resin.

$$\text{(Weak) } Li^+ < H^+ < Na^+ < NH_4^+ < K^+ < Rb^+ < Cs^+ < Cu^+ < Ag^+ \text{ (Strong)}$$

and for some divalent metal ions

$$\text{(Weak) } Mg^{2+} < Cu^{2+} < Cd^{2+} < Ni^{2+} < Ca^{2+} < Sr^{2+} < Pb^{2+} < Ba^{2+} \text{ (Strong)}$$

The approximate relative selectivity for strong cation-exchange is shown in Figure 6.5. Hydrogen ion is only weakly held compared to the majority of other cations. This fact is important in that the majority of the SPE cartridges

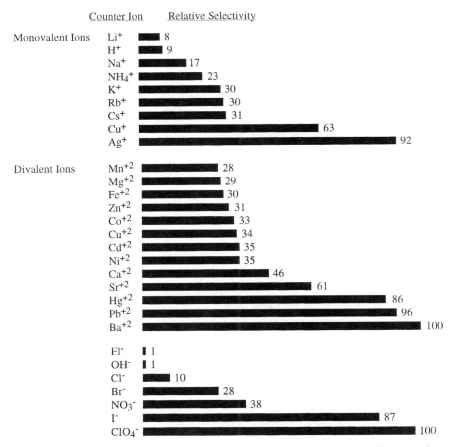

Figure 6.5. Relative selectivity for strong cation exchange of monovalent and divalent cations and for strong anion exchange.

and disks come in the hydrogen form of strong cation exchange, which means that they readily will exchange hydrogen for the metal ions in an aqueous sample. Sodium is held slightly more than hydrogen, and sodium is another common cation that is loaded on the ion-exchange sorbents. Thus, the sorbents are loaded with counter ions that are easily exchanged.

Likewise for anions, there is also a selectivity sequence in increasing order:

(Weak) $F^- <$ acetate $<$ formate $< Cl^- < Br^- < I^- < NO_3^- < SO_4^{2-} <$ citrate (Strong)

These selectivity sequences are for strong cation and anion exchangers. Weak cation and anion exchangers generally have a different selectivity with the H^+ preferred to the common cation, and for weak base resins, the H^+ ion also preferred, which gives rise to the positive charge of the weak-base resin (Fig. 6.2).

Resin selectivity is attributed to many factors, including charge, that is, monovalent or divalent, with preference given to the divalent ions. Trivalent ions are even more strongly held by the ion-exchange sorbent. In addition there is an inverse relationship between the ionic radius of the hydrated ion

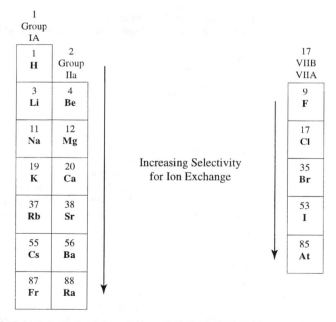

Figure 6.6. Selectivity follows the columns of the Periodic Table for many elements and their oxides.

and the selectivity to ion exchange. This statement means that small ions such as H^+ or Li^+ have a small size but a single charge so that they have a large hydrated radius; whereas Cs^+ is much bulkier and has, therefore, a smaller hydrated radius. Thus, the largest monovalent ions in the periodic table, such as Cs^+, would be the most highly sorbed by the ion-exchange sorbent among the singly charged ions. Thus, a column of the Periodic Table shows a selectivity sequence for ion exchange (Fig. 6.6).

Other factors that affect the selectivity include:

1. Van der Waals interaction of the counter ion with the resin or sorbent matrix, that is, an aromatic ring could interact with the hydrocarbon component of the sorbent.
2. A counter ion that is greatly polarized may bind more tightly than that recognized by a hydrated radius comparison.

6.5. DONNAN MEMBRANE THEORY

When an ion-exchange sorbent is equilibrated with an aqueous solution containing a different counter ion than that bound to the resin, an exchange reaction occurs. A combination of the Donnan membrane theory of equilibrium, selectivity coefficients, and the law of mass action describe the final equilibrium condition for the ratio of ions that would remain on the sorbent (Khym, 1974). The Donnan membrane theory states that when an ion-exchange gel membrane is brought to equilibrium with an electrolyte solution that the concentration of free electrolyte (mobile electrolyte) will be greater outside the ion exchanger than inside the exchanger. This concept is shown diagrammatically in Figure 6.7. The basic concept is that the fixed charge inside the membrane controls how much of the mobile ion of the same charge can enter inside the membrane or gel structure of the ion-exchange sorbent.

In Figure 6.7, the dashed line around the sphere represents the membrane surface. The ion-exchanger sphere contains fixed negatively charged sulfonate groups attached to the silica gel surface of the 40-μm particle. The negatively charged sulfonate is balanced by the mobile positively charged K^+ ions, called counter ions, on the inner side of the gel (I_s), and on the outside of the gel (O_s). At the beginning of equilibration, the K^+ and Cl^- ions are present in equimolar amounts of each on the outside of the gel. This fact is true because charge balance must be maintained.

In the beginning state when a new ionic solution is added to the resin, the concentration of new ions in the gel is not at equilibrium and there is more of the new ions outside the gel than inside the gel. In the final state, the concentration of charge (and new ions) in solution is equivalent on either side of the

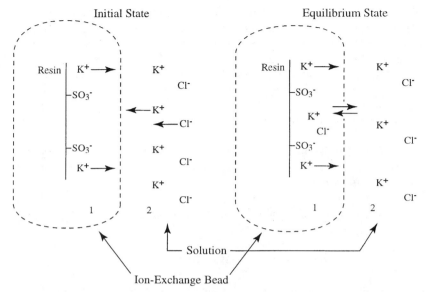

Figure 6.7. Donnan membrane equilibrium shown diagrammatically at initial and at equilibrium state. Note that inside the bead that $K^+ > Cl^-$. [after Khym (1974), published with permission of Prentice-Hall, Inc.]

gel surface. Because electroneutrality must be maintained, a certain amount of the K^+Cl^- outside the gel must enter the gel. Because of the fixed charge present inside the gel, the number of atoms that may enter the gel is much less than the number of atoms on the outside of the gel. Thus, the mass of charge inside the gel is equal or greater than the charge on the outside of the gel.

In the example shown in Figure 6.7, two atoms cross the boundary and five remain on the outside of the gel. It is intuitively obvious that the concentration of the electrolyte remains greater on the outside of the gel, because less electrolyte is free to enter the gel, where the ion exchange may take place. Thus, the combination of exchange capacity of the sorbent and the concentration of the electrolyte will affect the amount of electrolyte that may enter the ion-exchange gel.

This result is expressed mathematically in the Donnan equilibrium theory by the following relationships. First, the potassium and chloride ion concentrations outside the gel are equal in the starting condition by electroneutrality:

$$[K_o^+] = [Cl_o^-] \tag{6.3}$$

Second, the concentration of potassium inside the gel is equal to the sum of the mobile chloride sites and the fixed ion-exchange sites as shown in Eq. (6.4):

$$[K_i^+] = [Cl_i^-] + [RSO_3^-] \tag{6.4}$$

Because of this the Donnan membrane theory states that the activity quotient inside the gel and outside the gel are equal at equilibrium. In mathematical terms then, the following equation is true:

$$[K_i^+][Cl_i^-] = [K_o^+][Cl_o^-] \tag{6.5}$$

In other words, the total charge is equal inside and outside the gel. Equation (6.5) has hidden in it the charge due to the $[RSO_3^-]$ inside the gel. This may be derived as:

$$[K_i^+] = [Cl_i^-] + [RSO_3^-]_{inside} \tag{6.6}$$

Thus, the potassium inside the resin is equal to the sum of all the ion-exchange sites in the resin, $[RSO_3^-]$, plus the chloride that has migrated into the resin from the outside salt solution (KCl). Thus, intuitively it is obvious that the free KCl solution outside the resin is greater than the KCl inside the resin, or

$$[Cl]_{inside} < [Cl]_{outside} \tag{6.7}$$

This result—that chloride inside the resin is less than chloride outside the resin—may be shown mathematically by the following equations:

$$[K_o^+] = [Cl_o^-] \tag{6.3}$$

Equation (6.3) simply states that the potassium and chloride ions outside the gel are equal to one another in activity or concentration, which, of course, is due to the dissolution of the KCl solution that is passed through the resin. Thus, the product of potassium and chloride outside the resin is equal to the square of the chloride concentration outside the resin by substitution into Eq. (6.3):

$$[K_o^+][Cl_o^-] = [Cl_o^-]^2 = [K_i^+][Cl_i^-] \tag{6.8}$$

Thus, the product of the KCl inside is equal to the square of the chloride solution outside the resin.

$$[Cl_o]^2 = [Cl_i][[RSO_3^-]_{inside} + [Cl_i^-]] \tag{6.9}$$

In Eq. (6.9), the potassium inside the resin, $[K_i^+]$, is substituted by Eq. (6.6). If we then multiply out the terms in Eq. (6.9), we find:

$$[Cl_o]^2 = [Cl_i^-]^2 + [RSO_3^-]_{inside}[Cl_i^-] \qquad (6.10)$$

From the statement in Eq. (6.10), it shows that the square of the chloride outside the resin is equal to the square of the chloride inside the resin plus another term, the fixed charge inside the resin, $[RSO_3^-]_{inside}$, times the chloride inside the resin. Thus, the addition of this extra term in the equality of Eq. (6.10) demonstrates that the concentration of chloride inside the resin is less than outside the resin, which is the statement given in Eq. (6.7).

If one uses a value of 10N for the concentration of the RSO_3^- inside a strong cation exchanger (Khym, 1974) and if one assumes a concentration of KCl of 0.1 M that is passed through the resin, it is found that the difference in chloride concentrations may be calculated from Eq. (6.5):

$$[K_i^+][Cl_i^-] = [K_o^+][Cl_o^-] \qquad (6.5)$$

$$[10][x] = [0.1][0.1-x] \qquad (6.11)$$

$$10x = 0.01 - 0.1x \qquad (6.12)$$

$$10.1x = 0.01 \qquad (6.13)$$

$$x = 0.001 \qquad (6.14)$$

Thus, the amount of chloride going into the resin is 100 times less than that present on the outside of the resin, when the solution that is used for equilibration is 0.1M. The fact that the diffusable negatively charged electrolyte is excluded from the membrane does not mean that cation exchange does not occur; it is still possible for cations to migrate or diffuse out of the ion-exchange beads, but at a slower rate than if the total electrolyte is at a greater concentration.

The same reactions occur if an anion exchanger is considered in this reaction, except that there is an exclusion of the anions rather than cations. The Donnan principle does not apply to uncharged solutes. Neutral species are free to diffuse in and out of the exchanger. If one considers more than one cation in the example above, the Donnan principle still holds. By the Donnan hypothesis, stated in Eq. (6.5):

$$[K_i^+][Cl_i^-] = [K_o^+][Cl_o^-] \qquad (6.5)$$

and for sodium the same is also true:

$$[Na_i^+][Cl_i^-] = [Na_o^+][Cl_o^-] \qquad (6.15)$$

dividing Eqs. (6.5) and (6.15) yields:

$$[K_i^+]/[Na_i^+] = [K_o^+]/[Na_o^+] \tag{6.16}$$

Examination of this equation shows that exchange takes place rapidly to satisfy this equation. Thus, our example of the K^+ loaded exchanger would be quickly exchanged when a $NaCl$ solution is passed through the exchange resin. In the case of divalent or higher charged ions, the charge on the cation affects the Donnan calculations. For example, if one substitutes Sr^{2+} for K^+ in the example, then

$$[K_i^+][Cl_i^-] = [K_o^+][Cl_o^-] \tag{6.5}$$

$$[Sr_i^+][Cl_i^-]^2 = [Sr_o^+][Cl_o^-]^2 \tag{6.17}$$

Sr^{2+} is the cation that is exchanged; thus, there are 2 Cl^- ions in the mass balance equation and likewise there must be the release of 2 K^+ ions for each Sr^{2+} ion exchanged. Thus, the ratio equation becomes

$$[K_i^+]^2/[Sr_i^{2+}] = [K_o^+]^2/[Sr_o^{2+}] \tag{6.18}$$

The meaning of this equation is that for every Sr^{2+} ion that migrates into the resin 2 K^+ ions are displaced.

The conclusions of the Donnan principle and its effect on ion exchange are:

1. There is a high exclusion of total electrolyte in the solution of the ion exchanger.
2. The ion, whose sign is the same as the ion exchanger (i.e., cations in an anion-exchange sorbent), is excluded much more than the ion that is exchanged (in this case the anion).
3. The Donnan principle acts independently on each univalent ion present in solution.
4. The Donnan effect is reduced under high concentrations of electrolyte.
5. Nonelectrolytes are not affected by the Donnan membrane potential.

6.6. LAW OF CHEMICAL EQUILIBRIUM AND SELECTIVITY COEFFICIENTS

The last consideration is the law of chemical equilibrium. It simply states that the law of mass action (which states that the rate of a chemical reaction is proportional to the concentration of the reacting substances) is applicable to the state of equilibrium that exists in these reversible ion-exchange reactions. The mathematical forms of the law of chemical equilibrium as applied to ion exchange are written as for any chemical reaction. However, the activities of ions in the resin cannot be evaluated exactly; thus, the equilibrium quotient for the ion-exchange reaction is an apparent exchange constant rather than a true equilibrium constant (Khym, 1974).

$$RSO_3^- H^+ + K^+ = RSO_3^- K^+ + H^+ \tag{6.19}$$

$$K_{eq} = [RSO_3^- K^+][H^+]/[RSO_3^- H^+][K^+] \tag{6.20}$$

The same is true for the anion reaction. In this expression, concentrations are substituted for activity and activity coefficients are ignored. Generally, the concentrations measured are used uncorrected. Frequently normality is the unit chosen for the aqueous phase, while the counter ion per gram is used for the exchanger.

If divalent or trivalent ions are involved, then the mass-action equilibrium quotient is changed by the power of the number of ions involved. For example,

$$3\,RSO_3^- K^+ + Fe^{3+} = (RSO_3^-)_3\,Fe^{3+} + 3\,K^+ \tag{6.21}$$

$$K_{eq} = [(RSO_3^-)_3\,Fe^{3+}][K^+]^3/[RSO_3^- K^+]^3\,[Fe^{3+}] \tag{6.22}$$

The value of K_{eq} would depend on the units used to express the concentration of potassium and iron in both the resin phase and in the solution phase. The units do not completely cancel in this example, the way they do in the previous example that involve ions of equal charge.

The K_{eq} for an ion exchange reaction is a measure of the exchange power of a resin for a particular ion relative to another and is therefore also referred to as the selectivity coefficient. If the selectivity coefficients are determined for a series of ions, then one has the data to establish an affinity scale. A typical affinity scale for cations and anions on a strong exchange resin is shown in Table 6.2. The affinity scale simply means that the ions with the largest numerical value have more "power" to displace ions that are beneath it. Thus, Rb^+ has 2.2 times more power than Li^+ if the concentration of both are equal (Table 6.2). Or if Rb^+ were being replaced by Li^+, then it would take 2.2 times more Li^+ than Rb^+ to make the exchange. For the values in Table 6.2 an arbitrary value of 1.0 is assigned for the reference ion, which is the hydrogen ion for cations and chloride for anions. Typical ion-exchange SPE cartridges usually come in either the hydrogen or the chloride forms.

Many variables will affect the selectivity of the ion-exchange resin for the ions shown in Table 6.2, including capacity of the exchanger and crosslinked

**Table 6.2. Selectivity Coefficients for Cations and Anions on
Strong Exchange Resins**

Cation	Selectivity	Anion	Selectivity
Li^+	1	F^-	0.09
H^+	1.3	OH^-	0.09
Na^+	1.5	Cl^-	1.0
NH_4^+	1.8	Br^-	2.8
K^+	2.1	NO_3^-	3.8
Rb^+	2.2	ClO_4^-	10

After Khym (1974).

structure or silica structure. The silica matrix may also affect selectivity because of its weak cation-exchange capacity. Less is known and published about the silica-exchange resins as compared to the styrene resins, which have been used in ion exchange for many years.

6.7. KINETICS OF ION EXCHANGE

The exchange reaction on an ion-exchange resin is probably an instantaneous reaction; however, the time it takes to diffuse the ions to the point of contact with one another depends on the size and charge of the ions. Larger ions diffuse more slowly, and the pore diameter of the resin, temperature, and particle size of the exchanger all have an effect on the rate of ion exchange. Film diffusion is necessary at the layer immediately surrounding the ion-exchange bead, and diffusion is occurring with the migration of ions within the exchange bead itself. Because of these factors, the flow rate used in ion-exchange SPE for cartridges (large particle size) should be considerably slower than with reversed-phase or normal-phase SPE (Khym, 1974). The use of disks for ion exchange have an advantage over cartridges because of the smaller particle size used in disks (8–12 μm instead of 40–60 μm), which allows for more rapid kinetics and good retention of analytes.

6.8. METHODS DEVELOPMENT: GENERAL PRINCIPLES

The selection of an ion-exchange method involves first a consideration of the structure of the analyte and its pK_a in order to see if it can be either positively or negatively charged at a particular pH. Second is to consider the molecular weight and size of the solute, which can affect the movement of the molecule into the pores of the ion-exchange resin. A third consideration is what other charged solutes are present that may interfere with the ion-exchange reaction.

6.8.1.   Selecting Cation or Anion Exchange

The first selection in ion-exchange methods development is to choose either cation or anion exchange, which is decided by the charge on the analyte (Fig. 6.8). First, check the pK_a of the solute and determine whether it can be a positive, negative, or a neutral species. Examples of several different compounds are shown in Figure 6.8. Compounds such as paraquat or diquat (herbicides) are always protonated; thus, they will always bind by cation exchange. On the other hand, the triazine, prometryn, has a pK_a of ~ 4.8; thus, the molecule will be 50% protonated at pH 4.8, and totally protonated at pH 2.8. At pH 6.8 it is a neutral species. Thus, it is possible to isolate this compound effectively by cation exchange only when the pH is acidic.

An example of an organic anion is the pesticide 2,4-D with a pK_a of 4.9. In this case, however, the compound is an anion at pHs greater than the pK_a of 4.9 (50% anionic) and is totally anionic at pH 6.9. Amino acids, such as glycine, may be a cation, anion, or zwitterion, depending on the pH. Thus, this analyte could be isolated using either cation or anion exchange. Thus, the selection of the mechanism of ion exchange is rather straightforward to decide, as is the pH of the sample necessary for ion exchange to occur.

Having considered the molecule and its pK_a, one must consider the stability of the ion exchanger over the pH range necessary to provide the ionic form of the molecule. Ion exchangers built on a styrene–divinylbenzene structure can tolerate extremes of pH from 1 to 13. On the other hand, the silica-based exchangers, which are commonly used in SPE, have a much smaller range of pH (2–8). At pHs above or below this range, the silica will dissolve and hydrolyze the ion-exchange sorbent. Another consideration is the difference between the weak and strong exchangers. A strong exchanger is one that has its ion-exchange site in a charged, ionic form no matter what the pH (Fig. 6.2). A weak exchanger will loose its ionic exchange site as a function of pH.

For example, a weak cation exchanger with a pK_a of 4.8 will loose its ion-exchange site at pH 2 but will remain a charged site at pH 7.0. Likewise, a weak anion-exchange site will loose its ionic site above pH 7.0. Essentially the weak exchangers do not work well beyond their pK_a. The advantage of the weak exchanger in SPE, however, is that it is possible to remove the ionic exchange site with a pH adjustment. This is an important step because it is much easier to elute solutes if the charge from the resin is removed. Furthermore, the competing inorganic anions or cations in a sample do not compete as well for the ion-exchange sites of weak exchangers as do organic anions or cations; thus, there may be often much less interference with other analytes when using the weak exchangers. An example of when weak ion exchange would be useful is a case where the mechanism of ion exchange may be too strong for easy desorption of the analyte. In these cases, the use of the weak exchanger with some organic modifier, such as methanol, added to the mobile phase would make elution easier.

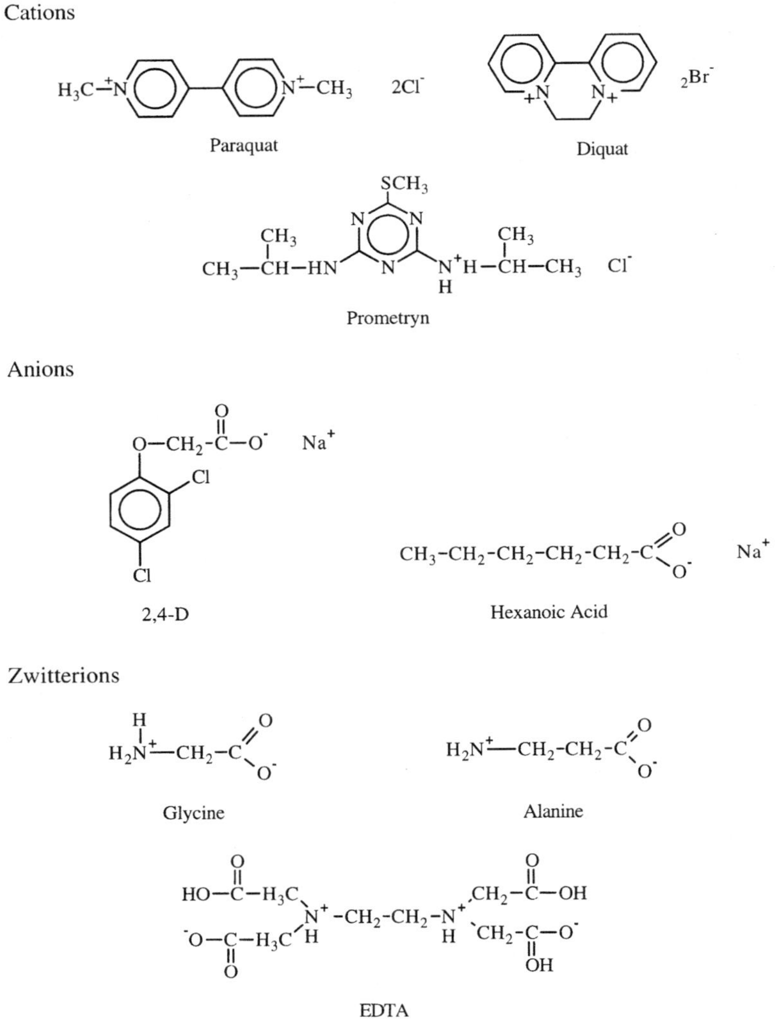

Cations

Paraquat

Diquat

Prometryn

Anions

2,4-D

Hexanoic Acid

Zwitterions

Glycine

Alanine

EDTA

Figure 6.8. Structure of various organic analytes that may form cations, anions, and zwitterions.

6.8.2. Molecular Size and Weight

The molecular size of the solute is the next factor to consider in methods development. As the size and charge of the molecule increase, the analyte becomes too large to effectively enter the pores of the exchanger (Fig. 6.9). Therefore, there are special ion exchangers made for large molecules. These are called wide-pore exchangers. They are effective for high-molecular-weight compounds, such as biological molecules, because size exclusion is not so likely to occur (Fig. 6.9). Typically, the large-pore exchangers have pores of 275 to 300 Å; whereas, normal exchangers have pores of only 60 Å. The smaller pore size will exclude molecules with molecular weights as low as 2000.

Another problem with the large molecules is that they frequently contain many charged sites that will bind them to the exchanger, sometimes irre-

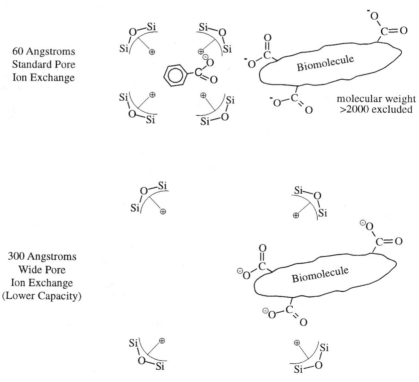

Figure 6.9. An example of a normal and wide-pore ion exchanger and the effect of size exclusion.

versibly to the ion-exchange sorbent. Thus, to elute them, all the ion-exchange sites must be removed by replacing them with other ions. For this reason, organic solutes with molecular weights greater than 500 do not undergo a true ion-exchange reaction (Khyme, 1972). By using the large-pore exchangers, the multiple sites can be greatly reduced, and an equilibrium sorption may be achieved. However, large-pore exchangers generally have considerably less exchange capacity (0.2–0.5 meq/g, see Table 6.1); furthermore, these exchangers use weak ion-exchange functional groups that are more easily eluted by pH adjustment.

6.8.3.   Interferences and Competing Ions

An aqueous sample will contain competing cations and anions that must be considered when carrying out the ion-exchange experiment. If the ionic strength of the sample (the concentration of ions in solution) is greater than $0.1 M$, then ion exchange will probably not be an effective method for the isolation of trace organic constituents in the sample. An ionic strength of $0.1 M$ is equal to 0.1 meq/mL of a monovalent cation. Given that an average sorbent will vary from 0.2 to 1 meq/cartridge or disk, then only 2 to 10 mL of sample could be passed through the sorbent before exceeding the exchange capacity of the sorbent and filling all the exchange sites.

In an aqueous sample where the total concentration of cations or anions may not be known, a simple procedure may be used to estimate the molarity of the sample. First, measure the specific conductance of the solution and express it in terms of microsiemens.

1. Let us say that a water sample has a specific conductance of 400 microsiemens (μs).
2. Multiply the conductance by 0.6 to obtain the total dissolved solids in milligrams per liter (Hem, 1970).
3. $0.6 \times 400 = 240$ mg/L as dissolved solids.
4. Assume that half of this value is cationic for charge balance, or 120 mg/L.
5. Assume that the major cation is sodium with an equivalent weight of 23.
6. The molarity of the sample is 120 mg/L divided by 23, which is equal to ~ 5.2 meq/L.

Thus, in this case the sample would contain approximately 0.5 meq/100 mL of sample, which is approximately the cation-exchange capacity of many of the cartridges and disks. Thus, 100 mL would be approximately the absolute maximum amount of sample that could be applied to a sorbent, and a volume that is half this amount would be a good volume (~ 50 mL) to start with in a methods development experiment. The final rule of thumb then is to take the ion-exchange capacity of the sorbent, such as 0.5 meq/cartridge, and divide by the conductance of the sample in microsiemens for the volume of sample:

$$V = 50 \, C/SC \tag{6.23}$$

where V is the volume (in mL), C is the ion-exchange capacity of the sorbent (in μeq/cartridge), and SC is the specific conductance (in μS). The example above gives:

$$V \text{ (in mL)} = 500 \, (\mu\text{eq/cartridge}) \times (50)/SC \text{ (in } \mu\text{S)} \tag{6.24}$$

$$V \text{ (in mL)} = 25,000/400 \tag{6.25}$$

$$V \text{ (in mL)} = 62.5 \tag{6.26}$$

Thus, a volume of approximately 60 mL would be a good starting point for methods development, and Eq. (6.23) is the rule-of-thumb equation to use for ion-exchange capacity.

6.9. METHODS DEVELOPMENT EXPERIMENT

Some issues to consider as one begins methods development for ion exchange follow:

1. The experiment begins with first deciding on a mechanism of exchange, as described earlier. Next consider the pK_a of the molecule and the pH of the sample to ensure that sorption by ion exchange will occur.
2. What is the molecular weight and size? Do I need a wide-pore exchanger or will the standard 60-Å pore work?
3. What are the interferences in the sample and what is the milliequivalence of exchange capacity needed for this sample? When these questions have been answered, then run the first step and check the capacity of the sorbent to bind the analyte and check for breakthrough. The breakthrough curve will look similar to that described in Chapter 4. If the sample is too concentrated for the sorbent, it may be possible to dilute with deionized water and to run the sample through the sorbent. This procedure presumes that the detection limit of the method is adequate with a smaller sample.

4. When the analyte is bound by the ion-exchange mechanism, then use the desorption eluent to displace the analyte and proceed with the method of identification. The two areas that take the most time in methods development are: first, the choice of retention and the problem of interferences with the analyte and second the elution of the analyte from the sorbent in an efficient manner.

6.10. GENERAL METHOD FOR ION-EXCHANGE SOLID-PHASE EXTRACTION

The general procedure for preparation and use of the ion-exchange sorbents begins with a prewash and conditioning of the sorbent.

1. Prewashing of the sorbent and glassware may be necessary, depending on the analytes that are being analyzed. If a polymer-based sorbent is being used for ion exchange, then a rinse with acetone and isopropanol is advised to clean up the styrene–divinylbenzene sorbent, which may hydrolyze or bleed unwanted materials into the analysis. This step is critical if trace organic analysis is involved. If the analysis is for metal ions or inorganic anions, this step is probably not needed.

2. Next, condition either the silica-based bonded phases or the styrene–divinylbenzene copolymers with methanol to wet and activate the sorbent for good mass transfer when the sample is passed through the sorbent. Rinse the methanol from the sorbent with deionized water. Several bed volumes should be sufficient to displace the majority of methanol and is comparable to that used in reversed-phase SPE.

3. Apply the sample to the ion-exchange sorbent. Generally, the void volume of the sorbent will be approximately 0.1 mL for each 100 mg of sorbent present. Thus, a typical 500-mg cartridge will contain about 0.5 mL void volume. Thus, after addition of the sample, wash the remaining sample through the sorbent with a 0.5-mL rinse with deionized water to displace any remaining sample.

4. Elution of the ion-exchange sorbent may be carried out by several strategies. First, if the sorbent is a cation-exchange sorbent, the cations may be displaced with the H^+ of a strong acid (i.e., 20% solution of HNO_3). This eluent will work well for metal cations. Another approach is to elute the cations with a more strongly held cation (Table 6.2); thus, sodium could be displaced by potassium on a

strong cation-exchange sorbent. This approach would be used for organic cations and combined with a methanol eluate (50 : 50 methanol/water) to help prevent sorption to the matrix of the sorbent by van der Waals interactions and to assist elution. Also the pH can be adjusted in the elution solvent such that the organic cation will loose its charge; that is, the pH should be two pH units above the pK_a. Sodium hydroxide or potassium hydroxide may be used here to raise the pH in combination with the methanol ($0.1 N$). Organic compounds that contain nitrogen (amines) work well by this mechanism.

5. Elution of anion-exchange sorbents consists of displacing the analyte with strong base (OH^- ions) or with a replacing anion. If the matrix of the sorbent is a styrene–divinylbenzene copolymer, then a $0.1 N$ solution of NaOH may be used as a displacing solvent. This solution is too strong for a silica-based bonded anion-exchange sorbent and would dissolve the silica matrix. Thus, the NaOH eluent would be best used with a styrene–divinylbenzene-based anion exchanger, especially in the elution of inorganic anions. For organic anions, a 50 : 50 mixture of $0.1 N$ NaOH and acetonitrile may be used. Caution should be exercised when using sodium hydroxide and methanol as an eluent because of its potential to esterify organic acids. For the silica-bonded phases, phosphate buffer may be used as a displacing ion at a concentration of $0.1 M$. For organic acids, a sodium phosphate buffer may be mixed, 1 : 1, with methanol as an eluent.

6.11. EXAMPLES USING ION-EXCHANGE SOLID-PHASE EXTRACTION

The types of examples that use ion exchange include general chemistry (salt conversions and preparation of deionized water), purification of organic compounds, environmental applications (trace enrichment and interference removal), food and beverage samples (sample clean-up), pharmaceutical and biological samples (trace enrichment), metal ions (trace enrichment), and humic substances (hydrogen saturation), to name but a few. These examples are explained in the following sections.

6.11.1. Salt Conversions

In this application of SPE, a salt solution is converted from one form to another. Let us say from a NaCl salt to a $NaNO_3$ salt. If one takes a strong anion exchanger in chloride form, it is first rinsed with HNO_3 acid to displace the chloride ion and to convert the exchanger to a nitrate form [Eq. (6.27)]. In this step, the nitric acid should be at least pH 2 ($< 0.01\,N$) or greater to prevent dissolution of the silica matrix.

$$R_1R_2RN^+Cl^- + HNO_3 = R_1R_2RN^+NO_3^- + HCl \qquad (6.27)$$

Typically, an excess of 2 to 5 times the exchange capacity is applied at a slow flow rate to convert the sorbent to a NO_3^- sorbent. In this case, 200 mL of $0.01\,N$ HNO_3 (2 meq) acid would be applied, which assumes 1 g of sorbent with 1 meq/g capacity. The excess acid is rinsed from the column with deionized water, and the resin is now in a nitrate form. To make the salt conversion, an NaCl solution is passed through the column and converted to an $NaNO_3$ salt [Eq. (6.28)].

$$R_1R_2RN^+NO_3^- + NaCl = R_1R_2RN^+Cl^- + NaNO_3 \qquad (6.28)$$

The solution leaving the exchanger is now converted to the sodium nitrate salt. The amount of NaCl solution required is a function of the exchange coefficient of the anion-exchange resin. In the case of a nitrate exchanger being changed to a chloride exchanger [Eq. (6.28)], the nitrate is favored by selectivity by a factor of approximately 4 to 1 (Table 6.2). Thus, the breakthrough of chloride may be more rapid than the volume predicted from exchange capacity because the nitrate is more tightly bound. A pilot study should be done to determine the correct volumes for clean recovery of the sodium nitrate.

6.11.2. Preparation of Deionized Water

A small sample of water may be rapidly converted to deionized water (DI) using a combination of cation and anion exchange. In this example, the columns consist of a hydrogen-saturated cation-exchange resin followed by a hydroxyl-saturated anion-exchange resin. The process consists of first exchanging all the cations to a hydrogen form in the water sample:

$$RSO_3^- H^+ + \text{major cations (e.g., } Na^+, K^+, Ca^{2+}, Mg^{2+})$$
$$= RSO_3^- (\text{major cations}) + H^+ \qquad (6.29)$$

The exchange leaves the hydrogen ion in place of the major cations; thus, one is left with HCl, H_2SO_4, and H_2CO_3. These acids are removed in the hydroxyl-loaded anion-exchange resin and release OH^-, which reacts with the H^+ ion to form H_2O. Thus, the final product of this ion-exchange reaction is water with the removal of the ions. The water is deionized. This process also removes weakly charged ions, such as phenols, natural dissolved organic carbon, silicic

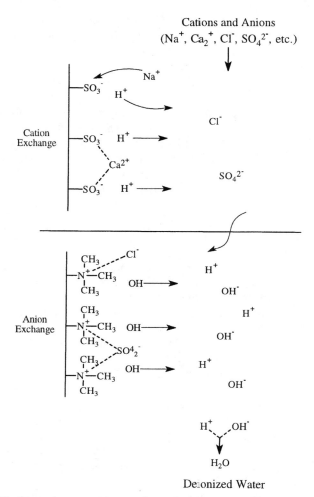

Figure 6.10. Ion-exchange reactions used to make salt conversions and to deionize water.

acid, boric acid, and carbonic acids. Figure 6.10 shows both the salt conversion reaction and the deionization reaction.

6.11.3. Isolation of Metal Cations

Metal cations may be isolated from water samples by ion-exchange SPE using Empore disk extraction. The major cations, calcium, sodium, magnesium, and

$$Ca^{2+} > Na^+ > Mg^{2+} > K^+$$

Figure 6.11. Typical abundance of major cations in natural fresh water (Hem, 1970).

potassium (Fig. 6.11) are removed from a water sample by cation exchange using a strong cation-exchange sorbent on a disk. The disk is the 47-mm disk that contains 0.35 meq/disk. For example, a water sample has a conductance of 250 μS; thus, using Eq. (6.23) one obtains:

$$V \text{ (in mL)} = 350 \times 50 / 250 \tag{6.30}$$

$$V \text{ (in mL)} = 70 \tag{6.31}$$

Thus, the methods development would begin with a volume of approximately 70 mL. Use the disk in its hydrogen form:

SPE Conditions Used

Sample	River water (50 mL) with conductance of 250 μS.
Solutes	Ca^{2+}, Na^+, K^+, and Mg^{2+}
Sorbent	Empore disk, ~500 mg of strong cation exchange resin, H^+ form, precondition with 10 mL each of acetone and DI water, condition with 10 mL of 20% HNO_3, then 2×10 mL DI water.
Eluent	Elute with 2×5 mL of 20% HNO_3.
Reference	3 M brochure on cation extraction. Published with permission of 3 M. See Appendix.

6.11.4. Isolation of Triazine Herbicides from Soil

The triazine herbicides, atrazine, propazine, and simazine, are isolated from soil extracts using strong cation exchange. The pK_a of these compounds is approximately 2. Thus, a strong cation-exchange resin, which has a pK_a of < 1, should capture these compounds by ion exchange. The herbicides contain nitrogen (Fig. 7.6) that occurs as secondary amines, which are capable of possessing a positive charge. The advantage of using cation exchange for these compounds in soil extracts is that typical methanol extracts of soil contain high concentrations of natural organic acids that will interfere with the extraction and analysis of the analytes, especially if high-pressure liquid chromatography (HPLC) is used. The natural organic acids will not be isolated on the cation exchange sorbent; thus, the extract will be purified during the extraction step. The advantage of sorbing the organic cations by ion exchange is that the soil interferences may be washed off with acetonitrile and water, and neither solvent will desorb the herbicides because of the sorption by ion exchange. The wash with 1 mL of 0.1M potassium phosphate is used to overcome the exchange capacity of the disk so that the elution is effective. The competing cations in the soil extract will not be very high given the fact that the extracting solvent is acetonitrile, which has a low solubility for cations.

SPE Conditions Used

Sample	Soil extract (20 g). Extract with 40 mL of acetonitrile/water (80 : 20). Acidify extract with acetic acid (1/100 with DI water), add 25 mL of acid to 5 mL of extract for a total of 30 mL.
Solutes	Atrazine, propazine, and simazine.
Sorbent	Empore disk, ~ 500 mg of strong cation exchange resin, H$^+$ form, pre-condition with 10 mL each of acetone and DI water, condition with 10 mL of 20% HNO$_3$, then 2 × 10 mL DI water.
Eluent	Wash disk with 1/100 acetic acid solution (5 mL), then 5 mL of acetonitrile, then 5 mL of water, then add 1 mL of 0.1 M potassium phosphate to disk. Elute triazines with 2 × 5 mL volumes of a 50 : 50 acetonitrile/0.1 M potassium phosphate buffer.
Reference	Modified Bakerbond procedure. Published with permission of J. T. Baker, Inc. (see Suggested Reading, Chapter 1).

6.11.5. Purification of Organic Compounds

Nonionic organic compounds may be removed from ionic organic compounds using ion exchange. The reactions will work equally well in methanol or other organic solvents as they do in water. An example is the separation of nonionic and anionic surfactants from one another. In this separation, a mixture of anionic and nonionic surfactants (Fig. 6.12) is loaded on a weak anionic exchange resin in the protonated form with methanol as the solvent. Both molecules are soluble in methanol, but the anionic surfactant is tightly bound on the weak anion-exchange resin by anion exchange while the nonionic surfactant passes through the exchange resin. The anionic surfactant may be effectively eluted from the weak anion-exchange resin by elution with dilute base in methanol (ammonium hydroxide in methanol).

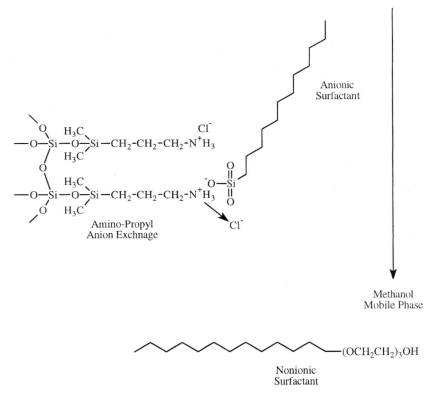

Figure 6.12. Example of a separation of anionic and nonionic surfactants using ion-exchange SPE.

SPE Conditions Used

Sample	Methanol solution of liquid soap, which contains surfactants
Solutes	Nonionic and anionic surfactants (see Fig. 6.12).
Sorbent	~ 500 mg of NH/NH$_2$, H$^+$Cl$^-$ form, precondition with 2 mL each of methanol and DI water, condition with 2 mL of 0.1 M HCl, then 2 mL DI water, then 2 mL of methanol.
Eluent	Add sample to column and slowly aspirate through. Anionic surfactant sorbs by anion exchange. Nonionic compounds pass through in methanol. Elute anionic surfactants with 50 : 50 methanol 0.1 N NaOH, 3 mL.
Reference	Modified from *Anal. Chem.*, **59**: 1798–1802 (1988).

6.11.6.   Hydroxyatrazine Metabolites from Water

The hydroxyatrazine metabolites (Fig. 6.13) are cationic, polar compounds that are cations at a pH near their pK_a (~ 5). They are polar compounds that are not isolated by C-18 or SDB. They may be effectively isolated from water samples by cation exchange and quantified by HPLC. They are the major soil metabolites of the triazine herbicides.

SPE Conditions Used

Sample	River water sample (250 mL) buffered to pH 5 with $0.1M$ KHPO$_4$ at pH 5.
Solutes	Hydroxyatrazine metabolites, $pK_a \sim 5$.
Sorbent	500 mg of strong cation exchange resin, precondition with 2 mL each of methanol and DI water.
Eluent	Add sample to cation exchange resin. Elute with acetonitrile phosphate buffer pH 7, 2 mL.
Reference	*J. Agricul. Food Chem.*, **42**: 922–927 (1994).

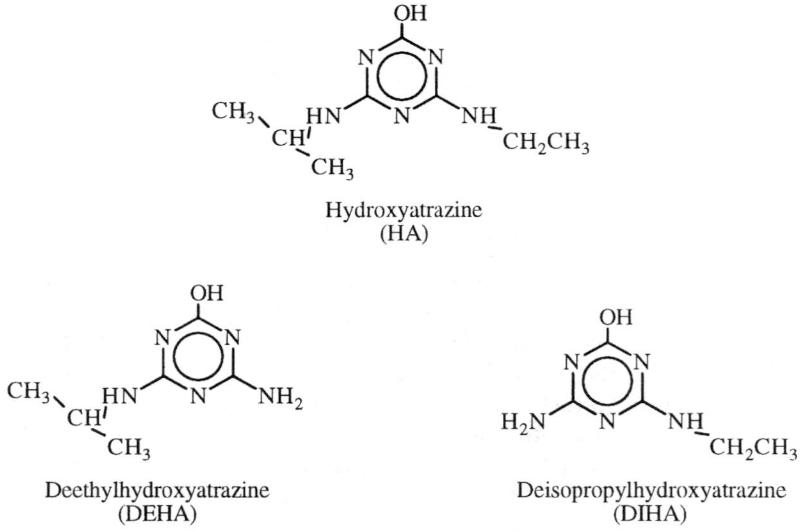

Hydroxyatrazine
(HA)

Deethylhydroxyatrazine
(DEHA)

Deisopropylhydroxyatrazine
(DIHA)

Figure 6.13.   Structures of hydroxyatrazine metabolites.

6.11.7.   Paraquat and Diquat in Water

Paraquat and diquat are permanently cationic and polar (see Fig. 6.8). In general, cationic organic compounds have a strong affinity for silica surfaces and are easily cation exchanged. Thus, they are only found in extremely trace levels in water because of their sorption by cation exchange onto plants and soils, as well as by the hydrophobic effect. They are concentrated by sorption onto a polar CN column and eluted with 1.5 N HCl. The mechanism is one of cation exchange onto the bonded sorbent and elution with acid removes the cationic herbicide.

SPE Conditions Used

Sample	River water (100 mL), no sample additions.
Solutes	Paraquat and diquat.
Sorbent	400 mg of CN, precondition with 2 mL each of methanol and DI water.
Eluent	Elute with 2 aliquots of 1.5 N HCl, 2 mL each.
Reference	Bakerbond Application Note EN-020 Paraquat and Diquat from Pond Water. Published with permission of J. T. Baker, Inc. (see Suggested Reading, Chapter 1).

6.11.8.   Food and Beverage

Additives to diet cola—caffeine, saccharin, and sodium benzoate—represent several different classes of compounds that may be extracted on anion-exchange sorbents simultaneously. Caffeine and saccharin are neutral compounds and sodium benzoate is an organic anion (Fig. 6.14). Caffeine and saccharin sorb by van der Waals interaction on the silica-bonded ion-exchange sorbent and the sodium benzoate by anion exchange. The sorbent chosen is a strong anion exchange in chloride form. The sorbent is first cleaned and

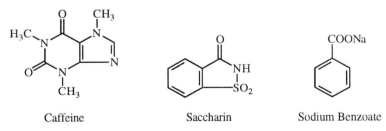

Caffeine   Saccharin   Sodium Benzoate

Figure 6.14. Structure of caffeine, saccharin, and sodium benzoate.

preconditioned with methanol, ammonium hydroxide, and then loaded with chloride using HCl. The eluent is a combination of $0.2\,M$ potassium phosphate and methanol.

SPE Conditions Used

Sample	Diet cola (10 mL), no sample additions, de-gas sample by shaking.
Solutes	Caffeine, saccharin, and sodium benzoate.
Sorbent	500 mg of strong anion exchange (SAX) in chloride form. Condition with 3 mL of methanol, 3 mL concentrated NH_4OH, 3 mL DI water, then 3 mL of $0.2\,N$ HCl.
Eluent	Elute with 2×2 mL of $0.2\,M$ K_3PO_4.
Reference	Bakerbond Application Note FF-006 Extraction of Caffeine, Saccharin, and Sodium Benzoate from Diet Cola. Published with the permission of J. T. Baker, Inc. (see Suggested Reading, Chapter 1).

6.11.9.   Food and Feeds

The extraction of methylimidazole from food is based on the cation exchange of the analyte onto the sorbent (Fig. 6.15). The food sample is prepared by homogenizing in methanol, filtering the sample, and evaporating the filtrate to dryness. The extract is reconstituted in water and filtered. The sample is now ready for application to the strong cation-exchange resin.

SPE Conditions Used

Sample	Aqueous extract of food sample, as described above.
Solutes	Methylimidazole from food.
Sorbent	500 mg of strong cation exchange in H^+ form
Eluent	Condition column with 3 mL of DI water. Aspirate 5 mL of sample through the sorbent. Wash with 2×3 mL of DI water. Elute with 2×1 mL of $0.5\,N$ HCl.
Reference	Bakerbond Application Note FF-010, Extraction of methylimidazole from food. Published with the permission of J. T. Baker, Inc. (see Suggested Reading, Chapter 1).

Figure 6.15. Structure of methylimidazole.

6.11.10. Pharmaceutical and Biological Applications

The extraction of amino acids (Fig. 6.8) from medical plant extracts is again the use of cation exchange to remove organic bases (amino acids) from extracts of plants. Homogenize and blend the plant material with methanol. Filter the extract and evaporate to dryness. Reconstitute the sample with 10% methanol/water and filter. The extract is prepared for cation exchange.

SPE Conditions Used

Sample	Aqueous/methanol extract of plant materials.
Solutes	Amino acids.
Sorbent	500 mg of strong cation exchange sorbent in H^+ form. Condition the sorbent with 3 mL methanol and 3 mL of DI water.
Eluent	Elute with 1 mL of 0.1 N HCl.
Reference	Bakerbond Application Note PH-015, Extraction of Amino Acids from Medical Plant Extracts. Published with the permission of J. T. Baker, Inc. (see Suggested Reading, Chapter 1).

6.11.11. Extraction of Cimetidine from Urine

Cimetidine is a drug used for the treatment of ulcers. The structure includes an imidazole ring (Fig. 6.16) that can be protonated and removed by cation exchange. The method uses a weak cation-exchange sorbent (COOH) for the isolation.

SPE Conditions Used

Sample	Urine sample (1 mL), add 25 μL of 14.8 M NH$_4$OH and mix.
Solutes	Cimetidine, an ulcer drug.
Sorbent	500 mg of weak cation-exchange sorbent in ammonium form ($-CH_2CH_2COONH_4$). Condition the column with 3 mL of methanol, followed by 3 mL of 0.15 M ammonium hydroxide.

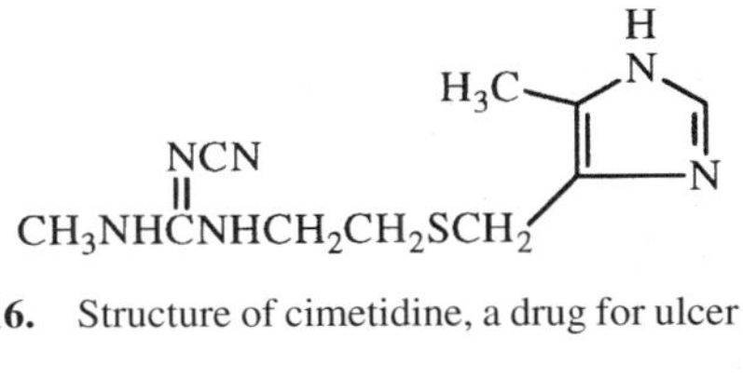

Figure 6.16.   Structure of cimetidine, a drug for ulcer treatment.

Eluent   Wash with 3 mL of 0.15 M ammonium hydroxide. (This step
                unlocks the cation-exchange mechanism.) Then elute with
                2×500 μL of methanol.
Reference   Bakerbond Application Note PH-034 Extraction of Cimetidine
                from Urine. Published with the permission of J. T. Baker, Inc.
                (see Suggested Reading, Chapter 1).

6.11.12.   Dacthal Isolation and Derivatization

Metabolites of the herbicide dacthal may be concentrated from groundwater
and then derivatized on ion-exchange disks (Field and Monohan, 1995). Trace
enrichment consisted of passing 100 mL of groundwater through a strong
anion-exchange disk (Empore) and isolating the metabolites by a combination

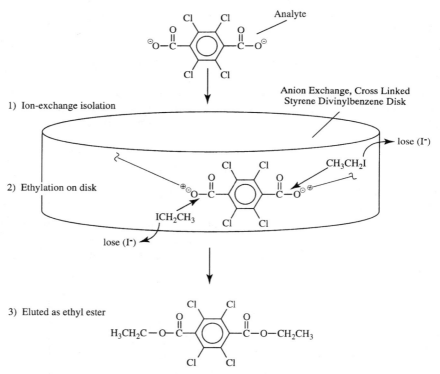

Figure 6.17. Isolation and derivatization of dacthal metabolites on a strong anion-exchange
disk. (After Field and Monohan 1995).

of ion exchange and matrix interaction onto the styrene–divinylbenzene copolymer. The aromatic ring of the dacthal metabolites are capable of π–π interactions to enhance the ion-exchange mechanism (Fig. 6.17). The carboxylic-acid metabolites are then simultaneously eluted and derivatized by placing the 13-mm disk in a 2-mL autosampler vial together with 140 μL of ethyl iodide and 1 mL of acetonitrile and heating for 1 h at 100 °C. The method has a detection limit of 0.02 μg/L and is reproducible.

SUGGESTED READING

Adams, B. A. and Holmes, E. L. 1935. Ion-exchange resins, *J. Soc. Chem. Ind.*, **54**: 1; British Patents 450308–9 (1936).

Kunin, R. and Myers, R. J. 1958. *Ion Exchange resins*, Wiley, New York.

McDonald, P. D. and Bouvier, E. S. P. 1995. *Solid Phase Extraction Applications Guide and Bibliography, A Resource for Sample Preparation Methods Development*, 6th ed., Waters, Milford, MA.

Pietrzyk, D. J. 1969. Ion-exchange resins in non-aqueous solvents—III, solvent-uptake properties of ion-exchange resins and related adsorbents, *Talanta*, **16**: 169–179.

Simpson, N. and Van Horne, K. C. Eds. 1993. *Varian Sorbent Extraction Technology Handbook*. Varian Sample Preparation Products, Harbor City, CA.

Walton, H. F. 1976. *Ion-Exchange Chromatography*. Dowden, Hutchinson & Ross, Stroudsburg, PA.

Zief, M and Kiser, R. 1988. *Sorbent Extraction for Sample Preparation*. J. T. Baker, Phillipsburg, NJ.

REFERENCES

Field, J. A. and Monohan, K. 1995. In-vial derivatization and Empore disk elution for the quantitative determination of the carboxylic acid metabolites of dacthal in groundwater, *Anal. Chem.*, **67**: 3357–3362.

Hem, J. D. 1970. Study and interpretation of the chemical characteristics of natural water, U.S. Geological Survey Water-Supply Paper 1473.

Khym, J. X. 1974. *Analytical Ion-Exchange Procedures in Chemistry and Biology*. Prentice-Hall, Englewood Cliffs, NJ.

ENVIRONMENTAL ANALYSIS

An understanding of simple methods development is crucial to developing effective environmental applications of solid-phase extraction (SPE). The four mechanisms outlined in Chapter 2 (reversed phase, normal phase, ion exchange, and mixed mode) are sufficient for the majority of SPE applications in environmental analysis. The molecule's structure and the sample matrix are the main factors that will determine which mechanism of isolation and separation will be the most appropriate. The fundamental approach to selection of sorbents will be a key topic and many examples are given. This chapter will also discuss applications of SPE to environmental matrices. These include water, soil, and air, for a variety of compounds.

7.1. STRUCTURE AND CHEMISTRY OF MICROPOLLUTANTS

The first considerations in determining the most appropriate SPE methodology are the structure and polarity of the analytes of interest. Table 7.1 shows a selection of environmentally important compounds as examples for SPE methods development from aqueous solution. The polarity range of environmentally important analytes is broad and stretches from nonpolar compounds, such as polychlorinated biphenyls (PCBs), dioxin, and 1,1,1-trichloro-2-2-bis(4-chlorophenyl)ethane (DDT), to moderately nonpolar compounds, such as polynuclear aromatic hydrocarbons (PAHs), to polar compounds such as the herbicides. The most polar compounds are those containing multiple polar functional groups or an ionic functional group, either anionic or cationic. The type of SPE cartridge and elution solvent that are used depends on the polarity of the compound.

For example, for the most nonpolar compounds, both the C-18 and C-8 columns will work well to sorb the analytes from aqueous solution by reversed phase. However, it has been found that C-8 gives equally good recoveries for the most nonpolar analytes, probably due to the fact that the nonpolar compounds elute somewhat easier from C-8 cartridges because of decreased van der Waals interaction. The elution solvent for the nonpolar compounds must be one that solubilizes the compounds, such as hexane/acetone, or hexane/ethyl acetate, or in many cases just ethyl acetate. Finally, the sorbent

Table 7.1. Classes of Organic Compounds Commonly Encountered in Environmental Analysis of Water by Trace Enrichment SPE

Compound	Compound Class	Polarity/SPE	Choice
PCB congeners	Polychlorinated biphenyls	Nonpolar	C-8/C-2
PAH	Polynuclear aromatic hydrocarbons	Nonpolar	C-18/C-8
DDT, DDE	Chlorinated insecticide	Nonpolar	C-8/C-2
Dieldrin, endrin	Organochlorine insecticides	Nonpolar	C-18/C-8
Diazinon	Organophosphorus insecticides	Nonpolar	C-18
Acridine	Azaarenes	Nonpolar	C-18 endcapped
Phthalate	Phthalate esters	Moderately nonpolar	C-18
Chlorsulfuron	Sulfourea herbicides	Moderately nonpolar	C-18
Metobromuron	Phenylurea herbicide	Moderately nonpolar	C-18
Atrazine	Triazine herbicides	Moderately nonpolar	C-18
Alachlor	Acetanilide herbicide	Moderately nonpolar	C-18
Carbofuran	Carbamate herbicide	Moderately nonpolar	C-18
2,4-Dichlorophenol	Chlorinated phenols	Moderately nonpolar	C-18
Chlorpyrifos	Phosphorothioate	Moderately nonpolar	C-18
LAS	Surfactants	Moderately nonpolar Ionic	C-8
Phenol	Phenols	Polar	SDB
Herbicide metabolites	Carboxylic acids	Polar and ionic	Acid/SDB
2,4-D	Phenoxyacid herbicide	Polar and ionic	Acid/SDB
Paraquat	Cationic herbicides	Polar and ionic	CN

must be thoroughly dried under vacuum, if hexane or methylene-chloride elution solvents are being used. If the sorbent is not properly dried, then the hexane solvent has a difficult time eluting the solutes from the sorbent because of water sorbed to the silica bonded phase.

For moderately polar compounds in aqueous solutions, the use of the C-18 sorbent is the most efficient. The sorbent has good capacity for these compounds and is easy to elute with a solvent such as ethyl acetate. For most polar compounds (such as phenol in Table 7.1), the C-18 sorbent is typically not hydrophobic enough to sorb the compounds from a water sample. Therefore, a nonpolar sorbent with greater surface area is required. Typically there are two choices, a styrene–divinylbenzene (SDB) sorbent and graphitized carbon. The SDB sorbents are easier to use, cleaner, and easier to elute. The graphitized-carbon sorbents contain active sites of anionic exchange, which give more capacity for many polar compounds, but they are subsequently more difficult to elute. The graphitized-carbon sorbents are discussed in more detail in Chapter 12.

Finally, ionic compounds are more readily isolated by reversed phase, if the pH of the sample is adjusted such that the molecule is not ionic or a lesser fraction is ionic. Typically, the pH of the sample should be lowered to two pH units below the pK_a for organic acids in order to protonate all of the carboxyl groups and to have the maximum sorption capacity (Thurman et al., 1978). If ionic compounds are still polar after pH adjustment, then it is typically necessary to isolate them using an ion-exchange method. This topic was discussed thoroughly in Chapter 6.

Normal-phase SPE is used mainly for the clean up of organic solvents that have been used to extract analytes from some matrices, such as soils, sediments, and biological samples (e.g., fish). Good examples include the clean-up procedures used for PCBs, dioxins, and PAHs from soils and sediments. In these instances, silica gel is used to retain polar organic substances that are released from soils and sediments during extraction. The nonpolar analytes are bound less tightly to the silica or alumina surface and are desorbed in hexane or hexane/chloroform solvent mixtures. Both silica and alumina gel are used to generate "clean extracts" for gas chromatography/mass spectrometry (GC/MS) analysis (see Chapter 5).

7.2. GENERIC APPROACH TO ENVIRONMENTAL SOLID-PHASE EXTRACTION

The environmental analysis of organic compounds in aqueous samples by SPE encompasses a wide variety of compounds and literally thousands of studies have been published using SPE (Baker bibliography, Varian bibliography, and Waters bibliography; see Suggested Reading, Chapter 1). In spite of the variety of literature, there is a straightforward approach to SPE that will generally solve the analysis problem for water samples. In fact, there could be described a generic approach to SPE in environmental analysis of water.

The generic approach to SPE of aqueous samples follows the following method (Fig. 7.1). First look at the polarity of the compound or its solubility and whether it can form an ion. Thus, is it ionic or not? If nonionic, then what is its polarity? Is the molecule nonpolar, moderately polar, or polar? If the molecule is nonpolar or moderately polar, then the generic C-18/C-8 method described below will work. Alternatively, either a less polar sorbent may be used (i.e. either a C-8 or even a CN sorbent) to isolate the most nonpolar analytes. If the compound is polar (i.e., aqueous solubility greater than 1000 mg/L), then either the SDB or graphitized carbon will work best as a sorbent.

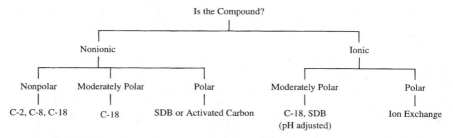

Figure 7.1. Generic approach to selection of an SPE method for environmental analysis.

If the molecule is ionic, and is moderately polar (i.e., as a rule of thumb, a molecule with 12 or more carbon atoms per ionic functional group is moderately polar), then there is a good chance that the molecule will still isolate on C-18, without pH adjustment to remove the ionic charge. If the molecule is anionic, because of a carboxyl group, the pH may be lowered such that the molecule becomes protonated and is no longer ionic. This adjustment of pH will greatly increase the hydrophobicity of the solute and allow isolation on C-18. It is important to choose a silica sorbent that is trifunctional when using acidified samples so that the lowering of pH does not hydrolyze the silica and the C-18 reversed phase. If the molecule is anionic because of a strong acid group, such as a sulfonic acid, it is not possible to lower the pH sufficiently to protonate the solute.

If the molecule is ionic and polar, so that even with pH adjustment, it cannot be made sufficiently hydrophobic to isolate on C-18, SDB, or graphitized carbon; at this point, the use of ion exchange is recommended. With ion exchange the analytes must compete with other anions and cations present in the sample for ion-exchange sites (Chapter 6). Generally, if the ionic strength of the water sample is less than 0.1 or a conductivity of 1000 μmhos/cm, then ion exchange may be used to effectively isolate the organic analytes. In water of high ionic strength, seawater, for example (0.7 ionic strength), the use of ion exchange for organic molecules is not effective because sodium or chloride will compete for the ion-exchange sites on the sorbent (see Chapter 6).

7.3. WATER SAMPLES AND TRACE ENRICHMENT

7.3.1. Introduction

Analysis of water samples for organic compounds involves the use of gas and liquid chromatography. In either case, the analytes must be concentrated and removed from the water phase and into an organic solvent suitable for instrumental analysis. Solid-phase extraction is one of the best sample-handling procedures for this task. The concentration step in the analysis of water is called trace enrichment (Little and Fallick, 1975). In all cases, some elementary clean up of the sample by SPE is required before instrumental analysis.

Generally, the task given to SPE is trace enrichment of the organic analytes in a water sample. The amount of water required for this task will vary with the detection limit required (i.e., µg/L or ng/L). Chapter 3 discussed how to calculate the mass of analyte required and the volume of sample needed. Generally speaking, a detection limit of 1 µg/L by GC/MS requires about 10 mL of water, 0.1 µg/L requires 100 mL of water, and 0.01 µg/L requires 1 L of water. Thus, the approach to SPE for trace enrichment is in many ways dictated by the level of detection.

For example, the majority of pesticides used in the United States have health advisories in the microgram-per-liter to submicrogram-per-liter range. Many of the methods recommended by the Environmental Protection Agency (EPA) require a 1-L sample. In Europe, the health standards for pesticides in groundwater is 0.1 µg/L per compound. For these reasons, the common sample for trace enrichment varies from 100 mL to 1 L.

The mechanism of sorption in trace enrichment is usually reversed phase, and the commonly used sorbent is C-18 or SDB. There are two formats of SPE for trace enrichment, cartridges and disks. Both formats may be automated, and the choices for off-line automation include several vendors for disks (see Chapter 10). For on-line automation of trace enrichment, there are several instruments available (Chapter 10). Because of the large volumes of water that must be pumped through the sorbents for trace enrichment, automation of the process is recommended. The disks have an advantage over cartridges in that a liter of water may be processed in approximately 15 min with no losses due to fast flow rates. Cartridges take considerably longer to process. This fact is due to the large surface area of the disk and the small particle size of the sorbent (~ 8 µm) compared to the 40-µm particle size of the cartridge. The slight advantage of the cartridge over the disk is that a small volume of solvent may be used to elute the cartridge, which simplifies the evaporation of the eluent prior to gas chromatography or liquid chromatography. Figure 7.2 shows a typical off-line approach to the trace enrichment process for organic analytes from a 1-L water sample prior to GC/MS. An example of on-line trace enrichment is shown in Figure 1.8.

Another use of trace enrichment for environmental samples, such as soil and sediment, is the concentration of analytes from organic solvents that have been used to extract the solid matrix, such as methanol, acetone, or acetonitrile (water-miscible solvents). Commonly, water-miscible organic solvents may be used to recover analytes from soils and sediments. These polar organic

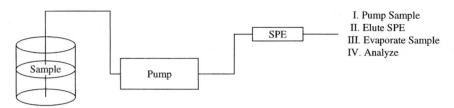

Figure 7.2. Off-line trace enrichment of pesticides from a 1-L water sample for analysis.

solvents work well to remove pesticides and other contaminants from soil, which often is polar because of low organic carbon contents and various mineral phases. Rather than try to evaporate these polar organic solvents, which evaporate slowly because of low vapor pressures, and alternative method is to dilute the solvent extract with water and to concentrate the analytes on C-18 using the trace enrichment process. This method is especially effective for nonpolar organic solutes, such as PCBs, DDT, many insecticides, and herbicides. Generally, if the aqueous solubility of the compound is less than 500 mg/L, then the analyte should be easily enriched on C-18 in the presence of 10% methanol, acetonitrile, or acetone. Thus, one needs only to dilute the organic extract with distilled water to approximately 10% organic solvent, and then to process the sample according to the trace enrichment procedure.

7.3.2. Removal of Interferences

The interferences that are collected in a typical trace-enrichment process include particulate matter that can accumulate on the head of the column or onto the filter disk itself. Next are natural organic substances, such as aquatic humic substances, which will be removed by sorption to the head of the column, and may interfere with the analyte sorption. Aquatic humic substances are natural, colored organic acids from 500 to 5000 molecular weight. They are formed in soil and aquatic environments and constitute the majority of the dissolved organic matter in natural waters (Thurman, 1985). Inorganic salts pass through the column during trace enrichment and are not part of the final eluate. If the sorbent is then eluted with a nonpolar solvent (ethyl acetate), the majority of the aquatic humic substances will remain sorbed to the C-18. If however, methanol is used, the humic substances will co-elute. Figure 7.3 shows the major interferences that may occur during trace enrichment. Several manufacturers sell SPE products to filter the sample. For example, IST offers a "foamlike" filter plug at the head of its SPE syringe columns. The filter is a depth filter and allows much faster flow rates through the column than sorbent

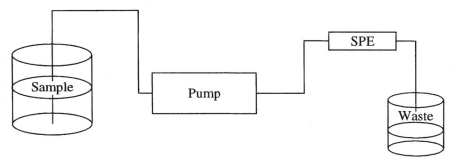

Figure 7.3. Removal of interferences during trace enrichment.

alone. Also the disks that are used in SPE (see Chapter 11) often have a glass-fiber filter above them when used in a syringe barrel. The free disks could also be used with a glass-fiber filter placed in front to remove particulates. It should be remembered that if the sample is not filtered, then the particulates are also partially to entirely eluted with the SPE sorbent during the elution step. Thus, the sample includes both a dissolved and particulate result when the sample is later analyzed.

7.3.3. Generic Eluents

The literature contains many choices for eluting solvents in trace enrichment by C-18 SPE. One of the most common eluents is ethyl acetate, which removes the majority of the hydrophobic substances sorbed to the resin. If this eluent is followed by methanol, then the combination makes an excellent trace enrichment combination for GC/MS (ethyl acetate eluent) and for high-pressure liquid chromatography (HPLC) (methanol eluent). A short drying step is required to expel water from the voids of the sorbent before elution with ethyl acetate. The explanation for this combined procedure was discussed in Figure 3.5 on selective elution. In summary, the ethyl acetate removes residual sorbed water present in the silica matrix and gives an effective elution of nonpolar organic solutes without the prior extensive drying of the sorbent commonly used for more hydrophobic solvents, such as chloroform and hexane/acetone. Methanol then follows the ethyl acetate and removes ionic organic analytes, or the most polar analytes, which were not effectively eluted by the ethyl acetate. See Figure 3.5 for more detailed discussion. If the solutes are very hydrophobic, such as DDT or PCBs, then ethyl acetate should be mixed 50 : 50 with methylene chloride as an eluting solvent. This combination will be a more effective eluent than ethyl acetate alone.

7.4. ENVIRONMENTAL PROTECTION AGENCY METHODS

7.4.1. Cartridge Methods

The EPA (United States) has standard methods for the analysis of pollutants in water. These methods cover approximately 100 organic compounds and toxic substances. The methods were originally developed using solvent extraction as the method to remove the analytes from water. The method involved the extraction of 1 L of water in a separatory funnel using methylene chloride. Now there are at least 12 methods that are approved for SPE rather than liquid extraction, one of which is Method 525.1.

There is one excellent study on the optimization of this method for 46 compounds including pesticides, PAHs, and other contaminants on the EPA list (Raisglid et al., 1993). Raisglid and co-workers optimized the 525.1 Method using an automated workstation made by Tekmar called the AutoTrace SPE Workstation (see Chapter 10). The automation parameters discussed in this study will be addressed later in Chapter 10.

Raisglid and co-workers (1993) found that methylene chloride did not give the best absolute recovery for the 46 compounds (63%), but a mixture of acetone/ethyl acetate (75 : 25) gave recoveries of 84%. These recoveries were based on a 1-L sample and 500 mg of C-18 sorbent. The ability of the acetone to solvate and remove water from the bonded silica columns (C-8 and C-18) improved the recoveries. Drying times were reduced from 10 min with methylene chloride to 1 min with the acetone/ethyl acetate. Furthermore, the removal of sodium sulfate from the drying of methylene chloride further increased the recoveries of the analytes.

The method of Raisglid and co-workers (1993) reduced the elution solvent use from 10 to 5 mL. Furthermore, they added a soaking step between the two elution solvents. The 5-mL volume was divided into 3 and 2 mL elution volumes with the soaking of the sorbent. Apparently diffusion from the silica pores is required for good elution. This was apparent because their study of flow rate showed that there was no difference in recovery during elution.

They further tested the C-2, C-8, and C-18 bonded-silica phases for the best overall recovery. They found that the C-2 was not hydrophobic enough for some of the 46 test compounds and that the C-18 was too strong for the most nonpolar analytes. Thus, the best overall phase was the C-8 with average absolute recoveries of 91 and 97% relative recovery compared to 84% absolute and 96% relative for the C-18 (Table 7.2).

A very interesting finding of this study (Raisglid et al., 1993) was the fact that the absolute recoveries were 10% higher when the sample was run at 4 vs. 25 °C. The improved recovery was attributed to less diffusion into the pores of the silica sorbent and elution was improved. Overall this study gives an

Table 7.2. Percent Recoveries of 46 EPA Priority Pollutants from a Modified Method of EPA 525.1

Compound	C-8 (EC)		C-18 (EC)	
	Absolute	Relative	Absolute	Relative
1 Hexachlorocyclopentadiene	78	81	77	75
2 Dimethylphthalate	93	103	86	91
3 Acenaphthalene	94	101	97	106
4 Acenaphthene-d10	IS	IS	IS	IS
5 2-Chlorobiphenyl	89	96	90	100
6 Diethylphthalate	104	115	102	111
7 Fluorene	90	97	91	102
8 2.3-Dichlorobiphenyl	79	92	79	97
9 Hexachlorobenzene	72	81	72	84
10 Simazine	84	95	74	75
11 Atrazine	99	111	94	104
12 Pentachlorophenol	ND	205	165	164
13 Iindane	99	118	98	116
14 Phenanthrene-d10	IS	IS	IS	IS
15 Phenanthrene	84	96	83	101
16 Anthracene	81	90	80	95
17 2.4.5-Trichlorobiphenyl	76	85	76	89
18 Atachlor	117	123	103	117
19 Heptachlor	91	107	84	96
20 di-*n*-Butylphthalate	113	119	98	116
21 2.2'.4.4'-Tetrachlorobiphenyl	83	88	78	89
22 Aldrin	85	90	76	86
23 Heptachlor epoxide	98	106	89	103
24 2.2'.3'.4.6-Pentachlorobiphenyl	88	92	82	91
25 Gamma-chlordane	98	104	86	98
26 Pyrene	86	89	80	95
27 Alpha-chlordane	97	102	86	98
28 Trans nonachlor	97	101	86	97
29 2.2'.4.4'.5.6'-Hexachlorobiphenyl	84	120	87	123
30 Endrin	101	127	86	127
31 Butylbenzylphthalate	119	150	88	124
32 di(2-Ethylhexyl) Adipate	128	155	87	116
33 Benz(*a*)anthracene	86	109	78	110
34 Chrysene d-12	IS	IS	IS	IS
35 Chrysene	85	112	79	112
36 2.2'.3.3'.4.4'.6-Heptachlorobiphenyl	93	119	86	121
37 Methoxychlor	115	113	78	112
38 2.2'.3.3'.4.5'.6.6'-Octachlorobenzene	88	120	88	124
39 di(2-Ethylhexyl)phthalate	129	157	92	124

(contd.)

Table 7.2. (*continued*)

Compound	C-8 (EC)		C-18 (EC)	
	Absolute	Relative	Absolute	Relative
40　Benzo(*b*)fluoranthene	90	111	75	101
41　Benzo(*k*)fluoranthene	90	114	75	101
42　Benzo(*a*)pyrene	86	106	68	91
43　Perylene-d12	104	97	92	85
44　Indeno(1.2.3.*c.d*) Pyrene	105	122	68	90
45　Dibenz(*a.h*)anthracene	113	130	71	93
46　Benzo(*g.h.t*)perylene	96	121	69	93
Avg:	91	97	84	96

Recoveries were based on a 1-L sample and a 500-mg cartridge of C-8 and C-18).
After Raisglid and co-workers (1993); published with permission.

excellent overview of how to analyze the priority pollutants in surface water, groundwater, and wastewater samples.

Finally, it should be noted that the official EPA method calls for the use of methylene chloride rather than water-miscible solvents. Methylene chloride is currently under consideration for removal in approved methods by EPA because of its toxicity and the possibility of damage to the ozone. This topic will be addressed again in the following section on disk methods.

7.4.2.　Disk Methods

The Empore extraction disks also have been approved for EPA Method 525.1. Early studies found that recoveries of the pollutants from the disk were low (40–70%) using methylene chloride as an eluting solvent (Varian sample product literature, 1995). These low recoveries most likely arise from the inability of methylene chloride to displace and penetrate the water saturating the internal porous silica sorbent. Dye studies by 3 M have shown this to be the case (3 M, Product Literature, Appendix). An elution strategy that has improved recoveries is presented. First rinse the sample bottle with 5 mL of ethyl acetate to remove any analyte sorbed to the glass walls, then elute the disk with this solvent. Repeat the procedure using methylene chloride, 5 mL. Finally, follow the last elution with 5 mL of 1 : 1 methylene chloride/ethyl acetate following the same procedure. The result is 85 to 95% recovery of analytes. Currently, the Empore extraction disks may be used to isolate the organic compounds from 1-L water samples for a variety of EPA methods (Table 7.3).

Table 7.3. EPA Methods That Also Use the Empore Extraction Disks

Method Number	Analytes	Method
506	Phthalate and adipate esters in drinking water	GC/PID
508.1	Chlorinated pesticides in water	GC/ECD
513	2,3,7,8-Tetrachlorodibenzo-*p*-dioxin (TCDD) in drinking water	GC/MS
515.2	Chlorinated acids in water	GC/ECD
525.2	Organic compounds in drinking water (Table 7.2)	GC/MS
549.1	Diquat and paraquat in drinking water	HPLC-UV
550.1	PAHs in drinking water	HPLC-UV & Fluorescence
552.1	Haloacetic acids and dalapon in drinking water	GC/ECD
3535	Test methods for evaluating solid waste	Appropriate Methods
1664	Proposed for oil and grease	Gravimetric IR

The majority of the methods in Table 7.3 use the C-18, 47-mm Empore extraction disk. Three methods though deviate from the C-18 protocols; the first is Method 515.2 for chlorinated acids. This method uses the 47-mm SDB disk. The increased capacity of the SDB disk is used to capture the more soluble chlorinated organic acids.

Next, Method 549.1 for diquat and paraquat uses an interesting combination of deactivating solution and ion-pair reagent. Because the diquat and paraquat are permanently fixed organic cations (Fig. 7.4), they have a high affinity for the free silanol groups on the silica surface. Thus, a concentrated solution of cetyltrimethylammonium bromide is applied to the disk to sorb and deactivate the cation-exchange sites of the C-8 disk. A solution of 1-hexane-sulfonic acid is also added to the C-8 disk, which turns the disk into a cation-exchange resin. The sample is now applied to the disk and held by ion exchange with the 1-hexanesulfonic acid. Finally the ion pair of the paraquat and 1-hexanesulfonic acid is eluted from the C-8 disk with an eluting solution containing methanol, acid, and diethylamine, which disrupts the ion pair. The sample is analyzed by HPLC.

The last method to consider is Method 552.1, which analyzes for haloacetic acids using an strong anion-exchange disk of 47-mm diameter. The soluble acids are effectively sorbed by anion exchange and recovered by an eluting solvent of methanol and 10% sulfuric acid. Thus, the elution consists of displacing the ions with sulfate and also protonating the acids to reduce their

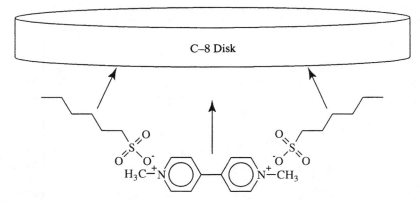

Figure 7.4. Structure of paraquat and diquat with ion-pair reagent sorbing to an extraction disk.

affinity for the ion-exchange resin. A summary of these methods may be obtained either through Varian Sample Preparation Products or through 3 M (see Appendix Product Guide). Also read the section in Chapter 11 on disk products for more ideas, applications, and hardware for SPE using disk technology. Finally, there is a comprehensive basic introduction to disk analysis of water samples by Markell and Hagen (1996) that is quite useful for the environmental analyst. A book by Barcelo and Hennion (1997) is quite useful to the environmental analyst. This book has an extensive section on on-line HPLC and GC/MS analysis of environmental compounds.

7.5. ON-LINE SOLID-PHASE EXTRACTION—HIGH-PRESSURE LIQUID CHROMATOGRAPHY: ENVIRONMENTAL APPLICATIONS

Environmental applications of on-line SPE have been expanding because of the availability of instruments to make the transition from sample pumping to a SPE column (a minicolumn or guard column for HPLC) and injecting the sample into an HPLC. The advantage of using an on-line approach is that small particle-size packing material may be used (10 μm) because of the high-pressure pumps of the HPLC system. Therefore, there is a large number of theoretical plates and good column efficiency (see Chapter 4 for discussion). Furthermore, the sample may be isolated on the SPE precolumn while the HPLC is running the previous sample, so there is no loss in time between isolation and analysis. A disadvantage to the system is that the number of SPE precolumns commercially available are generally less than that offered in off-line SPE. Finally, the cost of the on-line systems are considerably greater than the off-line systems (see discussions in Chapter 10).

For a good discussion of on-line SPE coupled to HPLC, the work of Hennion and Pichon (1994) and Hennion and co-workers (1990) discuss environmental applications. In their work, the styrene–divinylbenzene co-polymers are used as SPE columns with good capacity for many of the environmentally relevant contaminants and allow for direct analysis by HPLC. Hennion and Pichon (1994) discuss and review a number of studies on pesticides and related compounds by on-line SPE. There is also more dis-cussion of on-line SPE methods coupled to GC/MS in Chapter 10.

7.6. LARGE-VOLUME ANALYSIS BY SOLID-PHASE EXTRACTION

A specific application of environmental SPE is sample preparation of extra-large volumes (from 10 to 100 L). This work was pioneered in the early 1970s by Junk and co-workers (1974) for the analysis of trace organic compounds in water using styrene-divinylbenzene copolymers (XAD-2 resin from Rohm and Haas) and by Thurman and Malcolm (1981) and Leenheer and Stuber (1981) for the analysis of natural organic substances in water (humic sub-stances). One can obtain the XAD resins from Supelco (Appendix Products Guide) and still follow the protocol of this early work for the isolation of contaminants and humic substances from large volumes of water (10–1000 L of water).

The method is straightforward and entails packing columns of 1 L to 10 L of XAD resin (either the styrene–divinylbenzene or acrylic esters, XAD-2 and 8, respectively). The resins are cleaned extensively in Soxhlet (with 0.1 N NaOH, methanol, and acetonitrile) prior to their use to remove interferences and then packed in low-pressure chromatographic columns. Water is pumped through the columns and compounds are isolated and purified using the same protocol outlined in methods development (Chapter 3). The main disadvantage of the XAD resins continues to be the impurities present in the resins. Because the resins are synthesized for industrial purposes, the grades of solvents used in synthesis are not of the highest purity; thus, there is a considerable amount of impurities that are copolymerized or incorporated into the resin materials. These materials often end up in the final concentrate that is injected into the GC or GC/MS.

There is currently a company that has commercially made a device that both filters and passes a water sample through a series of columns of XAD resin (Axys, see Product Guide, Appendix). The device is specifically designed for the analysis of ultra-trace levels of PCBs, DDT, and dioxin in nat-ural waters. Literally, 100 L of water may be rapidly passed through the

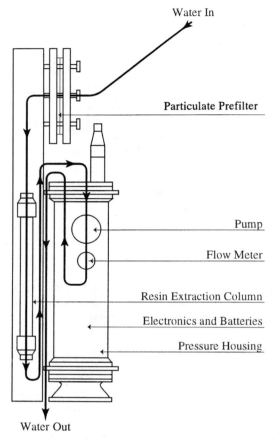

Figure 7.5. Large-volume device (Axys, Infiltrex II) for environmental SPE of PCBs and dioxin. (Published with permission of Axys Environmental Systems, Inc.)

columns for high concentration factors (detection limits of picograms per liter). Figure 7.5 shows an example of the device. The device has been used successfully by Ethier and Taylor (1994) for a study of DDT and PCBs in tributaries of the Great Lakes. Columns for the isolation of metals and radionuclides are also available. They also currently have shown the ability to use the device (shown in Fig. 7.5) for volumes as large as 1000 L for parts per quintillion (femtograms/liter) for dioxin. The Axys device has found usefulness for marine sampling where the device is submerged deep into the water column and then is activated. The device pumps the water directly onto the XAD-2 column at depth using a battery-powered pump and flow gauge.

Samples are filtered and then immediately passed through the XAD resin to sorb analytes.

Another concept for extra-large volume analysis that has been attempted only infrequently (McDonnelland et al., 1993), but that could be quite useful for large volume work, is the use of large Empore disks of SDB to isolate organic substances from large volumes very rapidly. Currently it is possible to obtain 90-mm disks that contain approximately 2 g of 8-µm packing material. These disks would be quite useful for isolation of large volumes of water for PCBs and dioxin because of high column efficiency due to small packing size (8–10 µm C-18). Given the high capacity factors for these compounds, the disk should be capable of passing 10 to 100 L of water and maintaining good capacity. Practically speaking, the disk may have plugging problems beyond 10 to 30 L depending on the water sample. Larger disks may be available in the future to extend this technology. Finally, some manufacturers do make cartridges that contain up to 10 g of C-18. These formats are not practical for large volume work because of plugging of the column and the low flow rates that are required. They have their usefulness for preparing large masses of analyte for analysis.

7.7. SOIL ANALYSIS AND SOLID-PHASE EXTRACTION

The analysis of soil by SPE involves first the extraction of the analytes from soil by an organic solvent, followed by cleanup and purification by SPE. Thus, SPE is used for clean-up as well as for trace enrichment. The strategy is different for clean-up compared to that of trace enrichment. Two approaches are possible. First is to sorb unwanted interferences on the SPE and allow the analyte to pass through for direct analysis or further concentration (SPE clean-up). Second is to isolate the analyte directly by SPE and to elute it free from the interferences or to wash the interferences off the SPE sorbent prior to elution (trace enrichment).

7.7.1. Solid-Phase Extraction Cleanup

In this procedure, the interferences are removed by SPE and the analytes pass through the sorbent. An example is the isolation of nonionic herbicides from soil, atrazine and alachlor. Because soil is commonly extracted with methanol or acetonitrile, there is a substantial amount of humic material extracted at the same time. Both methanol and acetonitrile are miscible with water and readily penetrate the pores and surfaces of soil organic matter, which is hydroscopic and contains sorbed water. Because soil organic matter is attached to silicate surfaces and iron and aluminum hydroxides, it readily binds water to its surface. The solvent must be capable of removing the water and leaching the herbicides from soil.

A typical example would be to extract 10 g of soil with 40 mL of methanol/water (80 : 20% by volume). The solution contains a mixture of analytes and humic substances, which are the major degradation products of soil organic matter. The humic substances are a major interference in the analysis, especially if HPLC is to be used. If this extract were evaporated and applied to an HPLC column, resolution and sensitivity would be compromised by the high concentrations of humic substances.

The humic substances are soluble because of the presence of carboxyl and phenolic hydroxyl groups, and these groups may be used to affect a clean-up procedure. The negatively charged humic substances may be removed by anion exchange from the methanol/water extraction solvent. The procedure involves passing the methanol/water (80 : 20% v/v) through an anion-exchange sorbent in OH form. The hydroxide ion displaced by ion exchange of the carboxyl groups of the humic substances will further dissociate the humic phenolic-hydroxyl groups and result in effective removal of these compounds from the methanol/water without removing either atrazine or alachlor. The methanol/water fraction can then be analyzed directly by HPLC without further cleanup. If the concentration were too low for the analysis method, trace enrichment may be used, which is the second approach to soil analysis by SPE.

7.7.2. Trace Enrichment

The methanol extract of soil is diluted with water to increase the affinity of the solute for the C-18 sorbent. Figure 7.6 shows the effect of methanol concentration on the sorption capacity of atrazine and other triazine herbicides for C-18. At concentrations of methanol less than 10%, the capacity of the sorbent is more than adequate to isolate the atrazine and alachlor from the methanol/water phase. This procedure is many times simpler than evaporating the methanol; however, methanol evaporation may be necessary if more water-soluble compounds are present.

For example, Figure 7.6 shows that although the capacity for atrazine is adequate for the diluted methanol solution (8% methanol), it is inadequate for the isolation of the more soluble metabolite, deethylatrazine. As little as 2% methanol begins to affect the recovery of the deethylatrazine on the C-18 sorbent. And at 5% methanol, the deethylatrazine is only 60% recovered. Deisopropylatrazine, which is more soluble than deethylatrazine, is even more affected by the addition of methanol to the sample for trace enrichment. The methanol–water mixture could be evaporated so that only water is present and the aqueous phase added directly to the C-18 sorbent. Further dilution of the

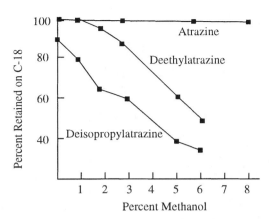

Figure 7.6. Effect of methanol concentration on capacity of atrazine, deethylatrazine, and deisopropylatrazine for a C-18 sorbent for a 100-mL sample. [Mills and Thurman (1992); published with permission.]

methanol–water mixture would result in a volume so large that breakthrough of the analytes would occur. Thus, an evaporation of methanol is followed by trace enrichment and the deethylatrazine eluted directly from the C-18 sorbent.

The concept of the deliberate addition of methanol to a water sample during trace enrichment entered the literature some years ago with early work by Andrews and Good (1982), who claimed that addition of methanol resulted in better solvation of the C-18 hydrophobic phase and more capacity when using large sample volumes (from 100 mL to 1 L). The data in Figure 7.6, and indeed the effect of methanol in reversed phase C-18, refutes this hypothesis. Reexamination of the data published by Andrews and Good (1982) shows that their data was for hydrophobic pesticides, such as dieldrin and heptachlor. These compounds have a tendency to sorb to containers and be removed from the aqueous sample prior to SPE. Furthermore, they are difficult to elute from C-18, depending on the elution solvent and drying time. Thus, the higher recoveries from the addition of methanol could easily have been caused by sample handling rather than by solvation of the C-18 sorbent. The addition of methanol to the mobile phase is not recommended for trace enrichment, if more hydrophilic compounds and metabolites are to be isolated. It may be used, however, for less soluble compounds, such as the nonionic nonpolar solutes (e.g., dieldrin, Section 7.10.2) as discussed later in this chapter.

7.7.3. Normal-Phase Solid-Phase Extraction of Soil Extracts

This involves the combination of the two ideas of trace enrichment and removal of interferences in soil extracts. In this procedure, the extract (a non-

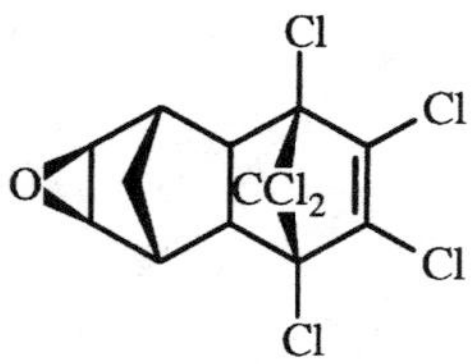

Figure 7.7. Structure of endrin.

polar solvent) is passed onto a normal-phase column and then analytes are eluted from the normal-phase sorbent in a series of fractions using solvents of increasing polarity. This method may be used when the soil has been extracted by a supercritical fluid and trapped in a hydrophobic solvent, Soxhlet extracted with a hydrophobic solvent, or extracted with a new procedure called accelerated solvent extraction (ASE). In these three procedures, the hydrophobic solvent is free of water, usually after drying over sodium sulfate.

An example is endrin in soil (Fig. 7.7). This highly chlorinated insecticide is extracted by Soxhlet extraction with hexane from a soil that has been mixed one to one with sodium sulfate to remove water. The hexane extract is further dried over sodium sulfate and the hexane is added to a hexane-loaded silica or CN SPE sorbent that has been prepared with hexane. The endrin is sorbed to the silica or CN by normal-phase mechanisms and may be eluted with a solvent that will overcome the interaction, such as methylene chloride or ethyl acetate. The more polar interferences are sorbed to the silica or CN sorbent and are not eluted with endrin. Thus, the SPE column performs both trace enrichment and SPE clean-up. See Chapter 5 for more examples.

7.8. AIR ANALYSIS AND SOLID-PHASE EXTRACTION

The typical method for air analysis involves the use of an air-sampling device, called a PUFF. The PUFF is an open-cell polyurethane foam material approximately 10 cm thick through which air is passed and onto which the compounds sorb. The air sample is usually filtered first through a 0.1-μm filter to remove particulates. Then the air passes through the PUFF, sucked through by a blower similar to that in a vacuum cleaner (Fig. 7.8). The PUFFs may be placed in tandem so any breakthrough of solute may be measured. The PUFF itself is extracted by Soxhlet for GC/MS analysis. If necessary, normal-phase SPE may be used to clean up the extract. The use of XAD resins and Tenax have also been applied to the isolation and analysis of volatile and semivolatile organic compounds in air (Wong et al., 1991). Furthermore, supercritical fluid

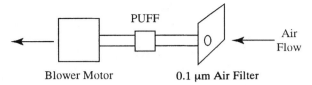

Figure 7.8. Air sampler for organic compounds using the principle of the foam PUFF.

extraction is used to improve recoveries of organic substances from these resins.

Formaldehyde and other aldehydes are receiving increasing attention both as toxic substances and as promoters in the photochemical formation of ozone in the atmosphere. They are released into residential buildings from plywood and particle board, insulation, combustion appliances, tobacco smoke, and various consumer products. Aldehydes are released into the atmosphere in the exhaust of motor vehicles and other equipment in which hydrocarbon fuels are incompletely burned. A sensitive method for analyzing aldehydes and ketones is based on the sorption of these compounds to an SPE sorbent and their subsequent reaction with 2,4-dinitrophenylhydrazine (DNPH) on the sorbent. They are then analyzed as their hydrazones by HPLC (Fig. 7.9). A gradient analysis by HPLC may separate as many as 17 components with detection by ultraviolet (UV) light.

The hydrazones may be detected by absorbance in the ultraviolet region, with maximum sensitivity obtained between 350 and 380 nm. The detection

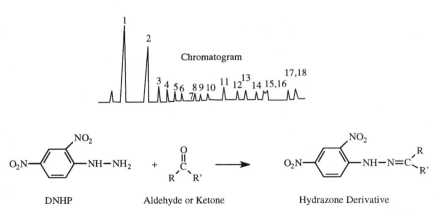

Figure 7.9. Formation of the hydrazone derivative of aldehydes from air and subsequent analysis by HPLC. (Published with permission of Waters.)

limit may be as low as 3 ppb by volume when a 100-L sample of air is taken. The SPE cartridges available for aldehydes in air are marketed by Waters (Product Guide, Appendix) and use the formation of a hydrazone derivative in the SPE cartridge—a type of on-cartridge derivatization. Waters also sells a small vacuum pump to pass air through the cartridge at a fixed rate.

7.9.　OIL AND GREASE ANALYSIS

Oil and grease involves the analysis of a second phase (an oil or grease phase) that is commonly present in wastewaters, such as effluents from paper mills, poultry plants, meat packing plants, manufacturing plants, and petroleum refineries. In the past, the method has used extraction by Freon, which is now considered hazardous for the ozone. The SPE method involves either the use of a silica C-18 cartridge or silica C-18 disk through which the sample is passed. The oil and grease are a complex mixture of hydrocarbons and waxes, which are not truly soluble but form a second phase and are removed by the sorbent. They may be eluted with methylene chloride and quantified gravimetrically (Varian Products Guide, Appendix).

The method involves the extraction of a 1 L sample through the special cartridge (Varian EnvirElut Oil and Grease) or the oil-and-grease disk by 3 M. Concentrations may be measured from 5 to 1000 mg/L. Blanks are typically less than 1 mg/L; all measurements are gravimetric. These methods are friendly to the ozone layer because of the use of nonchlorinated solvents.

7.10.　EXAMPLES OF ENVIRONMENTAL SOLID-PHASE EXTRACTION

7.10.1.　Organochlorine Pesticides in Water

These compounds are examples of nonpolar analytes with the structures shown in Figure 7.10. The organochlorine pesticides are no longer used in Europe and the United States but may be found in developing countries or may still persist in water and sediments of agricultural areas of Europe and the United States. The isolation method uses a reversed-phase mechanism but with a moderately polar phase, the CN sorbent, to enhance recovery during elution. The generic elution method will work well for these compounds, but care must be taken during solvent evaporation to prevent losses. A keeper solvent, such as acetone, may be added to prevent loss of analytes during evaporation.

Another approach is to use a disk with in-vial elution (this technique is discussed in Chapter 11. The method consists of using a filter disk and eluting the disk in the vial with ethyl acetate and immediately analyzing by GC/MS.

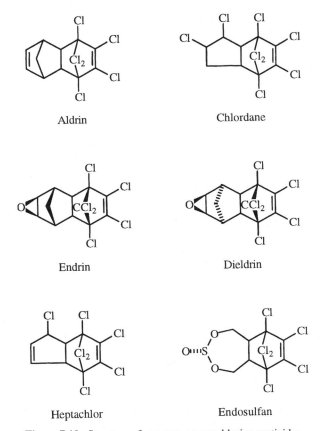

Figure 7.10. Structure of common organochlorine pesticides.

SPE Conditions Used

Sample	River water (100 mL), no sample additions.
Solutes	Endrin, dieldrin, chlordane, endosulfan.
Sorbent	360 mg of CN, precondition with 2 mL each of methanol, ethyl acetate, methanol, and distilled water.
Eluent	After sample addition, force residual water out of sorbent with air, and elute with ethyl acetate, 2 mL.
Reference	Modified from *Chromatographia*, **35**: 290–294 (1993).

7.10.2.   Organochlorine Pesticides in Soil and Sediment

Begin this method with the Soxhlet extraction of approximately 1 to 10 g of soil or sediment with 90% methanol/10% distilled water. The methanol will effectively wet the soil and sediments and remove the majority of the organochlorine pesticides. Alternatively, the soils and sediments may be extracted with heated solvent under pressure using the accelerated solvent extractor by Dionex (see Chapter 9 on the extraction of food and natural materials for details of analysis). After the extraction is complete, dilute the methanol extract with distilled water to a final concentration of 10% methanol. Process the extract through a C-18 cartridge as described in Section 7.10.1. In this case, the C-18 sorbent with greater hydrophobicity is used because of the 10% methanol present in the sample. Elute the cartridge with ethyl acetate and analyze by GC/MS.

SPE Conditions Used

Sample	Soil or sediment (20 g), no sample additions.
Solutes	Endrin, dieldrin, chlordane, endosulfan.
Extractant	90% methanol/10% distilled water, extract twice with 20 mL each time with Soxhlet, heated vial, or accelerated solvent extractor. Dilute to 360 mL total with distilled water and then process by SPE.
Sorbent	360 mg of C-18, precondition with 2 mL each of methanol, ethyl acetate, methanol, and distilled water.
Eluent	After sample addition, force residual water out of sorbent with air, and elute with ethyl acetate, 2 mL.
Reference	Modified from *Anal. Chem.*, **64**: 1985 (1992).

7.10.3.   Polyaromatic Hydrocarbons in Water and Soil

These compounds are another example of nonionic, nonpolar compounds. They are found in trace levels in water and result from combustion processes and hydrocarbon spills. They are trace enriched from water by sorption onto CN, C-8, or C-18 and elution with ethyl acetate/toluene. Toluene is added to the ethyl acetate eluent to increase solubility of the polyaromatic hydrocarbons (PAHs) and to enhance elution from the solid phase. The more hydrophobic PAHs, such as pyrene (Fig. 7.11), will recover more efficiently from a more polar reversed phase, such as CN or C-8 due to less strong van der Waals interactions between the analyte and the sorbent. The PAHs may be analyzed by either GC/MS or by HPLC. Soil samples may be processed as in Section 7.10.2 with 90% methanol extraction, followed by dilution with

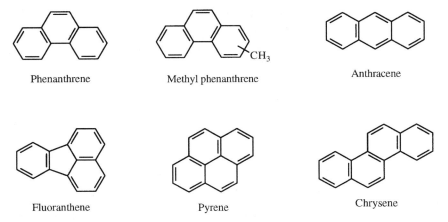

Phenanthrene   Methyl phenanthrene   Anthracene

Fluoranthene   Pyrene   Chrysene

Figure 7.11. Structures of polynuclear aromatic hydrocarbons.

water to 10% methanol and trace enrichment on C-18. Elute with ethyl acetate/toluene and proceed with GC/MS analysis.

SPE Conditions Used

Sample	River water (100 mL), no sample additions.
Solutes	PAHs.
Sorbent	360 mg of CN, C-8, or C-18, precondition with 2 mL each of methanol, ethyl acetate/toluene, methanol, and distilled water.
Eluent	Elute with ethyl acetate/toluene, 2 mL of 50/50. If a more hydrophobic solvent is chosen (hexane or methylene chloride), the sorbent must be carefully dried and free of water.
Reference	Modified from *Chromatographia*, **28**: 203–211 (1989) and Bakerbond Application Note EN-019 PAHs from Drinking Water.

SPE Conditions Used

Sample	Soil or sediment (20 g), no sample additions.
Solutes	PAHs.
Extractant	90% methanol/10% distilled water, extract twice with 20 mL each time with Soxhlet, heated vial, or accelerated solvent extractor. Dilute to 360 mL total with distilled water and then process by SPE.

Sorbent   360 mg of C-18, precondition with 2 mL each of methanol,
               ethyl acetate, methanol, and distilled water.
Eluent   After sample addition, force residual water out of sorbent with
               air, and elute with ethyl acetate, 2 mL.
Reference   Modified from *Anal. Chem.*, **64**: 1985 (1992).

7.10.4. Nonionic Herbicides in Water and Soil

This example is a nonionic, moderately polar suite of pesticides. The compounds shown in Figure 7.12 are commonly used in the United States and occur in water at micrograms per liter concentrations. They have water solubilities of ~10 to 500 mg/L, which makes them moderately polar substances and easily isolated by trace enrichment using C-18. Filtration of the water sample removes the particulate matter, but not the herbicides. They are present chiefly in the water phase, with less than 1% sorbed to the suspended sediments for most nonionic compounds (Squillace and Thurman, 1992).

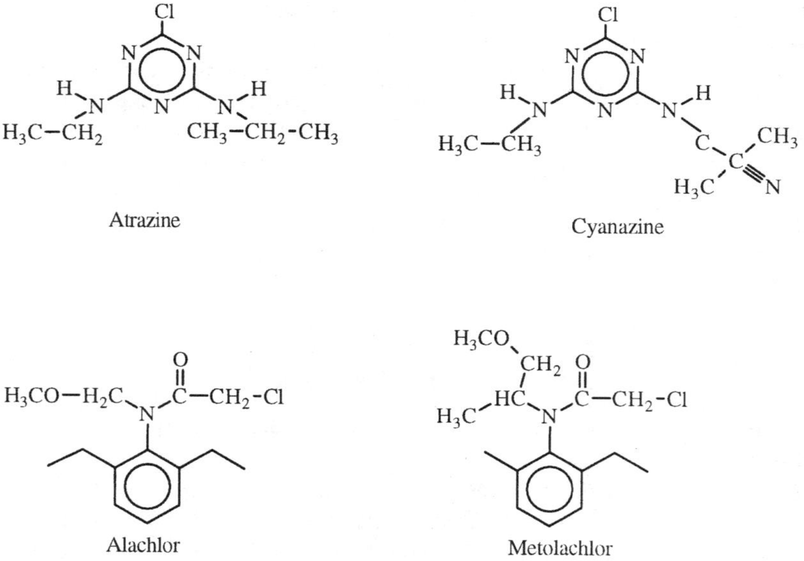

Figure 7.12. Structures of common corn herbicides.

SPE Conditions Used

Sample	Water (100 mL), no sample additions.
Sorbent	360 mg of C-18, 40-μm particles, precondition with 2 mL each of methanol, ethyl acetate, methanol, and distilled water.
Eluent	After sample addition, force residual water out of sorbent with air, and elute with ethyl acetate, 2 mL.
Reference	*Anal. Chem.*, **62**: 2043–2048 (1990).

SPE Conditions Used

Sample	Water (1000 mL), no sample additions.
Sorbent	500 mg of C-18 (3 M), 8-μm particles in 47-mm Teflon disk, precondition with 5–10 mL each of methanol, ethyl acetate, methanol, and distilled water.
Eluent	After addition of sample dry the sample with air for 10 min, then elute with ethyl acetate, 10 mL.
Reference	Modified procedure from the 3 M Brochure. Published with permission of 3 M.

SPE Conditions Used

Sample	Soil or sediment (20 g), no sample additions.
Solutes	Herbicides.
Extractant	90% methanol/10% distilled water. Extract twice with 20 mL each time with Soxhlet, heated vial, or accelerated solvent extractor. Dilute to 360 mL total with distilled water and then process by SPE.
Sorbent	360 mg of C-18, precondition with 2 mL each of methanol, ethyl acetate, methanol, and distilled water.
Eluent	After sample addition, force residual water out of sorbent with air, and elute with ethyl acetate, 2 mL.
Reference	Modified from *Anal. Chem.*, **64**: 1985 (1992).

7.10.5. Organophosphorus Pesticides in Water and Soil

Organophosphorus pesticides are another example of nonionic, moderately polar compounds (Fig. 7.13). These pesticides are commonly used today and include compounds such as methyl parathion, parathion, diazinon, and dimethoate. They are used as insecticides on crops such as cotton, as fungicides on fruits, and as nematocides on soil. They are moderately polar

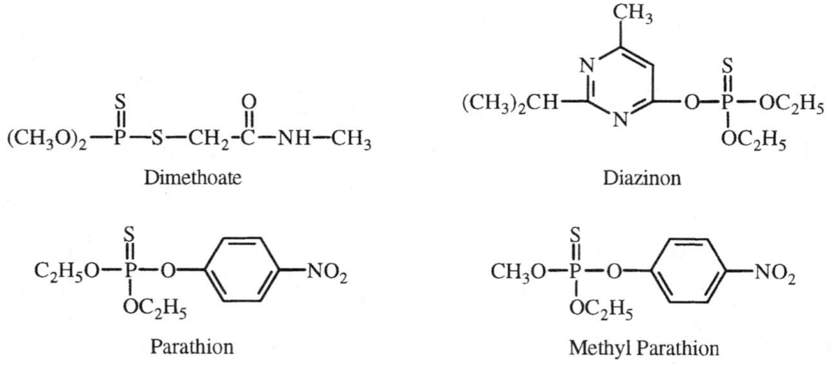

Figure 7.13. Structures of organophosphorus pesticides.

compounds, which means that they are readily isolated by trace enrichment on C-18. They may be eluted with a generic solvent such as ethyl acetate and analyzed directly by GC/MS.

SPE Conditions Used

Sample	River water (100 mL), no sample additions.
Solutes	Methyl parathion and other organophosphorus pesticides.
Sorbent	360 mg of C-18, precondition with 2 mL each of methanol, ethyl acetate, methanol, and distilled water.
Eluent	After sample addition, force residual water out of sorbent with air, and elute with ethyl acetate, 2 mL.
Reference	Modified from *J. Chromatog.*, **644**: 49 (1993).

SPE Conditions Used

Sample:	Soil or sediment (20 g) no sample additions.
Solutes:	Methyl parathion and other organophosphorus pesticides.
Extractant:	90% methanol/10% distilled water, extract twice with 20 mL each time with Soxhlet, heated vial, or accelerated solvent extractor. Dilute to 360 mL total with distilled water and then process by SPE.
Sorbent:	360 mg of C-18, precondition with 2 mL each of methanol, ethyl acetate, methanol, and distilled water.
Eluent:	After sample addition, force residual water out of sorbent with air, and elute with ethyl acetate, 2 mL.
Reference:	Modified from *Anal. Chem.*, **64**: 1985 (1992).

7.10.6. Nonvolatile Organic Compounds: Carbamates, Substituted Amides, and Phenylureas in Water and Soil

These compounds are nonionic and moderately polar to polar herbicides (Fig. 7.14). They are analyzed by liquid chromatography rather than by gas chromatography. Their more labile nature and lack of volatility lend these compounds to an approach using C-18 followed by elution with a polar solvent, such as methanol. Because of the aqueous solubility of some of these compounds, SDB may be substituted for the C-18 sorbent to increase the sorption capacity.

SPE Conditions Used

Sample	River water (100 mL), no sample additions.
Solutes	Carbamate, phenylureas, and substituted amides.
Sorbent	360 mg of C-18, precondition with 2 mL each of methanol and distilled water.
Eluent	Elute with methanol, 2 mL.
Reference	Modified from *Int. J. Environ. Anal. Chem.*, **42**: 15 (1990) and Bakerbond Application Note EN-025 Extraction of Nonvolatile Organic Compounds from Water. Published with permission of J. T. Baker, Inc. (See Suggested Reading, Chapter 1).

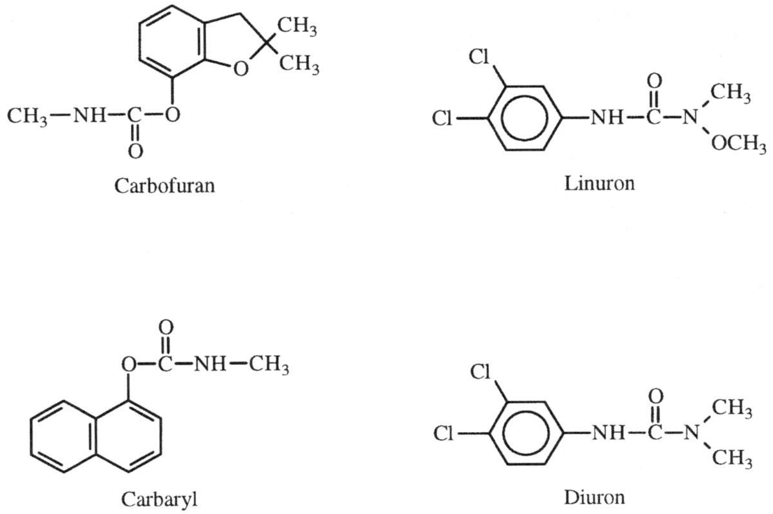

Carbofuran   Linuron

Carbaryl   Diuron

Figure 7.14. Structures of carbamate, phenylureas, and substituted amide pesticides.

SPE Conditions Used

Sample	Soil or sediment (20 g), no sample additions.
Solutes	Carbamate, phenylureas, and substituted amides.
Extractant	90% methanol/10% distilled water, extract twice with 20 mL each time with Soxhlet, heated vial, or accelerated solvent extractor. Dilute to 360 mL total with distilled water and then process by SPE.
Sorbent	360 mg of C-18, precondition with 2 mL each of methanol, ethyl acetate, methanol, and distilled water.
Eluent	After sample addition, force residual water out of sorbent with air, and elute with ethyl acetate, 2 mL.
Reference	Modified from *Anal. Chem.*, **64**: 1985 (1992).

7.10.7. Herbicide Metabolites in Water and Soil

These compounds are examples of nonionic, polar herbicide metabolites (Fig. 7.15). The examples shown below are degradation products of the triazine compounds shown in Section 7.10.4. They are nonionic compounds that are formed in soil by the dealkylation of the parent compounds. They are considerably more soluble than the parent compounds with aqueous solubilities equal to or greater than 1000 mg/L. Therefore, the sorbent to choose in this case is the styrene–divinylbenzene or graphitized carbon. The compounds are soluble in ethyl acetate or methanol, either of which may be used as an elution solvent. The C-18 sorbent may also be used with good recovery of either deethylatrazine or deisopropylatrazine, but with no recovery of the didealkylatrazine.

SPE Conditions Used

Sample	River water (100 mL), no sample additions.
Solutes	Deethylatrazine, deisopropylatrazine, and didealkylatrazine.
Sorbent	1 g of SDB, precondition with 2 mL each of methanol, ethyl acetate, methanol, and distilled water.

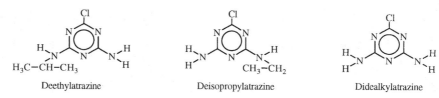

Deethylatrazine   Deisopropylatrazine   Didealkylatrazine

Figure 7.15. Structure of triazine metabolites.

Eluent	After sample addition, force residual water out of sorbent with air, and elute with ethyl acetate, 2 mL.
Reference	Unpublished data.

SPE Conditions Used

Sample	River water (100 mL), no sample additions.
Solutes	Deethylatrazine, deisopropylatrazine, and didealkylatrazine.
Sorbent	360 mg of C-18, precondition with 2 mL each of methanol, ethyl acetate, methanol, and distilled water.
Eluent	After sample addition, force residual water out of sorbent with air, and elute with ethyl acetate, 2 mL.
Reference	*Anal. Chem.*, **62**: 2043 (1990).
Notes	Low recovery of didealkylatrazine (60%). Substitution of SDB sorbent, 500 mg would bring up recovery to near 100 %. See in Section 7.10.8.

SPE Conditions Used

Sample	Soil or sediment (20 g), no sample additions.
Solutes	Deethylatrazine, deisopropylatrazine, and didealkylatrazine.
Extractant	90% methanol/10% distilled water, extract twice with 20 mL each time with Soxhlet, heated vial, or accelerated solvent extractor. Evaporate the majority of the methanol off sample under nitrogen, then process aqueous extract (~ 4 mL) by SPE.
Sorbent	1 g of SDB, precondition with 2 mL each of methanol, ethyl acetate, methanol, and distilled water.
Eluent	After sample addition, force residual water out of sorbent with air, and elute with ethyl acetate, 2 mL.
Reference	Modified from *Anal. Chem.*, **64**: 1985 (1992).

7.10.8. Phenols and Chlorinated Phenols in Water and Soil

These compounds are examples of ionic moderately polar compounds (Fig. 7.16). They are ionic only at certain pH ranges. Phenol itself is actually a polar compound, while the chlorinated phenols are moderately polar. Thus, phenol would best be isolated by an SDB sorbent, while the chlorinated phenols may be isolated effectively on C-18 or SDB. The pK_a of these compounds varies from 8 to 10. Thus, it is critical that the pH of the sample be approximately 7, or less, for good retention of the nonionic species.

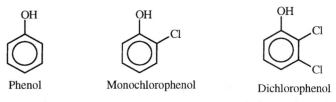

Figure 7.16. Structure of phenol and chlorinated phenols.

These compounds are important industrial chemicals, and phenol itself is listed on the EPA's contaminant list. The elution solvents that are miscible with water work best: acetone, acetonitrile, and methanol. There may be the possibility of methylation if methanol is used, and that should be considered during the analysis.

SPE Conditions Used

Sample	River water (100 mL), no sample additions.
Solutes	Chlorinated phenols.
Sorbent	1 g of C-18 or SDB, precondition with 2 mL of acetonitrile and distilled water.
Eluent	Elute with acetonitrile, 2 mL (methanol or acetone could be substituted).
Reference	Modified from *Chromatographia*, **18**: 323–325 (1984) and Bakerbond Application Note EN-012 Phenols. Published with permission of J. T. Baker, Inc.

SPE Conditions Used

Sample	Soil or sediment (20 g), no sample additions.
Solutes	Chlorinated phenols.
Extractant	90% methanol/10% distilled water, extract twice with 20 mL each time with Soxhlet, heated vial, or accelerated solvent extractor. Dilute to 360 mL total with distilled water and then process by SPE.
Sorbent	1 g of SDB, precondition with 2 mL each of methanol, ethyl acetate, methanol, and distilled water.
Eluent	After sample addition, force residual water out of sorbent with air, and elute with ethyl acetate, 2 mL.
Reference	Modified from *Anal. Chem.*, **64**: 1985 (1992).

7.10.9.   Acid Herbicides (2,4-D, 2,4,5-T, Dicamba) and Sulfourea Herbicides in Water and Soil

These herbicides are anionic compounds that are moderately polar (Fig. 7.17). The acid-herbicide family consists of compounds that are ionic or partially ionic at the pH of a typical water sample (pH = 7). Thus, the capacity of these compounds for C-18 is quite low without acidification of the water sample. Typically, the pH must be lowered to 2 pH units below the pK_a for effective recovery, usually not to be lower than pH 2.0. At pH units below this value, damage may occur to silica sorbent. Furthermore, the silica sorbents require trifunctional silica rather than monofunctional silica in order to prevent hydrolysis of the sorbent. Sorbents of SDB are more stable and can be used as low as pH 1. Methanol is a better eluent than ethyl acetate because of the ionic character of the analytes.

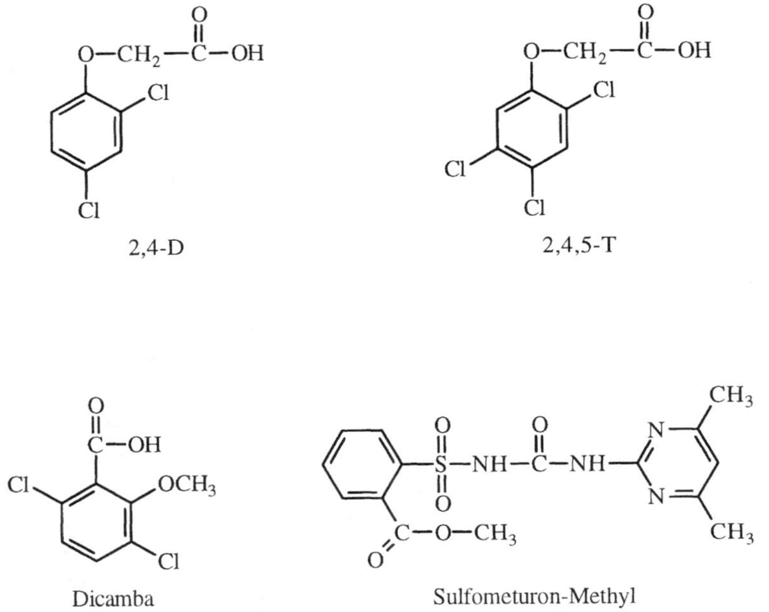

Figure 7.17. Structures of acid herbicides.

SPE Conditions Used

Sample	River water (100 mL), pH adjustment to 2.0 or 2 pH units below pK_a.
Solutes	Acid herbicides (2,4-D, 2,4,5-T, dicamba) and sulfourea herbicides.
Sorbent	400 mg of C-18 (trifunctional) or SDB, precondition with 2 mL each of methanol and distilled water (pH 2).
Eluent	Elute with methanol, 2 mL.
Reference	Modified from Journal of Chromatography, v. 25 (1987) 345–350, Analytical Chemistry, v. 59 (1987) 1739–1742, and Bakerbond Application Note EN-024 Chlorophenoxy Herbicides. Published with permission. (See Suggested Reading, Chapter 1.)

SPE Conditions Used

Sample	Soil or sediment (20 g), no sample additions.
Solutes	Acid herbicides, (2,4-D, 2,4,5-T, dicamba) and sulfourea herbicides.
Extractant	90% methanol/10% distilled water, Extract twice with 20 mL each time with Soxhlet, heated vial, or accelerated solvent extractor. Dilute to 360 mL total with distilled water and then process by SPE.
Sorbent	1 g of SDB, precondition with 2 mL each of methanol, ethyl acetate, methanol, and distilled water.
Eluent	After sample addition, force residual water out of sorbent with air, and elute with ethyl acetate, 2 mL.
Reference	Modified from Anal. Chem., **64**: 1985 (1992).

7.10.10.   Anionic Surfactants in Water and Soil

These compounds are permanently anionic and are moderately polar (surfactants are organic molecules that are surface active). This means that they concentrate on the surface of a liquid in which they are dissolved. Generally, these types of analytes contain both a hydrophobic and a hydrophilic segment. There are anionic, cationic, neutral, and amphoteric surfactants. They may be readily sorbed from water by reversed-phase SPE. Elution requires methanol or acetonitrile rather than ethyl acetate because of their polar, ionic functional groups, which are typically sulfate esters or sulfonic acids (Fig. 7.18).

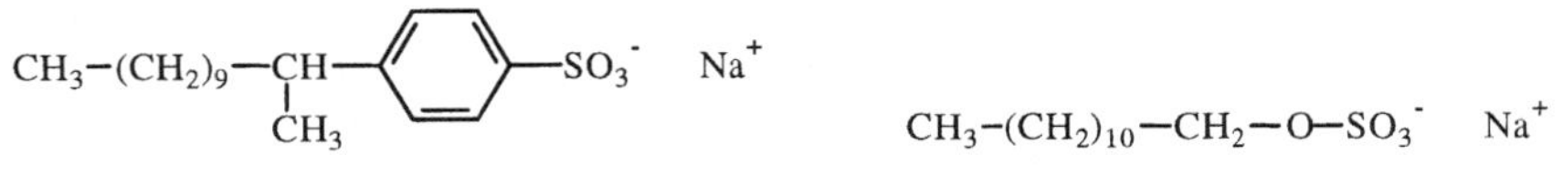

$$CH_3-(CH_2)_9-\underset{\underset{CH_3}{|}}{CH}-\text{⟨benzene⟩}-SO_3^- \quad Na^+$$

$$CH_3-(CH_2)_{10}-CH_2-O-SO_3^- \quad Na^+$$

Isomer of LAS   SDS

Figure 7.18. Structures of anionic surfactants.

SPE Conditions Used

Sample	Waste water (100 mL), no sample additions.
Solutes	Linear alkylsulfonate surfactants.
Sorbent	360 mg of C-8, but C-18 would work equally well. Precondition with 2 mL each of elution solvent acetonitrile or methanol, and distilled water.
Eluent	Elute with acetonitrile, but methanol works equally well, 2 mL.
Reference	*Anal. Chem.*, **62**: 2581–2586 (1990).

SPE Conditions Used

Sample	Soil or sediment (20 g), no sample additions.
Solutes	Anionic surfactants.
Extractant	90% methanol/10% distilled water, extract twice with 20 mL each time with Soxhlet, heated vial, or accelerated solvent extractor. Dilute to 360 mL total with distilled water and then process by SPE.
Sorbent	360 mg of C-18, precondition with 2 mL each of methanol, ethyl acetate, methanol, and distilled water.
Eluent	Elute with methanol, 2 mL.
Reference	Modified from *Anal. Chem.*, **62**: 2581 (1990).

SUGGESTED READING

Brouwer, E. R., Liska, I., Geerdink, R. B., Frintrop, P. C. M., Mulder, W. H., Lingeman, H., and Brinkman, U. A. Th. 1991. Determination of polar pollutants in water using an on-line liquid chromatographic preconcentration system, *Chromatographia*, **32**: 445–452.

Font, G., Manes, J., Molto, J. C., and Pico, Y. 1993. Solid-phase extractions in multiresidue pesticide analysis of water, *J. Chromatog.*, **642**: 135–161.

van der Hoff, G. R., Gort, S. M., Baumann, R. A., van Zoonen, P., and Brinkman, U. A. Th. 1991. Clean-up of some organochlorine and pyrethroid insecticides by automated solid-phase extraction cartridges coupled to capillary GC-ECD, *J. High Resol. Chromatog.*, **14**: 465–470.

Junk, G. A. and Richard, J. J. 1988. Organics in water: Solid-phase extraction on a small scale, *Anal. Chem.*, **60**: 451–454.

Loconto, P. R. 1991. Solid-phase extraction in trace environmental analysis, part I—current research, *LC-GC*, **9**: 460–465.

Loconto, P. R. 1991. Solid-phase extraction in trace environmental analysis, current research—Part II, *LC-GC*, **9**: 752–760.

McDonald, P. D. and Bouvier, E. S. P. 1995. *Solid Phase Extraction Applications Guide and Bibliography, A Resource for Sample Preparation Methods Development*, 6th ed., Waters, Milford, MA.

Raisglid, M., Burke, M. F., and Van Horne, K. C. 1993. Factors affecting the reliability of solid-phase extraction for environmental analysis, International Symposium on Laboratory Automation and Robotics, Amsterdam, pp. 1–7.

Snyder, J. L., Grob, R. L., McNally, M. E., Oostdyk, T. S. 1994. A different approach—using solid-phase disks and cartridges to extract organochlorine and organophosphate pesticides from soils, *LC-GC*, **12**: 230–242.

Simpson, N. and Van Horne, K.C., Eds., 1993. *Varian Sorbent Extraction Technology Handbook*. Varian Sample Preparation Products, Harbor City, CA.

Wells, M. J. M. and Michael, J. L. 1987. Reversed-phase solid-phase extraction for aqueous environmental sample preparation in herbicide residue analysis, *J. Chromatog. Sci.*, **25**: 345–350.

Zhou, S. W., Malaiyandi, M. and Benoit, F. M. 1990. An investigation on the concentration efficiencies of some macroreticular and ambersorb resins using radiolabeled organic contaminants commonly encountered in water, *J. Environ. Anal. Chem.*, **38**: 439–471.

REFERENCES

Andrews, J. S. and Good, T. J. 1982. Trace enrichment of pesticides using bonded-phase sorbents, *Am. Lab.*, April, 70–75.

Barcelo, D. and Hennion, M.-C. 1997. *Trace Determination of Pesticides and Their Degradation Products in Water*. Elsevier, Amsterdam.

Etheir, A. G. and Taylor, S. 1994. Intercomparison of *In Situ* Solid-Phase Extraction (SPE) and Liquid-Liquid Extraction (LLE) Techniques for the determination of trace organic contaminants in fresh water. Axys, Inc. internal publication.

Hennion, M. C. and Pichon, V. 1994. Solid-phase extraction of polar organic solutes from water, *Environ. Sci. Tech.*, **28**: 576A–583A.

Hennion, M. C., Subra, P., Rosset, R., Lamacq, J., Scribe, P., and Saliot, A. 1990. Off-line and on-line preconcentration techniques for the determination of phenylureas in freshwaters, *Int. J. Environ. Anal. Chem.*, **42**: 15–33.

Junk, G. A., Richard, J. J., Grieser, M. D., Witiak, D., Witiak, J. L., Arguello, M. D., Vick, R., Svec, H. J., Fritz, J. S., and Calder, G. V. 1974. Use of macroreticular resins in the analysis of water for trace organic compounds, *J. Chromatog.*, **99**: 745–762.

Leenheer, J. A. and Stuber, H. A. 1981. Migration through soil of organic solutes in an oil-shale process water, *Environ. Sci. Tech.*, **15**: 1467–1475.

Little, J. N. and Fallick, G. J. 1975. New considerations in detector-application relationships, *J. Chromatog.*, **112**: 389–397.

Markell, C. and Hagen, D.F. 1996. Solid-phase extraction basics for water analysis. In *Principles of Environmental Sampling*, 2nd ed. (Keith, L. H., Ed.), Chapter 7, American Chemical Society, Washington, D.C.

McDonnelland, T., Rosenfeld, J., and Rais-Firouz, A. 1993. Solid-phase sample preparation of natural waters with reversed-phase disks, *J. Chromatog.*, **629**: 41–53.

Mills, M. S. and Thurman, E. M. 1992. Mixed-mode isolation of triazine metabolites from soil and aquifer sediments using automated solid-phase extraction, *Anal. Chem.*, **64**: 1985–1990.

Raisglid, M., Burke, M. F., and Van Horne, K. C. 1993. Factors affecting the reliability of solid-phase extraction for environmental analysis: International Symposium on Laboratory Automation and Robotics, Amsterdam, pp. 1–7.

Squillace, P. J. and Thurman, E. M. 1992. Herbicide transport in rivers: Importance of hydrology and geochemistry in nonpoint-source contamination, *Environ. Sci. Tech.*, **26**: 538–545.

Thurman, E. M. 1985. *Organic Geochemistry of Natural Waters*. Martinus Nihjoff/Dr. W. Junk, Dordrechdt.

Thurman, E. M. and Malcolm, R. L. 1981. Preparative isolation of aquatic humic substances, *Environ. Sci. Tech.*, **15**: 463–466.

Thurman, E. M., Malcolm, R. L., and Aiken, G. R. 1978. Prediction of capacity factors for aqueous organic solutes adsorbed on a porous acrylic resin, *Anal. Chem.*, **50**: 775–779.

Wong, J. M., Kado, N. Y., Kuzmicky, P. A., Ning, H. S., Woodrow, J. E., Hsieh, D. P. H., and Seiber, J. N. 1991. Determination of volatile and semivolatile mutagens in air using solid adsorbents and supercritical fluid extraction, *Anal. Chem.*, **63**: 1644–1650.

CHAPTER

8

DRUGS AND PHARMACEUTICALS

8.1. INTRODUCTION

The use of solid-phase extraction (SPE) for the sample preparation of drugs and pharmaceuticals has increased over the last 15 years because of ease of operation, increased selectivity with many new phases, and interfacing of automation and robotics. A simple strategy now exists for SPE methods development of drugs and pharmaceuticals, which makes sample preparation extremely straightforward. This strategy is the use of generic mixed-mode SPE and is discussed in detail in this chapter.

There are three considerations in effective SPE of drugs from matrices such as urine, blood, and plasma. First is the chemistry of the drug and whether it is an acid, base, or neutral compound. Second is the matrix from which the drug is being removed. Third is the use of sorbents that will effectively retain the analyte while interferences are washed from the column. Then the drug may be eluted into a clean fraction that can either be analyzed directly by high-pressure liquid chromatography (HPLC), enzyme-linked immunosorbent assay (ELISA), or derivatized for analysis by gas chromatography/mass spectrometry (GC/MS). These three points are considered in this chapter, as well as some examples of simple methods for drugs of abuse by the National Institute for Drug Abuse (NIDA). The analytes include amphetamines, cocaine, marijuana, opiates, and phencyclidine (PCP). Furthermore, there is a discussion of general drug analysis, veterinary applications of SPE, and SPE applications by disk methods.

8.2. STRUCTURE AND CHEMISTRY OF DRUGS

Drugs may be classified into four categories: basic, acidic, neutral, and amphoteric compounds. The concept is based on the ionic state of the compound at various pH values and has been the basis of separation in liquid chromatography and rapid screening methods, such as those using Toxi-Lab (Chia and Gere, 1987) and JETUBE (Nickel, 1977; Brooks and Smith, 1991). A drug that contains a basic functional group, such as an amine, is capable of accepting a proton at low pH and becoming a cation. Thus, basic drugs are

197

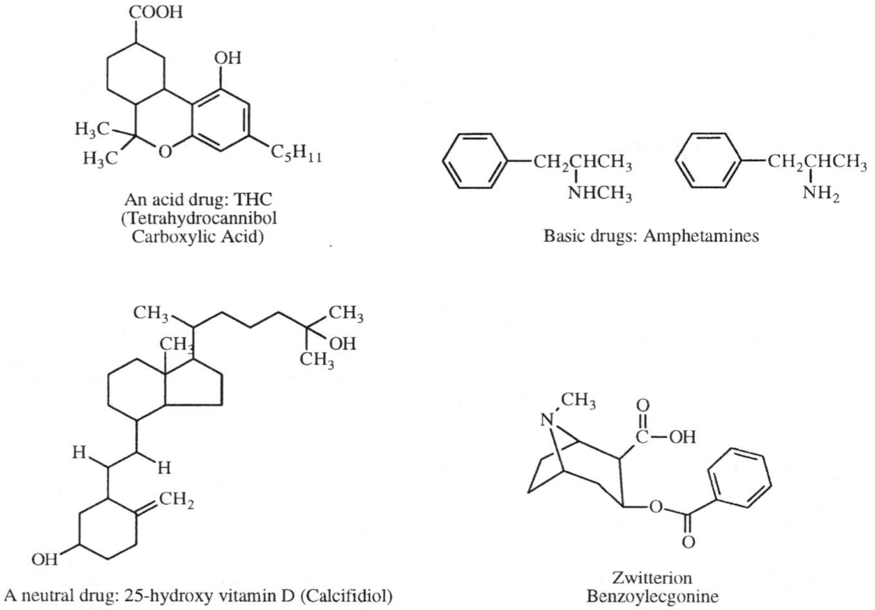

An acid drug: THC
(Tetrahydrocannibol
Carboxylic Acid)

Basic drugs: Amphetamines

A neutral drug: 25-hydroxy vitamin D (Calcifidiol)

Zwitterion
Benzoylecgonine

Figure 8.1. Four classes of drugs or pharmaceuticals, based on ionic state.

cations at a certain pH range. A drug that contains an acidic functional group, such as a carboxylic acid, is capable of donating a proton at pHs above 5 and becoming an anion. These are acidic drugs. Drugs that contain neither acidic nor basic functional groups are neutral drugs, because they are not ionic at any pH. Finally, the fourth category are drugs that contain both acidic and basic functional groups; they are called amphoteric drugs. They may be either cationic, anionic, or zwitterions (both positively and negatively charged at the same time), depending on the pH. Figure 8.1 shows examples of each of the four classes of drugs. A feature of over 80% of drugs is that they contain nitrogen (Engelhardt and Jungheim, 1990), often in a form that may be readily protonated, depending on the pH, to create a cation. This feature of drugs will be exploited by SPE, especially by a method of SPE called *mixed mode*, to concentrate and purify the drug for chromatographic analysis.

8.3. SAMPLE MATRIX

Matrices commonly encountered in drug analysis consist of body fluids, such as plasma, serum, lysed blood, urine, or sputum (saliva). These fluids are often

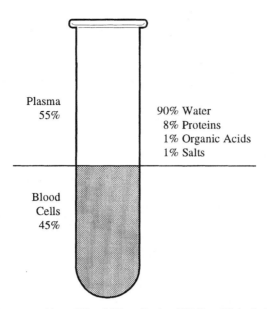

Plasma
55%

90% Water
8% Proteins
1% Organic Acids
1% Salts

Blood
Cells
45%

Figure 8.2. The composition of blood. [From Jordon (1993), published with permission.]

viscous and may need some prior treatment before SPE effectively can be applied. For example, a blood sample should be free of particulates, and several methods for sample preparation prior to SPE may be used. First is centrifugation, which may be applied to blood to separate the red blood cells from the plasma. Figure 8.2 shows that a centrifuged blood sample that has anticoagulant will consist of 55% plasma and 45% red blood cells. The plasma consists of approximately 90% water, 8% proteins, 1% organic acids, and approximately 1% salts.

If this sample of blood did not contain anticoagulant, it would have clotted. If the clotted-blood sample were again centrifuged, the two separated phases would now be red blood cells and serum. The serum is the clear liquid above the cells, similar to plasma. Where serum and plasma are different is in their protein concentration. Because the blood sample was allowed to clot, many of the proteins present in the serum have been used in the clot formation. The end result is that serum has approximately half the protein content of plasma and is less viscous. Typically, pH 7.0 phosphate buffer ($0.01\,M$ K_2HPO_4) must be added to a plasma sample at a 1:1 dilution prior to SPE because plasma is so viscous; whereas, it is not necessary to add buffer to serum samples. Although a buffer is often added to control pH. A plasma sample may show protein clots if the sample has been thawed and frozen several times. These characteristics make biological samples more difficult to work with than most environmental samples.

Urine samples may need addition of phosphate buffer prior to SPE for pH adjustment. If particulate matter is present, it is sometimes also necessary to filter or centrifuge. Furthermore, after storage in the refrigerator, urine samples may have precipitates, which may be dissolved by warming the sample prior to SPE. Sputum or saliva samples are quite viscous because they consist of high-molecular-weight proteins. These may be diluted 1 : 1 or 1 : 2 with potassium phosphate buffer. All biological samples should be free of clots and precipitates before SPE analysis. Finally, it is important to take more care in cleaning of the SPE apparatus when using biological samples because of the sticky nature of the proteinaceous substances being analyzed. Useful cleaning agents include 1% sodium dodecyl sulfate, 5% Liqui-Nox (Alconox, Inc.), $0.1N$ NaOH, and denatured alcohols.

One sorbent that has been commonly used for drug isolation and purification is reversed phase. In the reversed-phase approach, many organic components of body fluids may also be isolated. These interferences are sorbed to the C-18 column and co-elute with the drug. Because of these interferences, a new method of SPE was invented that would combine several sorption mechanisms. Out of the need for a more effective clean-up procedure for SPE of drugs from biological fluids, the mixed-mode sorbent was invented.

8.4. MIXED-MODE SORBENTS

The concept of mixed mode consists in using combined modes of sorption, usually reversed phase and cation exchange. Because the body fluids contain high concentrations of organic compounds that interfere with one mode of sorption, the mode of isolation has to be changed such that the drug is held by one mode, and the interferences are not held but are removed. Then the other mode of interaction is used to sorb the drug and the remaining interferences are eluted from the sample. Finally, both modes of interaction are canceled and the drug is eluted from the SPE tube. Figure 8.3 shows an example of how a metabolite of cocaine, called benzoylecgonine, is sorbed, isolated, and purified from urine using a mixed-mode sorbent.

The example sorbent for the mixed-mode isolation of benzoylecgonine is a combination of reversed phase using C-8 and cation exchange using a benzene sulfonic-acid cation-exchange resin in hydrogen form. Figure 8.4 shows how the mixed-mode separation works and is summarized from a study by Dixit and Dixit (1990), Chen and co-workers (1992), and Mills and co-workers (1993). In step 1 of Figure 8.4 a urine sample is added to the mixed-mode

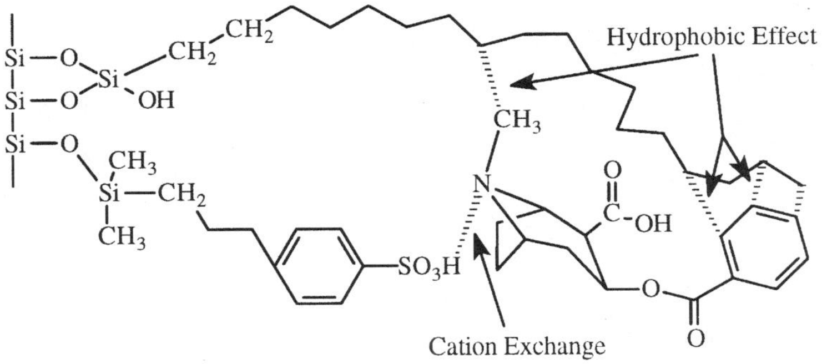

Figure 8.3. An example of benzoylecgonine sorption by a mixed-mode sorbent. (After Mills et al., 1993.)

sorbent. A typical volume of sample is 5 mL per column, which may contain approximately 500 mg of sorbent. The pH of the urine sample is 7.5; therefore, the analyte is negatively charged at the carboxyl site, and there is no charge at the nitrogen site. The analyte is an amphoteric compound capable of both cationic and anionic charge. The urine contains many salts, which immediately exchange with the hydrogen present on the cation-exchange site and convert it to a sodium form. However, also in step 1, the sample is loaded on the sorbent with a potassium phosphate buffer, and the resin is converted to the potassium form by addition of this buffer. The urine may contain a mixture of cations, but the buffer is potassium. The pH of the urine is 7.5, which ionizes the carboxyl group of the benzoylecgonine metabolite but leaves the amine group neutral; thus, the metabolite is sorbed in its ionic form by reversed phase, with a potassium ion acting as a counter ion. Because the volume of sample is small, only 5 mL, the sorbent is capable of removing the benzoylecgonine from urine by reversed phase, in spite of it being an anion.

Next (still step 1), the sorbent is washed with deionized water to remove excess salts that are present in the void volume of the sorbent and to remove water-soluble organic compounds, such as urea. The carboxyl group on the benzoylecgonine is present as a sodium or potassium salt. Step 2 is the acid-lock step, which involves a rinse of the sorbent with 0.1N HCl. The acid displaces the cations present on the cation-exchange resin and converts the carboxyl anion to a carboxylic acid. Furthermore, the amine nitrogen is protonated and the benzoylecgonine is bound by cation exchange, as well as by reversed phase. Thus, the analyte is "locked on" by two binding mechanisms simultaneously. Step 3 removes the excess acid with deionized water and then a rinse with methanol to remove organic interferences that do not contain

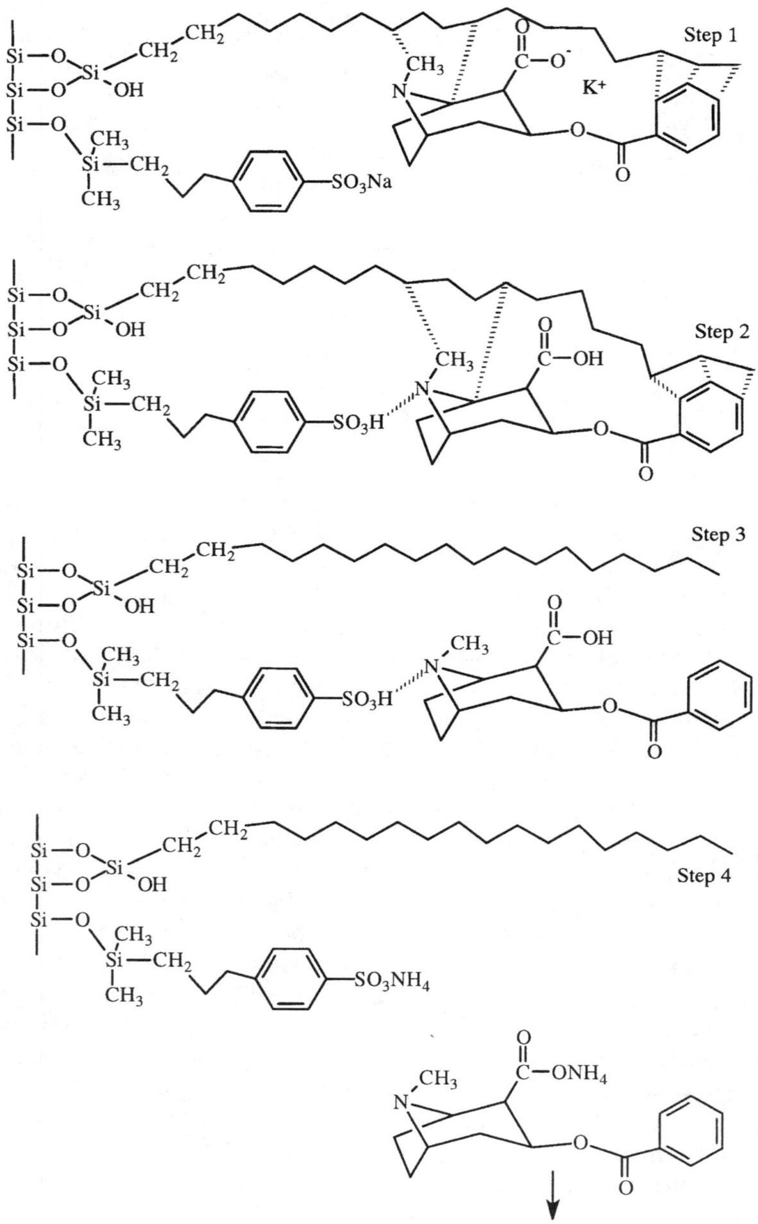

Figure 8.4. Four-step process for the isolation of basic and amphoteric drugs using mixed-mode sorbents.

nitrogen. Thus, the methanol rinse breaks the hydrophobic bond but not the cation-exchange bond. This is the critical step that cleans up the extract for good chromatography. Step 4 elutes the benzoylecgonine with a mixture of ethyl acetate, methanol, and 2% ammonium hydroxide (80 : 18 : 2 by volume). This eluent breaks both mechanisms of sorption and elutes the benzoylecgonine for analysis by gas or liquid chromatography. The 2% ammonium hydroxide removes the proton and "unlocks" the ion-exchange step of sorption. The methanol allows for excess water to be removed during the elution step, some of which is held by ion-exchange sites. Also the methanol breaks up the reversed-phase interactions and facilitates the addition of the ammonium hydroxide to the ethyl acetate. The ethyl acetate elutes the drug and prevents sorption by reversed phase.

8.5. APPLICATION OF MIXED MODE

The mixed-mode application may be applied to nearly all basic and amphoteric drugs. Table 8.1 shows a list of drugs that may be separated and isolated by mixed-mode sorbents. Acidic and neutral drugs cannot be isolated by a mixed-mode method, rather they should be isolated by the traditional mechanism of reversed phase. Examples will be shown in a later section. A good guide to the use of mixed mode is found in the catalog of World Wide Monitoring, now called United Chemical Technologies (see Appendix and Suggested Reading in this chapter).

There have been reports of alkylation of drugs at ion-exchange sites on the surface of some solid-phase extraction columns (Rymut et al., 1996). The authors report that alkyl sulfonates may exist in trace amounts on some of the mixed-mode sorbents, and this can lead to alkylation of basic nitrogen that is primary or secondary (i.e., contains a basic proton that is easily exchangeable).

Table 8.1. Examples of Drugs Isolated by Mixed Mode

Acepromazine	Ephedrine	Oxycodone
Acetaminophen	Fentamyl	Phenethylamine
Amphetamine	Hydromorphone	Procaine
Barbiturates	Ketamine	PCP
Benzoylecgonine	Lidocaine	Propoxyphene
Caffeine	Methadone	Phenylpropanolamine
Codeine	Morphine	Quinine
Cotinine	Nalorphine	Tetracaine
Diazepam	Nicotene	Theophyline

Table 8.2. Selected Examples of Specialized SPE Products for Drug Analysis

Trade Name	Manufacturer	Sorbent Chemistry
BondElut Certify	Varian-Analytichem	C-8 and cation exchange
SPEC II (disk)	Toxi-Lab	C-8 and cation exchange
NARC-2	J.T. Baker	C-8 and cation exchange
CleanScreen DAU	World Wide Monitoring	C-8 and cation exchange

8.6. SORBENTS FOR DRUG ANALYSIS

The XAD-2 resin (styrene–divinylbenzene copolymer) was one of the first sorbents to be used for sample preparation in drug analysis (Fujimoto and Wang, 1970). It is a copolymer of styrene–divinylbenzene with 50-Å pores and a surface area of approximately 200 to 300 m^2/g. The small pore size excluded much of the high-molecular-weight proteinaceous material of urine and blood samples; thus, it was a useful sorbent. It was followed by modern SPE in the late 1970s by the SepPak by Waters Associates, who introduced C-18 sorbents. The SepPak cartridge was disposable and easily used for SPE. Varian Sample Preparation Products marketed SPE columns in 1978 called ClinElut, which were silica sorbents. Since then a variety of mixed-mode sorbents and specialized drug products have entered the market (Table 8.2). The Appendix gives a complete list of product manufacturers for SPE products for drug analysis.

8.7. NIDA DRUGS OF ABUSE

The NIDA drugs of abuse include five classes of compounds: amphetamines, opiates, cocaine, phencyclidine, and tetrahydrocannibinol (marijuana). The analysis of these drugs includes both parent and degradation products. The key compounds for GC/MS analysis are discussed with the methods required for each of these compounds.

8.7.1. Marijuana

Tetrahydrocannabinol (marijuana) is rapidly metabolized in the body to the tetrahydrocannibinol carboxylic acid or THC (Fig. 8.5). It is the THC metabolite that is commonly monitored in urine for drug analysis of marijuana. Because the THC structure contains both a phenolic hydroxyl group and a car-

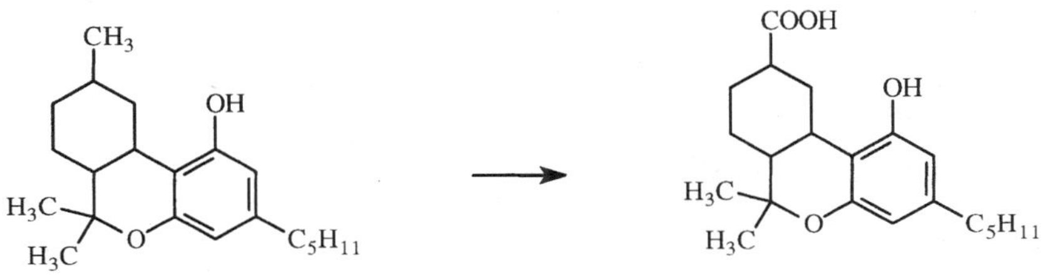

Tetrahydrocannabinol → Tetrahydrocannibolcarboxylic acid "THC"

Figure 8.5. Metabolism of tetrahydrocannibinol (marijuana) to THC in the body.

boxylic-acid group but no nitrogen atoms, the mixed-mode mechanism will not work. The use of mixed mode of an anion exchange and reversed phase will also not work because the majority of the interferences in an urine sample are negatively charged, and these interferences will be isolated with the drug. Thus, the method for THC involves the isolation of the drug by reversed phase and the washing of interferences off the sorbent with a weak polar solvent (20% acetonitrile) while retaining the THC. Finally, the THC is eluted with a stronger organic solvent (ethyl acetate).

THC Method

1. Prepare C-8 resin for sample by activating the column with methanol and phosphate buffer (pH 7.0, $0.01M$ K_2HPO_4).
2. Apply 5 mL of urine to the column at the normal pH of urine. Sorption occurs by reversed phase. The THC is hydrophobic and will sorb effectively without pH adjustment.
3. Rinse column with 2 mL of deionized water to rid sample of salts and urea.
4. Wash column with 20% acetonitrile in water, 5 mL. This wash solution will remove urine impurities without removing THC from the column. The strength of the wash solvent is quite important as well as the volume. The strength of the solvent is kept low enough not to remove the analyte.
5. Elute column with 2 mL of ethyl acetate.
6. Evaporate the ethyl acetate to dryness under nitrogen at 50 °C. THC is ionic and remains in vial.
7. Derivatize and analyze by GC/MS.

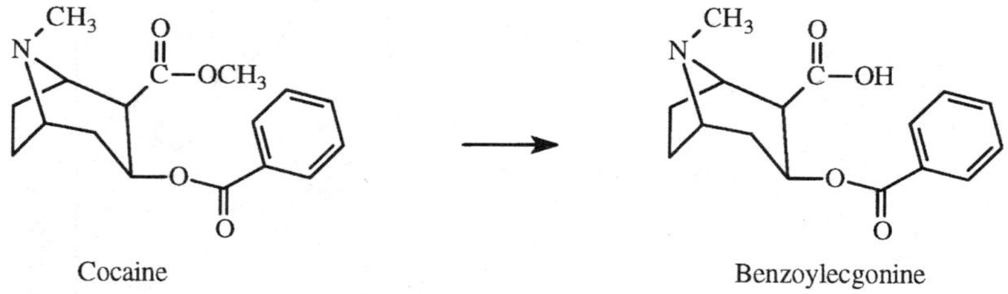

Figure 8.6. Metabolism of cocaine to benzoylecgonine in the body.

8.7.2. Cocaine

Cocaine metabolizes in the body to several compounds. The NIDA drug confirmation is based on the metabolite, benzoylecgonine. Benzoylecgonine is an amphoteric compound (contains both an amine and a carboxylic acid) that will isolate by mixed-mode SPE (Fig. 8.6).

Cocaine Method

1. Prepare the mixed-mode sorbent (e.g., Cleanscreen by World-Wide Monitoring) by washing with 2 mL of methanol and 2 mL of phosphate buffer (pH 7.0, 0.01 M K$_2$HPO$_4$).
2. Apply 2 mL of urine to the sorbent.
3. Rinse column with 2 mL of deionized water to rid sample of salts and urea.
4. Wash the column with 2 mL of 0.1 N HCl. This is the acid-lock step.
5. Wash the column with 3 mL of methanol to remove urine interferences.
6. Elute column with ethyl acetate, methanol, and ammonium hydroxide (80 : 18 : 2% v/v).
7. Evaporate to dryness under nitrogen.
8. Derivatize and analyze by GC/MS.

8.7.3. Phencyclidine

Phencyclidine, or PCP, is a hallucinogenic drug. Its structure is shown in Figure 8.7, and it is analyzed as the parent compound in urine. Because PCP contains a basic nitrogen, it is easily isolated by a mixed-mode approach. It may be analyzed directly by GC/MS without derivatization.

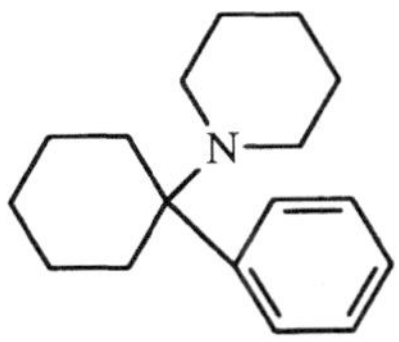

Figure 8.7. Structure of PCP.

PCP Method

1. Prepare the mixed-mode sorbent (e.g., Cleanscreen by World-Wide Monitoring) by washing with 2 mL of methanol and 2 mL of phosphate buffer (pH 7.0, 0.01 M K_2HPO_4).
2. Apply 2 mL of urine to the sorbent.
3. Rinse column with 2 mL of deionized water to rid sample of salts and urea.
4. Wash the column with 2 mL of 0.1 N HCl. This is the acid-lock step.
5. Wash the column with 3 mL of methanol to remove urine interferences.
6. Elute column with ethyl acetate, methanol, and ammonium hydroxide (80 : 18 : 2% v/v).
7. Evaporate to dryness under nitrogen.
8. Analyze by GC/MS.

8.7.4. Amphetamines

Methamphetamine and amphetamine are used as stimulants and their structure is shown in Figure 8.8. Both are basic compounds and may be isolated and purified by mixed-mode sorption. Both compounds are analyzed as parent compound in urine or blood.

Methamphetamine   Amphetamine

Figure 8.8. Structure of amphetamine and methamphetamine.

Amphetamine Method

1. Prepare the mixed-mode sorbent (e.g., Cleanscreen by World-Wide Monitoring) by washing with 2 mL of methanol and 2 mL of phosphate buffer (pH 7.0, 0.01 M K_2HPO_4).
2. Apply 2 mL of urine to the sorbent.
3. Rinse column with 2 mL of deionized water to rid sample of salts and urea.
4. Wash the column with 2 mL of 0.1 N HCl. This is the acid-lock step.
5. Wash the column with 3 mL of methanol to remove urine interferences.
6. Elute column with ethylacetate, methanol, and ammonium hydroxide (80 : 18 : 2% v/v).
7. Evaporate to dryness under nitrogen.
8. Derivatize and analyze by GC/MS.

8.7.5.   Opiates

NIDA specifications call for the analysis of two opiates, codeine and morphine, whose structures are shown in Figure 8.9. Both compounds are amphoteric compounds because of the phenolic hydroxyl group and the amine group. Therefore, they are easily concentrated and purified by mixed-mode sorption. The compounds are present in urine as glycoside conjugates through the phenolic hydroxyl group. Thus, they must first be hydrolyzed in acid before isolation by SPE.

Opiate Method

1. Prepare the mixed-mode sorbent (e.g., Cleanscreen by World-Wide Monitoring) by washing with 2 mL of methanol and 2 mL of phosphate buffer (pH 7.0, 0.01 M K_2HPO_4).
2. Apply 2 mL of hydrolyzed urine to the sorbent.
3. Rinse column with 2 mL of deionized water to rid sample of salts, acid, and urea.
4. Wash the column with 2 mL of 0.1 N HCl. This is the acid-lock step.
5. Wash the column with 3 mL of methanol to remove urine interferences.
6. Elute column with ethyl acetate, methanol, and ammonium hydroxide (80 : 18 : 2% v/v).
7. Evaporate to dryness under nitrogen.
8. Derivatize and analyze by GC/MS.

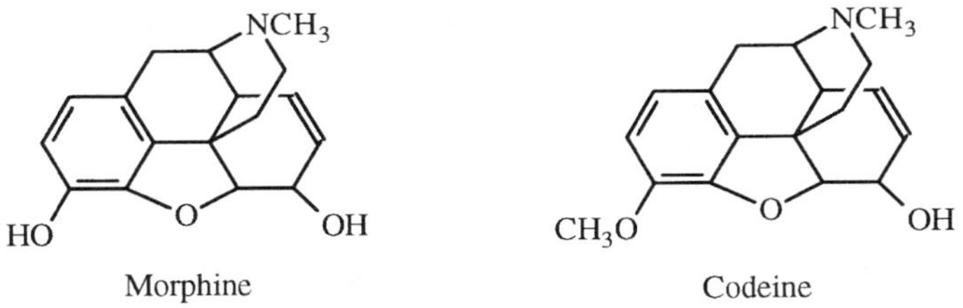

Figure 8.9. Structure of opiates.

8.8. GENERAL DRUG ANALYSIS AND METHODS

There are a number of drug classes that have been analyzed by SPE. Many of them are published and reviewed in the bibliographies of Waters and Varian (see Suggested Reading, Chapter 1). Furthermore, J. T. Baker publishes their Bakerbond application notes for drug analysis. Also, there are several review articles on drug analysis for biological samples and tissues (Krishnan and Ibraham, 1994; Scheurer and Moore, 1992; McDowall, 1989). The review by Scheurer and Moore (1992) reviews the extraction of drugs from solid tissues, such as bone, liver, brain, intestine, kidney, and adipose tissue. The review of Krishnan and Ibraham (1994) deals chiefly with drugs in urine and plasma. There are a number of published studies that deal with many types of drugs including tricyclic antidepressants, local anesthetics, steroids, anticonvulsants, anticancer drugs, antiepileptic drugs, blood pressure medications, and heart medications (Krishnan and Ibraham, (1994).

8.9. VETERINARY APPLICATIONS

Solid-phase extraction has been applied in different veterinary applications. One important one has been the use of SPE to concentrate and isolate drugs of abuse in horse racing (Phillips et al., 1990). There are also uses for medical applications in dogs for antidiabetic agents (Zhong and Lakings, 1989), in cattle for growth agents (Gaspar and Maghuin-Rogister, 1985), and in swine for steroids (Prasad et al., 1986).

8.10. EXAMPLES OF METHODS FOR PHARMACEUTICAL AND BIOLOGICAL ANALYSIS

Various drugs and pharmaceuticals are discussed in this section to show other types of drug sample preparation methods using SPE. Neutral and acidic drugs may be approached as any neutral or acidic organic compound, and the methods development strategy discussed in Chapter 3 is appropriate.

8.10.1. Cyclosporine in Blood

The cyclosporine method uses a hydrophobic or reversed-phase interaction on a CN sorbent. The structure of cyclosporine is shown in Figure 8.10.

Cyclosporine Method

1. To 1 mL of blood (anticoagulant added as heparin) add 2 mL of 70 : 30 water/acetonitrile; allow sample to hemolyze for 5 min.
2. Centrifuge and pass diluted sample through the CN sorbent.
3. Wash sorbent with 80 : 20 0.5 M acetic acid/acetonitrile, and 60 : 40 0.15 M acetic acid/acetonitrile.
4. Elute with acetonitrile, 2 mL.
5. Because the oligopeptide is hydrophobic and high molecular weight, a CN sorbent is used in a reversed-phase mechanism for this isolation.
6. Reference is *Clin. Chem.*, **31**: 196–201 (1985).

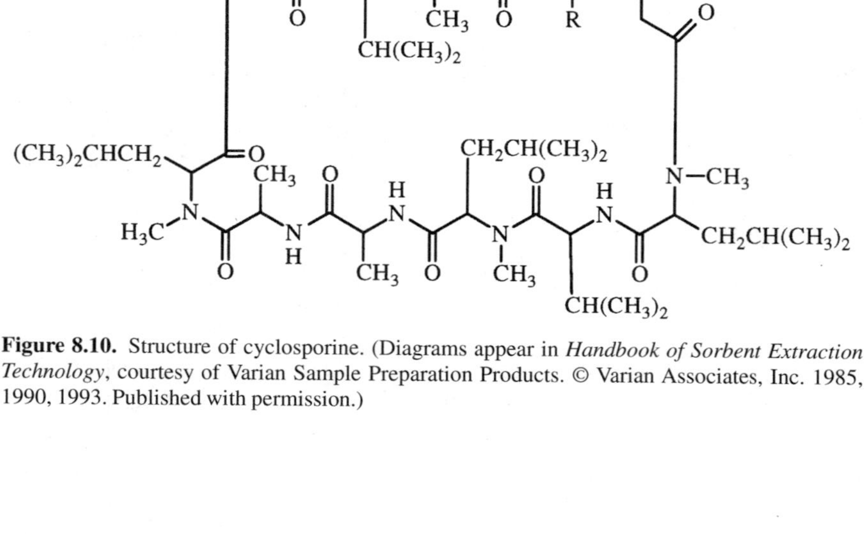

Figure 8.10. Structure of cyclosporine. (Diagrams appear in *Handbook of Sorbent Extraction Technology*, courtesy of Varian Sample Preparation Products. © Varian Associates, Inc. 1985, 1990, 1993. Published with permission.)

8.10.2.   Vitamin D in Serum

Two forms of vitamin D are shown in Figure 8.11. The molecule is hydrophobic and may be isolated from serum using a reversed-phase mechanism on a C-18 sorbent.

Vitamin D Method

1. Condition the C-18 sorbent with methanol and water, 2 mL of each.
2. To 3 mL of serum add 4 mL of 0.1 N HCl, then 8.5 mL of methanol and pass through the C-18 sorbent.
3. Wash sorbent with 70 : 30 methanol/water to remove interferences.
4. Elute vitamin D metabolites with 90 : 10 methanol/water, 3 mL.
5. Condition the NH_2 sorbent with hexane.
6. Pass eluent from C-18 sorbent through NH_2 sorbent.
7. Wash column with 0.5% isopropanol in hexane.
8. Elute 25-hydroxy vitamin D with 4 : 96 isopropanol/hexane.
9. Elute 1,25-dihydroxy vitamin D with 25 : 75 isopropanol/hexane.
10. Reference is *Clin. Chem.*, **30**: 56–61 (1984).

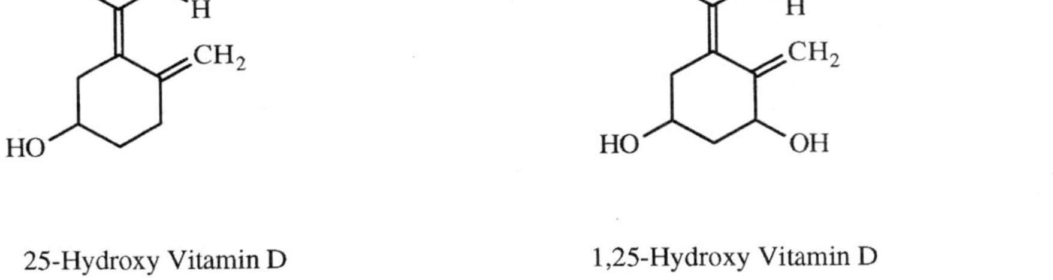

25-Hydroxy Vitamin D   1,25-Hydroxy Vitamin D

Figure 8.11. Structure of vitamin D and metabolite. (Diagrams appear in *Handbook of Sorbent Extraction Technology*, courtesy of Varian Sample Preparation Products. © Varian Associates, Inc., 1985, 1990, 1993. Published with permission.)

8.10.3. Lipids from Serum and Tissue

Seven different classes of lipids, including fatty acids, phospholipids, cholesteryl ester, cholesterol, triglycerides, diglycerides, and monoglycerides are isolated and separated (Fig. 8.12). Fatty acids and phospholipids are both

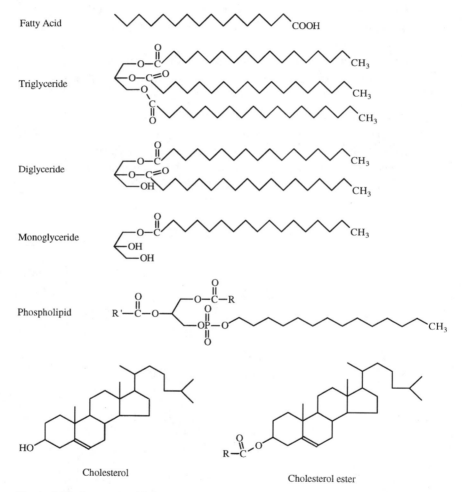

Figure 8.12. Structures of lipid classes. (Diagrams appear in *Handbook of Sorbent Extraction Technology*, courtesy of Varian Sample Preparation Products. © Varian Associates, Inc., 1985, 1990, 1993. Published with permission.)

anionic, cholesterol and cholesteryl esters are extremely nonpolar, and mono-, di-, and triglycerides that all contain polar ester and hydroxy groups. The matrix for this analysis is a chloroform extract of serum or adipose tissue. The sorbent used is an amino sorbent, which is used in normal phase to interact though hydrogen bonding and through weak ion-exchange mechanisms with the polar functional groups of the analytes. The method is referenced from Kaluzny and co-workers (1985).

Lipid Method

1. Condition the amino column with 2 mL of hexane.
2. Pass the chloroform extract through the NH_2 column.
3. Elute the first fraction (neutral lipids) with 2 : 1 chloroform/propanol.
4. Elute second fraction (fatty acids) with 2% acetic acid in diethyl ether.
5. Elute phospholipids with methanol.
6. The neutral lipids, fraction 1, is refractionated with a second amino column after evaporation and reconstitution with hexane.
7. The hexane extract is applied to the second amino column. The cholesteryl esters are eluted with hexane, with the second column attached to the first to trap cholesterol from sorbent 1, which elutes with triglycerides.
8. Triglycerides are eluted with hexane containing 1% diethyl ether and 10% methylene chloride.
9. The two columns are separated and cholesterol is eluted from both; the di- and monoglycerides are eluted from the upper NH_2 column.
10. Cholesterol is eluted with 5% ethyl acetate in hexane.
11. Diglycerides are eluted with 15% ethyl acetate in hexane.
12. Monoglycerides are eluted with 2 : 1 chloroform/methanol.

8.10.4.   Vitamin B_{12} in Urine or Aqueous Solution

The structure of vitamin B_{12}, called cyanocobalamin, is shown in Figure 8.13 and is a water-soluble vitamin. It has a molecular weight of 1355 and has a complex chemical structure. The molecule is hydrophilic and is not easily extractable into organic solvents. Thus, SPE is an excellent way to isolate it from aqueous solutions followed by HPLC analysis. It is isolated by reversed-phase sorption onto a C-18 sorbent.

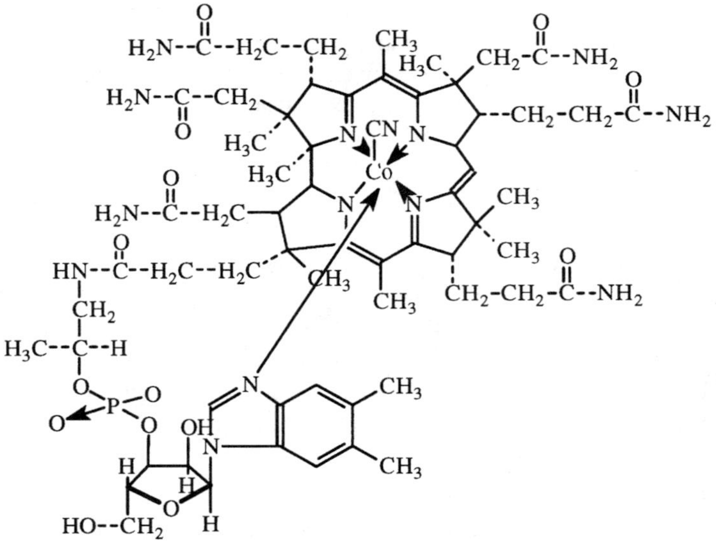

Figure 8.13. Structure of vitamin B_{12}.

Vitamin B_{12} Method

1. Condition the C-18 sorbent with methanol and water, 2 mL of each, followed by one column volume of 0.05 M sodium dihydrogenphosphate.

2. Ten milliliters of urine sample or aqueous sample is added to a 1-g column of C-18.

3. Wash column with one volume of distilled water; do not allow the column to dry.

4. Elute vitamin B_{12} with 2 mL of 50% ethanol–0.05M sodium dihydrogenphosphate followed by 3 mL of distilled water.

5. Reference is *J. Chromatog. Sci.*, **28**: 42–45 (1990).

8.11. AUTOMATION AND ON-LINE SOLID-PHASE EXTRACTION FOR DRUG ANALYSIS

McDowall and co-workers (1986, 1989) gave an overview of SPE and automation of SPE. They also discuss the general automation of drugs for SPE analysis. Chapter 10 will address this in detail with examples of the latest equipment in automation. Also Werkhoven-Goewie and co-workers (1983) show the on-line SPE and HPLC analysis of anti-tumoragenic drugs, which is relatively straightforward approach with 1 to 2 mL of sample being used for the analysis.

8.12.   DISK TECHNOLOGY AND DRUG ANALYSIS

The use of disk technology in drug analysis was first introduced by Ansys, Inc (SPEC disc). They have embedded irregularly shaped 30-µm diameter, porous silica and bonded-silica particles into glass-fiber filters (Blevins and Henry, 1995). Thus, the flow characteristics of the SPEC disc are faster than the flow characteristics of the Empore disk, which is described in Chapters 1 and 11, although 3M is introducing a new disk in 1996 with faster flow rates). The SPEC disc flows with a much faster flow rate under the same pressure (note that SPEC uses the spelling *disc* and Empore uses *disk*).

The SPEC disc (solid-phase extraction concentrator) construction is based on an adsorbent bed that is uniformly distributed throughout a glass-microfiber filter. Because of the woven nature of the glass fiber, depth filtration occurs, so that particulates do not easily plug the filter. Figure 8.14 shows the apparatus used for SPEC discs. The cartridge consists of a filter holder, with a sample reservoir (syringe) above, and the vacuum is applied below. The SPEC discs may be stacked in the cartridge for different applications on the same sample. This includes filters as well as different sorption discs.

The disc has a depth-filter region from 100 to 200 µm, followed by the major extraction region that is 800 to 900 µm in thickness. Thus, the disc is both filter and sorbent. Several advantages of the disc over standard SPE cartridges include ease of processing cloudy samples (important in automated procedures where plugging may cause serious losses of samples), fast processing times, and much decreased volumes of solvent (Blevins and Schultheis, 1994).

The types of sorbents available include polar, nonpolar, cation and anion exchange, and mixed mode (see Product Guide, Appendix). The polar phases such as primary and secondary amines, aminopropyl, silica, and cyanopropyl may be used on phenols, carbohydrates, drug metabolites, and vitamin D. The nonpolar sorbents, such as C-8, C-18, and a C-18AR (acid resistant, trifunctional) are the classic sorbents used in drug analysis. The mixed-mode phase comes in two polarities, nonpolar strong cation-exchange and slightly polar strong cation exchange. These sorbents are the classical mixed-mode applications described earlier in this chapter.

The SPEC approach also is available in a classic syringe mode, with both the filter and the SPEC disc in the syringe (Fig. 8.15). The phases are available

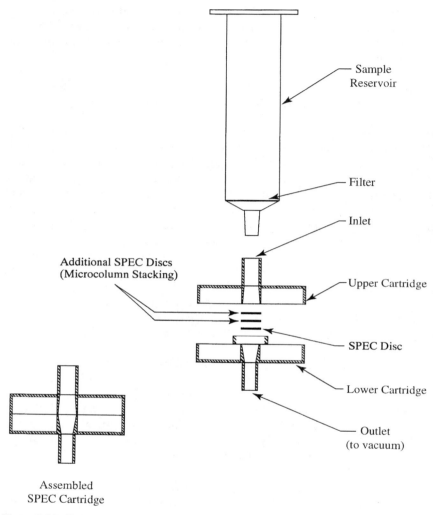

Figure 8.14. SPEC-Solid Phase Extraction Concentrator. (Published with permission of Ansys, Inc.)

in a 3-mL syringe and have 15 mg of sorbent material available on the cartridge. Methods are available from several manufacturers of SPE products for many drugs of abuse, such as 11-nor-delta 9-tetrahydrocannabinol-9-carboxylic acid (THC), which is the major metabolite of marijuana that is analyzed by GC/MS for NIDA. A C-18 disc is used for THC. SPEC have

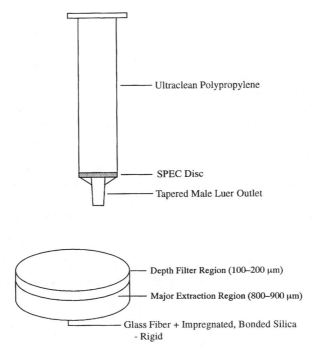

Ultraclean Polypropylene

SPEC Disc

Tapered Male Luer Outlet

Depth Filter Region (100–200 μm)

Major Extraction Region (800–900 μm)

Glass Fiber + Impregnated, Bonded Silica
- Rigid

Figure 8.15. Structure of the syringe model of the SPEC disc. (Published with permission of Ansys, Inc.)

methods for benzodiazepines and antidepressants that use a C-18 disc. The method for benzoylecgonine, amphetamines, opiates, and phencyclidine use strong cation-exchange discs. A study by Blevins and Henry (1995) describe the use of the SPEC discs for pharmaceutical applications. They show good recoveries for different drugs, such as amobarbital, hydrocortisone, and mepivacaine with recoveries from 92 to 100%.

Another interesting advantage of the disk technology, which is addressed by Hearne and Hall (1993) is the on-disk derivatization of THC. They found that the extract was considerably cleaner than the traditional method that uses elution into a test tube, followed by evaporation, and derivatization in a test tube. With the on-disk derivatization, the analyte is derivatized on the disk and eluted in the same step.

The Empore extraction disks have also been incorporated into syringes, especially for use in drug analysis. The syringe consists of the Empore disk, which is a particle-loaded membrane that is approximately 0.5-mm thick and contains 8 to 12 µm silica or silica-bonded particles that are immobilized within an inert matrix of polytetrafluoroethylene (PFTE) fibrils (see Chapter 1). The particle-loaded membranes are mounted into standard 1-, 3-, and 6-mL polypropylene syringe barrels and effective membrane diameters are 4, 7, and 10 mm (see Fig. 1.7). There is a prefilter before the membrane to remove particles before sorption onto the membrane. Several major differences between the SPEC disc and the Empore disk are the difference in flow of the two membranes, with the SPEC disc having a slightly faster rate because of the fiberglass matrix of the disc, and the increased capacity and kinetic rate of the Empore disk because of a smaller particle-size sorbent. The makers of the Empore disk (3M) have recently introduced a new holder that contains their disks in a microtiter-plate well system for use with standard liquid-handling equipment and for use in pharmaceutical and drug applications (Chapter 11). Both disks are effective in the isolation of drugs and their metabolites depending on the sorbent chosen.

8.13.   INNOVATIONS IN SOLID-PHASE EXTRACTION FOR DRUG ANALYSIS

There are a number of recent sorbents that may be useful for drug analysis that are currently being developed (see Chapter 12). They include restricted access packings that prevent the access of matrix components such as proteins, while selectively retaining the drug components and their metabolites in the interior of the sorbent. There is affinity chromatography, where the surface is modified with antibodies that bind specific molecules. There is a mixture of C-18 sorbents with hydrophilic endcapping for the isolation and separation of protein substances. Finally, matrix solid-phase dispersion (MSPD) has been used on tissue samples to remove drugs from solid samples. Matrix dispersion is the blending of a C-18 packing with a solid material in a mortar and pestle, which is followed by packing the sorbent in a syringe and eluting the drug from the C-18 sorbent (see Chapter 9).

SUGGESTED READING

Analytichem International, *Bond Elut Certify™ Instruction Manual for Drug Analysis*, Varian, Harbor City, CA.

Anderson, W. H. and Fuller, D. C. 1987. A simplified procedure for the isolation, characterization, and identification of weak acid and neutral drugs from whole blood, *J. Anal. Toxicol.*, **11**: 198–204.

Ford, B., Vine, J., and Watson, T. R. 1983. A rapid extraction method for acidic drugs in hemolyzed blood, *J. Anal. Toxicol.*, **7**: 116–118.

Furton, K. G. and Rein, J. 1990. Trends in techniques for the extraction of drugs and pesticides from biological specimens prior to chromatographic separation and detection, *Anal. Chim. Acta*, **236**: 99–114.

Johnson, E. L., Reynolds, D. L., Wright, D. S., and Pachla, L. A. 1988. Biological sample preparation and data reduction concepts in pharmaceutical analysis, *J. Chromatog. Sci.*, **26**: 372–379.

Jordan, L. 1993. *Handling Biological Samples with the BenchMate™ Workstation*. Zymark Publication, Hopkinton, MA.

King, J. W. and King, L. J. 1996. Solid-phase extraction and on-disc derivatization of the major benzodiazepines in urine using enzyme hydrolysis and Toxi-Lab Vc-Mp3 column, *J. Anal. Toxicol.*, **20**: 262–265.

Krishnan, T. R. and Ibraham, I. 1994. Solid-phase extraction technique for the analysis of biological samples, *J. Pharmaceut. Biomed. Anal.*, **12**: 287–294.

Lensmeyer, G. L., Wiebe, D. A., and Darcey, B. A. 1991. Application of a novel form of solid-phase sorbent (Empore® Membrane) to the isolation of tricyclic antidepressant drugs from blood, *J. Chromatog. Sci.*, **29**: 444–449.

Logan, B. K., Stafford, D. T., Tebbett, I. R., and Moore, C. M. 1990. Rapid screening of 100 basic drugs and metabolites in urine using cation exchange solid-phase extraction and high-performance liquid chromatography with diode array detection, *J. Anal. Toxicol.*, **14**: 154–159.

Marko, V., Soltes, L., and Radova, K. 1990. Polar interactions in solid-phase extraction of basic drugs by octadecylsilanized silica, *J. Chromatog. Sci.*, **28**: 403–406.

McDowall, R. D. 1989. Sample preparation for biomedical analysis, *J. Chromatog.*, **492**: 3–58.

Musch, G. and Massart, D. L. 1988. Isolation of basic drugs from plasma using solid-phase extraction with a cyanopropyl-bonded phase, *J. Chromatog.*, **432**: 209–222.

Patel, R. M., Benson, J. R. and Hometchko, D. 1990. Solid-phase extraction of amphetamine and methamphetamine using polymeric supports and hexanesulfonic acid as the ion-pairing reagent, *LC-GC*, **8**: 153–158.

Patel, R. M., Jagodzinski, J. J., Benson, J. R., and Hometchko, D. 1990. Mixed-mode sorbent for sample preparation, *LC-GC*, **8**: 874–878.

Parker, T. D., Wright, D. S., and Rossi, D. T. 1996. Design and evaluation of an automated solid-phase extraction method development system for use with biological fluids, *Anal. Chem.*, **68**: 2437–2441.

Platoff, Jr., G. E. and Gere, J. A. 1991. Solid phase extraction of abused drugs from urine, *Forensic Sci. Rev.*, **3**: 117–133.

Scheurer, J. and Moore, C. M. 1992. Solid-phase extraction of drugs from biological tissues—a review, *J. Anal. Toxicol.*, **16**: 264–269.

United Chemical Technologies, Inc. Catalogue, 98 pages with references, Bristol, PA.

REFERENCES

Blevins, D. D. and Henry, M. P. 1995. Pharmaceutical applications of extraction disk technology: *Am. Lab.*, May, 32–35.

Blevins, D. D. and Schultheis, S. K. 1994. Comparison of extraction disk and packed-bed cartridge technology in SPE, *LC-GC*, **12**: 12–16.

Brooks, K. E. and Smith, N. B. 1991. Efficient extraction of basic, neutral, and weakly acidic drugs from plasma for analysis by gas chromatography-mass spectrometry, *Clin. Chem.*, **37**: 1975–1978.

Chen, X. H., Franke, J. P., Wijsbeek, J., and de Zeeuw, R. A. 1992. Isolation of acidic, neutral, and basic drugs from whole blood using a single mixed-mode solid-phase extraction column, *J. Anal. Toxicol.*, **16**: 351–355.

Chia, D. T. and Gere, J. A. 1987. Rapid drug screening using Toxi-Lab® extraction followed by capillary gas chromatography/mass spectrometry, *Clin. Biochem.*, **20**: 303–306.

Dixit, V. and Dixit, V. M. 1990. A solid-phase extraction technique for preparation of drugs-of-abuse samples: *Am. Lab.*, May/June, 47–51.

Engelhardt, H. and Jungheim, M. 1990. Comparison and characterization of reversed phases, *Chromatographia*, **29**: 59–68.

Fujimoto, J. M. and Wang, R. I. H. 1970. Analgesics in human urine, *Toxicol. Appl. Pharmacol.*, **16**: 186–189.

Gaspar, P. and Maghuin-Rogister, G. 1985. Rapid extraction and purification of diethylstibosterol in bovine urine hydrolysates using reversed-phase C_{18} columns before determination by radioimmunoassay, *J. Chromatog.*, **328**: 413–416.

Hearne, G. M. and Hall, D. O. 1993. Advances in solid-phase extraction technology, *Am. Lab.*, January, 28H–28M.

Kaluzny, B. D. 1985. Proceedings of the Second International Symposium on Sample Preparation and Isolation using Bonded Silicas, Analytichem International, Amsterdam.

Krishnan, T. R. and Ibraham, I. 1994. Solid-phase extraction technique for the analysis of biological samples: *J. Pharmaceut. Biomed. Anal.*, **12**: 287–294.

McDowall, R. D. 1989. Sample preparation for biomedical analysis, *J. Chromatog.*, **492**: 3–58.

McDowall, R. D., Doyle, E., Murkitt, G. S., and Picot, V. S. 1989. Sample preparation for the HPLC analysis of drugs in biological fluids,: *J. Pharmaceut. Biomed. Anal.*, **7**: 1087–1096.

McDowall, R. D., Pearce, J. C., and Murkitt, G. S. 1986. Liquid-solid sample preparation in drug analysis, *J. Pharmaceut. Biomed. Anal.*, **4**: 3–21.

Mills, M. S. Thurman, E. M. and Pedersen, M. J. 1993. Mixed-mode solid-phase extraction: Combining mechanisms of interaction; hydrogen bonding, cation exchange, and reverse phase, *J. Chromatog.*, **629**: 11–21.

Nickel, K. L. 1977. Improved extraction of aldosterone from plasma and urine before radioimmunoassay, *Clin. Chem.*, **23**: 885–886.

Phillips, D., Tebbett, I. R., and Kalita, S. 1990. Solid phase extraction and analysis of horse urine for aminocaproic acid, *Chromatographia*, **30**: 309–310.

Prasad, V. K., Ho, B., and Haneke, C. 1986. Simultaneous determination of prednisolone acetate, prednisolonek cortisone and hydrocortisone in swine plasma using solid-phase and liquid-liquid extraction techniques, *J. Chromatog.*, **378**: 305–316.

Rymut, K., Dombrowski, L., Chaney, G., Telepchak, M., and O'Dell, L. 1996. The alkylation of drugs at ion exchange sites on the surface of solid phase extraction columns, United Chemical Technologies, Inc., product literature.

Scheurer, J. and Moore, C. M. 1992. Solid-phase extraction of drugs from biological tissues—a review, *J. Anal. Toxicol.*, **16**: 264–269.

Werkhoven-Goewie, C. E., Brinkman, U. A. Th., Frei, R. W., De Ruiter, C., and De Vries, J. 1983. Automated liquid chromatographic analysis of the anti-tumorigenic drugs etoposide (VP 16-213) and teniposide (VM 26), *J. Chromatog.*, **276**: 349–357.

Zhong, W. Z. and Lakings, D. B. 1989. Determination of pioglitazone in dog serum using solid-phase extraction and high-performance liquid chromatography with ultraviolet (229 nm) detection, *J. Chromatog.*, **490**: 377–385.

FOOD AND NATURAL PRODUCTS

The analyses of specific compounds in food products requires extracting analytes often from a complex matrix, either solid or liquid. Solids, such as food, must be homogenized and extracted with organic solvents or aqueous buffers before solid-phase extraction (SPE). Food products often consist of matrices that contain salts, carbohydrates, proteins, fats, and oils. Interferences from the food products must be removed either during extraction or during the SPE isolation. This chapter will address the extraction and liquefaction of food products prior to SPE analysis. Examples of extraction of specific compounds from foods using SPE also will be given. The chapter consists of three major sections: liquefying and extraction of food products, removing matrix interferences in foods, and examples of isolation using SPE.

9.1. LIQUEFYING AND EXTRACTION OF FOOD PRODUCTS

9.1.1. Sample Preparation

In order to pass a sample through an SPE cartridge, the sample generally should be in a liquid form. Liquefaction of the sample is accomplished by homogenizing and dissolving the analyte in an appropriate organic solvent. This process can be accomplished by use of a blender to grind the food sample (Fig. 9.1) and then addition of an organic solvent, such as methanol, followed by filtration of the sample. This method is simply called homogenization. Dry ice or diatomaceous earth can be added to the blender to make the sample flow more freely as a solid.

A second approach is to homogenize the sample in some traditional fashion (blender or mortar and pestle) followed by Soxhlet extraction with an appropriate solvent. In this procedure, the solvent is boiled and clean solvent re-extracts the food product each time. This method is effective; however, it is quite slow often taking 24 h per sample in order to obtain good recoveries. The analyte must boil at a higher temperature than the solvent for good recovery. Recently, there have been innovations in solvent extraction that combine hot solvent leaching and traditional Soxhlet extraction (Majors, 1996). The sample is first immersed in boiling solvent and then the thimble is raised for

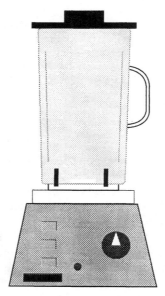

Figure 9.1. Blender for extraction of food and natural products.

Soxhlet extraction with solvent refluxing. Both manual and automated versions are available. The advantage is in reduced time of extraction over conventional extraction.

Other manual methods include a solid–liquid extraction, which consists of placing the solvent in a tube with sample and shaking the sample and filtering. This method is also called the shake–filter method. The solvent may be heated to improve extraction efficiency. Several other simple methods include sonication of the sample with an ultrasonic probe and heated solvent, and dissolution of the sample directly with solvent (Majors, 1996).

If food products consist of an aqueous matrix that must be removed before extraction, freeze drying may be considered. This method works best when the analyte is nonvolatile. The process of freeze drying gives a structure to the food product that is readily extracted by various solvents.

A recent advance is accelerated solvent extraction (ASE), a method that uses high temperature and pressure to push an organic solvent through a solid material and to collect the eluent in a vial (Fig. 9.2). The instrument, made by Dionex, is automated and may run 30 samples at once. The instrument can process one sample every 15 min with extraction efficiencies equal to that produced by Soxhlet extraction in 12 h (Dionex, Product Literature, Appendix). The extraction does not use supercritical fluids but consists of using elevated

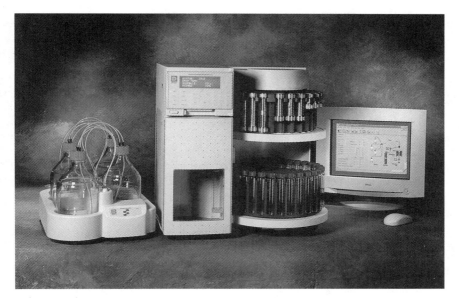

Figure 9.2. Photograph of an accelerated solvent extractor (ASE). (Copyright and courtesy of Dionex Corporation, Sunnyvale, California.)

temperatures of approximately 100 to 200 °C and pressure of 1500 to 3000 psi to achieve analyte recoveries equivalent to Soxhlet extraction with considerably less solvent (less than 15 mL for 10 g of sample) in 15 min. Compounds that have been tested include polynuclear aromatic hydrocarbons (PAHs), organophosphorus compounds, organochlorine pesticides, herbicides, and polychlorinated biphenyls (PCBs). Samples should be dry and ground before filling the extraction cells. Samples that contain water (greater than 10%) should be mixed in equal proportions with sodium sulfate. A cellulose disk is placed at the outlet of the extraction cell and approximately 10 g is extracted. The method has been used mainly on soils and sediments, but should also work well on food products.

Supercritical fluids may also be used for extraction of food. This method involves the use of supercritical carbon dioxide to remove analytes from food without dissolving the matrix (Fig. 9.3). Methanol is commonly added to the supercritical CO_2 in order to enhance the solubility of a compound.

The advantages of supercritical fluid extraction over conventional solvents is that the supercritical fluids have a higher diffusion rate than liquids, lower viscosity than liquids, and higher vapor pressure that allows easy evaporation of solvent. The lower viscosities and higher diffusion rates of supercritical

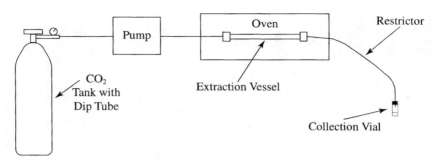

Figure 9.3. Concept of a supercritical fluid extractor.

fluids compared with liquids make them suitable for extraction of diffusion-controlled matrices, such as plant tissues and food. There are three review sttudies on the topic of supercritical fluid extraction of natural products that may be useful (Smith, 1995; Bevan and Marshall, 1994; Castioni et al., 1995).

Carbon dioxide is pumped from the bottom of the liquid tank and compressed, heated, and then the samples are extracted (Fig. 9.3). The analytes are trapped in an organic solvent as the carbon dioxide is vented from the extractor. Supercritical fluid extraction (SFE) involves some experimentation with the proper pressure to achieve the extraction of the analyte of interest. Several vendors sell the automated supercritical fluid extractors. Both the accelerated solvent extractor and the supercritical fluid extraction are expensive methods, in the price range of $50,000 each. Several general reviews of SFE include Gere and Derrico (1994), Smith (1995), and a general sample preparation review by Majors (1995).

Another extraction method that has been used in soil analysis (Fish and Revesz, 1996) and works for food analysis is microwave solvent extraction. It consists of placing a sample in an open or closed container capable of high pressure and heated by microwave energy, causing extraction of the analyte (Fish and Revesz, 1996). This method is yet another form of solvent extraction and differs in how the sample is heated. Its advantages lie in the fact that it can replace traditional solvent extraction and sonication with a fast and safe method of liquid–solid extraction.

There is also a new form of SPE, called matrix dispersion solid-phase extraction, which uses the solid phase (silica and C-18) to grind the samples and prepare the analyte for extraction. The food product is incorporated into the pores of the 40-μm packing material in the grinding process, which is done in a mortar and pestle, and the packing material is then packed into a

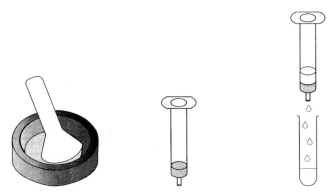

Figure 9.4. Matrix solid-phase dispersion (MSPD).

column for elution directly with organic solvent (Fig. 9.4). Basically the method consists of adding approximately 0.5 g of sample to 2 g of C-18 sorbent and grinding the sample until the pulverized material (milk, fat, liver, kidney, muscle, etc.) is incorporated into the packing material. The packing material is then transferred by funnel into a syringe-barrel column and plugged with a filter-paper disk. The column head is covered with a second disk and the contents are compressed by a plunger to a volume of ~4.5 mL. The column may then be eluted with a solvent or series of solvents to elute the analyte of interest (Barker et al., 1993). This method is discussed in detail in Chapter 12.

9.1.2. Choosing an Extraction Solvent

The extraction solvent should be chosen such that the analyte is extracted with as little co-extraction of interfering substances as possible. Furthermore, choose a solvent that is compatible with the solid-phase extraction method that is being used. For example, samples that are high in fat are best homogenized with nonpolar solvents, such as hexane. The use of hexane as an extractant dictates normal-phase sorbents with silica, alumina, Florisil, or a CN sorbent. Samples with high water content are best homogenized with acids, bases, or polar solvents, such as methanol, acetonitrile, or acetone. The use of methanol or one of the water-miscible solvents suggests the use of reversed phase (C-18) and dilution of the extract with water or buffer. Thus, the choice of the extraction solvent will dictate the type of sorbent that will be used in SPE.

9.2. REMOVING MATRIX INTERFERENCES IN FOODS

9.2.1.   Fats, Oils, and Lipids

Fats, oils, and lipids are common components of meats, nuts, and dairy products and manufactured goods, such as potato chips, cookies, and chocolate. They are soluble in nonpolar solvents, such as hexane and methylene chloride. The analyte, of course, should also be soluble in the extraction solvent. Typically normal-phase SPE would be used to retain a compound from this extraction solvent. A solid fat may be homogenized in a blender with hexane, filtered or centrifuged, then the solvent would be passed through a normal-phase column for retention of the solute. Another approach is the use of matrix solid-phase dispersion, where the solid would be ground into the silica and C-18 directly, then the analyte eluted directly from the ground mixture with either hexane or methylene chloride. The hexane or methylene chloride extract could then be applied directly to a normal-phase sorbent for separation. Liquid oils may be directly diluted with hexane or methylene chloride and applied to the normal-phase sorbent. Other lipid substances may be handled either as solids or liquids depending on their form.

9.2.2.   Proteins and Meat

Proteins are generally high molecular weight relative to the analyte that is being analyzed. Because the pores of most of the silica gels is 40 to 60 Å, proteins are poorly retained. Thus, water or aqueous buffer will commonly wash proteins from a C-18 sorbent, if they are not strongly sorbed. Proteins will also precipitate if salt or acid is added. Care should be taken not to lose analyte if precipitation is used. Meat is an obvious example of a proteinaceous material. Generally, it is homogenized in a blender and may be extracted by an organic solvent, such as hexane or methylene chloride as a nonpolar solvent, or by methanol if a polar solvent is required. Often the nonpolar solvent will work best because the fats present in the meat must also be solubilized. If emulsions occur in the extraction of meat and fat, addition of large amounts of sodium chloride will help in removing water and prevention of emulsions. Meat has been a food product that has been used extensively in solid-phase matrix dispersion. The use of the sorbent to grind the meat is a rapid and relatively simple process that prevents emulsions during the recovery of the analyte (Fig. 9.4).

9.2.3.   Carbohydrates, Polysaccharides, and Salts

Carbohydrates and polysaccharides are common constituents in food and beverages. These are polar substances that, if soluble, are not retained by nonpolar sorbents, such as C-18. The combination of high molecular weight and high polarity reduce sorption efficiency dramatically. These substances are often removed by aqueous extraction or by acidic or basic buffers. Examples of food products that contain carbohydrates and polysaccharides include grain products and confectioneries. Grain products may first be homogenized for the appropriate solute of interest. Fat must be extracted first with hexane before extraction of the solute and dispersion of the carbohydrate or polysaccharide.

Confectioneries, such as molasses and other sugar products, may be diluted with water or aqueous buffer prior to SPE. Chocolate, which has a high fat content, will need extraction with hexane or methylene chloride to remove the fat; then an aqueous solvent or polar solvent such as methanol may be used to extract the carbohydrate or polysaccharide. The choice of sorbents to recover specific analytes from this matrix would be reversed phase, such as C-8 or C-18.

Produce and vegetables are another group of foods that contain carbohydrates and polysaccharides. They may be homogenized with aqueous buffers, or a polar organic solvent, such as methanol. Acid digestion may also be a useful approach. Salts may be removed by aqueous extraction. They are not retained by nonpolar sorbents, such as C-18, and may be de-salted or washed from the extract. Alternatively, if the extractant is extremely nonpolar, such as hexane, then the salts are not extracted by the solvent.

9.2.4. Beverages

Beverages may often be treated as aqueous samples; thus, nonpolar and ion-exchange mechanisms will work for analyte recovery. Both of these mechanisms are effective with natural product analysis in wine, soft drinks, tea, and coffee. Milk, which commonly contains fats and proteins, may need dilution and filtration before application to a sorbent.

9.3. EXAMPLES OF ANALYTE ISOLATION FROM FOOD USING SOLID-PHASE EXTRACTION

9.3.1. Atrazine from Corn Oil

Atrazine is the major herbicide applied to corn fields in the midwestern United States. The purpose of this isolation is to determine if the corn oil contains traces of the herbicide atrazine.

Atrazine Method for Corn Oil

1. The corn oil should be diluted 1 : 10 with hexane and passed through a diol column. The oil is essentially not retained and passes through

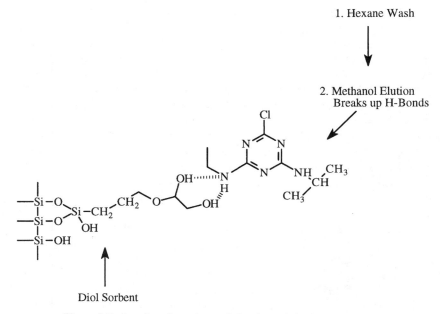

Figure 9.5. Sorption of atrazine to diol sorbent via hydrogen bonding.

the sorbent with the hexane eluent. Atrazine is sorbed to the diol through polar forces of hydrogen bonding between the amine groups of the atrazine and the hydroxyl groups of the diol.

2. Interferences are eluted with hexane, 2 mL.

3. Atrazine is eluted with a solvent that is capable of breaking hydrogen bonds, such as 2 mL of methanol. Figure 9.5 shows the mechanism of hydrogen bonding between atrazine and the diol hydroxyl groups.

9.3.2. Beverages: Organic Acids and Anthocyanins from Wine

Wine is a polar liquid suitable for direct application to SPE sorbents without sample preparation. Because the polarity of organic acids and anthocyanins differs widely, two SPE sorbents are used. Tartaric, malic, and fumaric acids are polar (Fig. 9.6) and not isolated by C-18. Rather, ion exchange is the method of isolation for these compounds. Anthocyanins, on the other hand, are nonpolar and may be retained by a reversed-phase mechanism. Thus, two types of sorbents are required for this method.

The method consists of a cyclohexyl column for the anthocyanins and an ion-exchange column for the organic acids. The cyclohexyl column operates

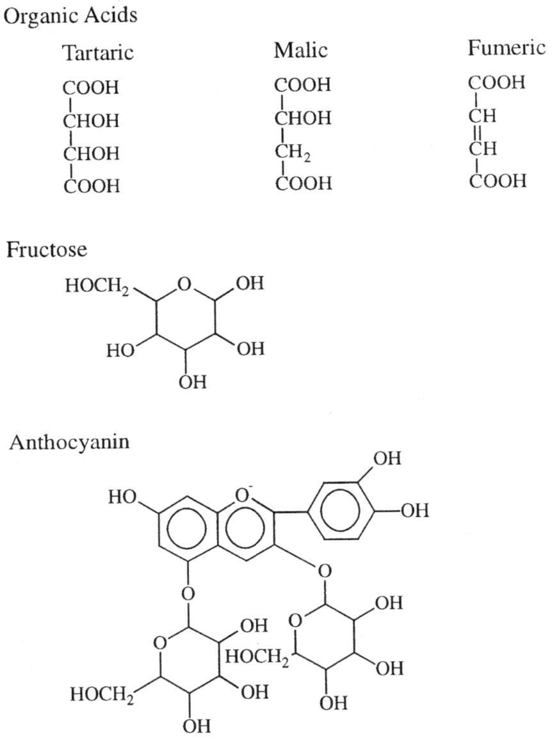
Organic Acids
Tartaric
COOH
CHOH
CHOH
COOH
Malic
COOH
CHOH
CH2
COOH
Fumeric
COOH
CH
CH
COOH
Fructose
HOCH2
O
OH
HO
OH
OH
Anthocyanin
OH
HO
O
OH
O
O
OH
OH
O
O
HOCH2
OH
HOCH2
OH
OH
OH
OH

Figure 9.6. Structure of organic acids, fructose, and anthocyanin found in wine. (Diagrams from *Handbook of Sorbent Extraction Technology*, courtesy of Varian Sample Preparation Products. © Varian Associates, Inc. 1985, 1990, 1993. Published with permission.)

in reversed-phase mode, and the anion-exchange column isolates by anion exchange, with the chloride of the anion-exchange resin being replaced by the organic acids in the wine (Fig. 9.7). The elution consists of strong acid, which protonates the organic acids and elutes the ion-exchange sorbent with the chloride of the HCl eluent.

Organic Acid and Anthocyanin Method for Wine

1. Condition the cyclohexyl and anion-exchange resin with methanol and water, 2 mL each.
2. Pass 50 mL sample through both sorbents in line, with the cyclohexyl column first followed by the anion-exchange column.
3. Wash cyclohexyl sorbent with de-ionized water.

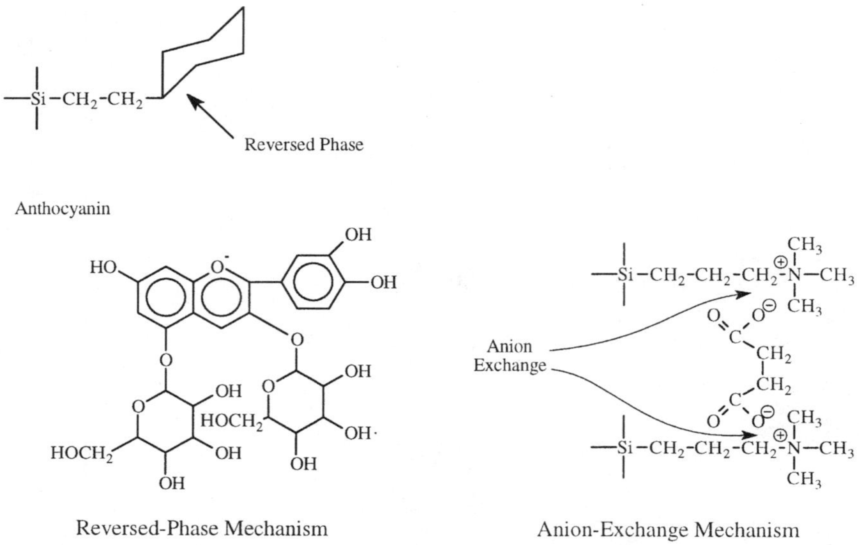

Anthocyanin

Reversed-Phase Mechanism   Anion-Exchange Mechanism

Figure 9.7. Mechanisms of sorption for anthocyanin and organic acids from wine.

4. Elute anthocyanins from cyclohexyl column with 2 mL of methanol.
5. Elute organic acids from anion-exchange resin with 1 mL of 0.1 N HCl.

9.3.3. Tetracycline Antibiotics in Meat

Tetracycline antibiotics (Fig. 9.8) are widely used in modern agriculture for veterinary treatment or growth production of food-producing animals. The determination of residual compounds in slaughtered animals is an important health safety issue (Oka et al., 1993). The method developed by Oka and co-workers (1993) uses C-18 cartridge clean up followed by thin layer chromatography (TLC) and fast atom bombardment (FAB) mass spectrometry.

Tetracycline Antibiotics in Meat

1. A sample (5g) was blended three times with 20, 20, and 10 mL of $0.1\,M$ Na_2EDTA-McIlvaine buffer (pH 4.0) using a high-speed blender and centrifuged each time. The supernatants were combined, centrifuged again, and filtered.
2. The filtrate was applied to a C-18 sorbent that had been first activated with 2 mL of methanol and then washed with 5 mL of the Na_2EDTA buffer solution.

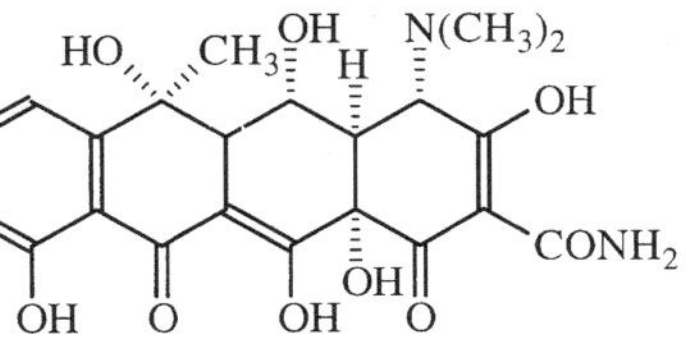

Oxytetracycline

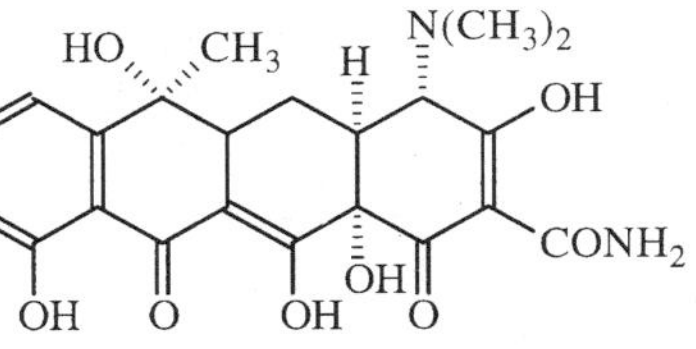

Tetracycline

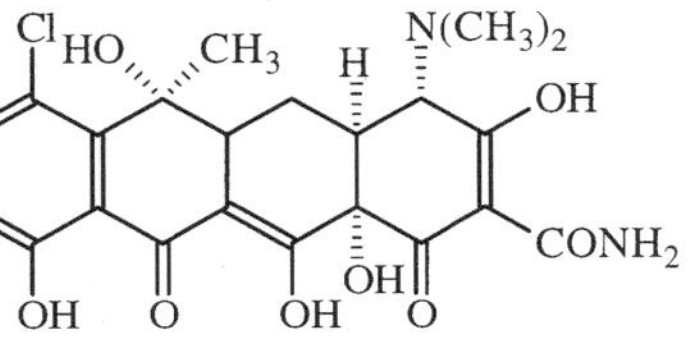

Chlortetracycline

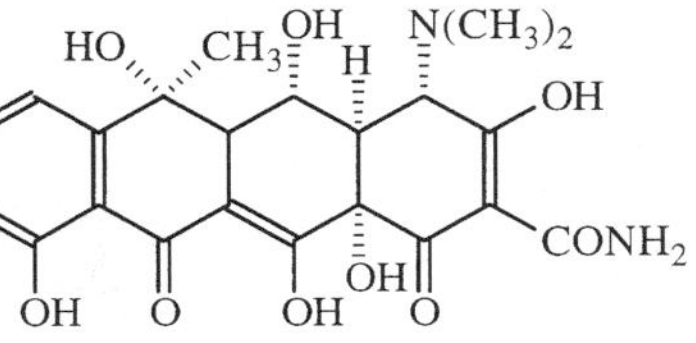

Doxycycline

Figure 9.8. Structure of the tetracycline antibiotics.

3. The cartridge was washed with 20 mL of distilled water and air-dried by aspiration for 5 min.
4. Tetracycline antibiotics were eluted with 10 mL of ethyl acetate, followed by 20 mL of methanol-ethyl acetate (5 : 95) by volume, and the eluate was evaporated to dryness under reduced pressure at 30 °C. The residue was redissolved in methanol and analyzed by TLC or high-pressure liquid chromatography (HPLC).
5. The reference for the method is Oka and co-workers (1993). The mixed-mode method discussed in the previous chapter would also work well for these compounds because of the amine groups. The addition of base would improve recovery with smaller amounts of eluent.

9.3.4. Fumonisin B_1 in Rodent Feed

Fumonisin B_1 (Fig. 9.9) is the major fumonisin metabolite produced by the fungus, *Fusarium moniliforme*. The metabolite has been implicated as a carcinogen in animals (Holcomb et al., 1993). The structure is shown in Figure 9.9, with both amine and carboxylic acid functionality. The assay for this compound in rodent feed was developed in order to determine that the research animals were not being exposed to fumonisin B_1 in their feed. The method uses both C-18 and anion exchange. In the C-18 reverse-phase mode, the sample is salted out with a 1% solution of KCl to increase the ionic strength

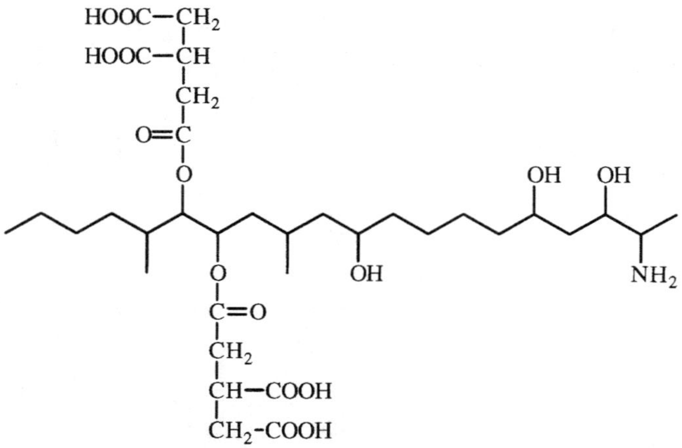

Fumonisin B$_1$

Figure 9.9. Structure of fumonisin B$_1$.

and to prevent ion-exchange sorption of the amine group to the free silanols. The C-18 is eluted with methanol/water (70 : 30), and the anion exchange is carried out in methanol. Acetic acid is used to elute the anion-exchange sorbent. A series of wash steps are carried out to purify the extract for HPLC.

Fumonisin B$_1$ Method for Rodent Feed

1. A 50-g sample of rodent feed is extracted with 200 mL of acetonitrile/water (50 : 50) in a 500-mL blender for 2 min. A 20-mL portion of this extract is centrifuged and 2 mL of the sample is diluted with 5 mL of aqueous KCl (1%) and is passed through a C-18 column. The column is conditioned with 5 mL of methanol and 5 mL of 1% KCl.

2. The column is then washed with 3 mL of 1% aqueous KCl, followed by 2 mL of acetonitrile/KCl (20 : 80). The wash cleans up the extract without eluting the fumonisin B$_1$.

3. Elute the fumonisin B$_1$ with 2 mL of methanol/water (70 : 30).

4. The eluate is diluted with 4 mL of the methanol/water (70 : 30) and is passed through a SAX (strong anion-exchange column). The SAX column is conditioned with 8 mL of methanol/water (70 : 30).

5. Wash the SAX sorbent with 8 mL of methanol/water (70 : 30) and 3 mL of methanol.

6. Elute the SAX sorbent with 14 mL of 1% acetic acid in methanol. The eluate is evaporated to dryness and is brought up in acetonitrile/water (50:50) for HPLC analysis.

9.3.5.   Vitamin K_1 in Food

Vitamin K_1 (Fig. 9.10) is primarily known for its role in the synthesis of coagulation proteins, and recent work has shown that it has a much wider role for other proteins in the body; thus, methods for its analysis in food are important (Booth et al., 1994). Concentrations of vitamin K_1 in foods have not been frequently reported; thus dietary intake is not well understood. This recent method by Booth and co-workers (1994) addresses the need for methods for vitamin K_1 and uses a silica-gel clean-up followed by HPLC analysis.

Vitamin K_1 Method for Food

1. Hexane was used to extract milk and juice samples and sonication was used to mix the samples. The hexane layer was separated and passed through a silica-gel column. Bread and vegetable samples were pulverized and extracted with hexane then cleaned in a silica-gel sorbent.

2. Precondition the silica gel (J. T. Baker, 3 mL) with 8 mL of hexane/diethyl ether (93:3 v/v), followed by 8 mL of hexane. Apply the 2-mL sample directly to the column followed by a wash with 8 mL of hexane. The vitamin K_1 was eluted with 8 mL of hexane/diethyl ether (93:3 v/v).

3. The samples were taken to dryness then re-dissolved in 20 μL of methylene chloride, followed by 180 μL of methanol (containing 10 mM $ZnCl_2$, 5 mM acetic acid, and 5 mM sodium acetate). The analysis was then carried out by HPLC.

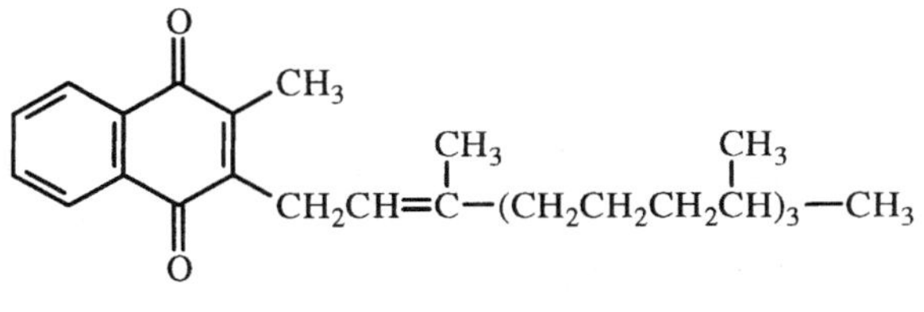

Figure 9.10. Structure of vitamin K_1.

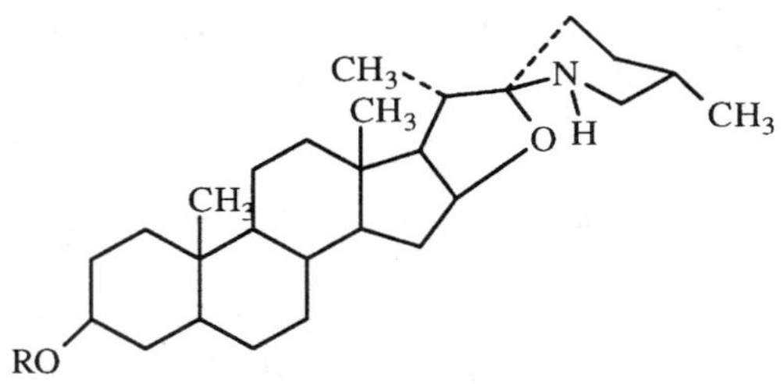

Tomatidine   R = OH

α-Tomatine   R = O-galactose-glucose-xylose
                                    |
                                    |
                                 glucose

Figure 9.11. Structure of α-tomatine.

9.3.6.   α-Tomatine in Tomatoes

The α-tomatine (Fig. 9.11) is the major glycoalkaloid present in the leaves, stems, and immature fruit of tomato plants. It is reported to be potentially toxic and also is reported to exert antifungal activity and to inhibit the growth of fruitworm and other insects (Friedman et al., 1994). The method consists of lyophilization of the tomatoes followed by grinding and sieving. The extraction is aqueous and reversed-phase C-18 is the sorbent used for the isolation of the α-tomatine.

α-Tomatine Method for Tomatoes

1. Fresh tomatoes were cubed and immediately frozen in liquid nitrogen. Samples were lyophilized. The dried tomatoes were then ground and sieved through a 0.5-mm screen.
2. One gram of tomato was extracted with 20 mL of 1% acetic acid for 2 hr, centrifuged, and filtered.
3. The C-18 sorbent was conditioned with 5 mL of methanol followed by 5 mL of water. The sample was applied in a volume of 30 mL.
4. The C-18 sorbent was washed with 10 mL of water, followed by 5 mL of 30 : 70 acetonitrile/water with 1% ammonium hydroxide, and then 5 mL of water.
5. The alkaloids were eluted with 10 mL of 70–30 acetonitrile/aqueous citric-acid phosphate buffer at pH 3.

6. The eluate was evaporated, re-extracted into butanol, evaporated, and then reconstituted in a methanol/acetic acid buffer for HPLC.

9.3.7. Wine Flavor Components

Aromatic substances are important in wine as they make a major contribution to the quality of the final product. Several hundred chemically different flavor compounds such as alcohols, esters, organic acids, aldehydes, ketones, and monoterpenes have been found in wines (Zhou et al., 1996). It is the combined contribution of these compounds that forms the character of a wine. The ability to determine these substances in wine may be used to help understand their role in flavoring. The method described (Zhou et al., 1996) uses a reversed-phase mechanism with XAD-2 to isolate the volatiles followed by Freon 11 extraction of the XAD-2 column.

Wine Flavor Components Method

1. A 12-cm column was packed with XAD-2 (styrene–divinylbenzene copolymer) with methanol and rinsed clean of methanol with distilled water. A 20-mL sample was applied to the column and washed with 50 mL of water to rid the column of carbohydrates.
2. With ice water circulating in the water jacket to cool the column, the column was eluted with Freon-11. The eluent was evaporated to ~ 100 µL and reconstituted to 200 µL with acetone.
3. Analyses were carried out by gas chromatography (GC). This method uses Freon-11, which is a compound harmful to the ozone layer. Freon-11 is used to remove the solvent peak in the GC run so that it does not co-elute with the volatile flavor compounds. The problem of solvent peak for volatile compounds may also be addressed with the use of a different technique, such as solid-phase microextraction. This method is shown in the following example.

9.3.8. Solid-Phase Microextraction for Flavor Analysis

Sample analysis based on sorption onto SPE sorbents followed by elution with organic solvent is the basis of the majority of SPE methods. However, the use of fibers to sorb organic compounds directly from solution was developed by Pawliszyn and co-workers (Arthur and Pawliszyn, 1990), and devices are commercially available from Supelco (see Chapter 12 for a description of the method). The adsorbed compounds may be directly desorbed into the GC without a solvent peak by simply heating the fiber in the inlet of the gas chromatograph. Solid-phase microextraction (SPHE) has mainly been used for environmental analysis; however, Yang and Peppard (1994) applied this method to the analysis of flavors in food and beverages. In their study, 25 compounds were tested for recovery from food and drinks.

Flavor Analysis by SPME

1. For liquid sampling, the SPME device was inserted into a 1.3-mL vial containing 0.5 mL of sample. The fiber remained in the liquid for 10 min under stirring.
2. The fiber was introduced into the GC injector in splitless mode and analyzed.
3. Compounds analyzed included a suite of volatiles. The authors found that the recoveries by SPME were variable and the use of an internal standard was necessary.

9.3.9.   Geosmin in Catfish

Aquaculture is the newest and one of the fastest growing segments of agriculture. In the United States, the business is approaching $1 billion per year and approximately half of this is channel catfish. The fish are grown in ponds with phytoplankton being one of the important members of the pond ecosystem. Some of the phytoplankton may produce compounds that give an off-flavor to the catfish. Geosmin is one of these substances, which gives a musty taste and odor to the fish. The method described here was taken from the work of Conte and co-workers (1996) and uses a combination of microwave-assisted distillation and solid-phase sorbent trapping of the geosmin for subsequent analysis by gas chromatography/mass spectrometry (GC/MS).

Geosmin Method for Catfish

1. The microwave distillation SPE apparatus is shown in Figure 9.12. A filet of catfish is placed in the microwave oven and heated, which drives the geosmin out of the tissue and is swept through the solid-phase extraction sorbent with argon gas. The condenser unit cools the gas, which increases the retention of geosmin, onto the 500 mg of C-18 in Figure 9.12. The sorbent is present in the bottom of the condenser unit. Charcoal, XAD-2, and C-18 were all tried for recovery of geosmin, and C-18 had the best recovery.
2. The C-18 is conditioned with 2 mL of ethyl acetate, methanol, and distilled water in that sequence.
3. A 20-g sample of catfish is homogenized with internal standard and microwaved at 40% power for 10 min.

4. The sorbent is rinsed with distilled water, 2 mL, and then eluted with three 1-mL aliquots of ethyl acetate.
5. The sample is dried with sodium sulfate and injected directly into the GC/MS.

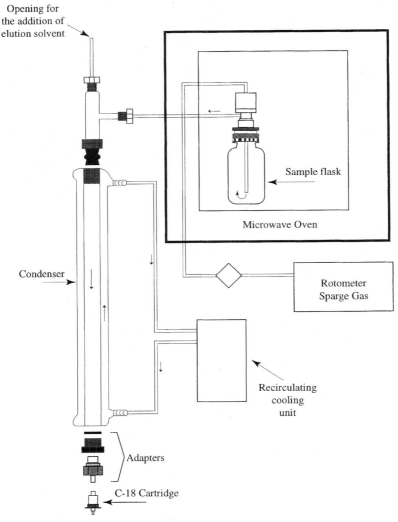

Figure 9.12. Microwave distillation solid-phase extractor for geosmin. [After Conte and co-workers (1996), published with permission.]

9.3.10.   Folic Acid in Cabbage

Folates are generally present in low concentration in foods; thus, any method of analysis requires isolation and sensitive detection (Vahteristo et. al., 1996). The method described here uses extraction of the folic acid (Fig. 9.13) with rapid microwave heating and enzymatic deconjugation with hog-kidney conjugase and chicken pancreatic enzymes. The extracts are purified with strong anion-exchange SPE. The final analysis is with HPLC.

Folic Acid Method in Cabbage

1. Samples, 2 to 5 g, of white cabbage were extracted with 5 to 6 volumes of 75 mM phosphate buffer at pH 6.0 (52 mM ascorbic acid/ascorbate mixture and 0.1% v/v 2-mercaptoethanol) under nitrogen. The sample was extracted by microwave for 1 min at 75% power. The samples were extracted for another 10 in a boiling-water bath.

2. Samples were centrifuged and the extract was adjusted to pH 4.9 with acetic acid and hog-kidney conjugase and chicken pancreas were added and incubated at 37 °C for 2 h.

3. The sample extracts were purified and concentrated by strong anion-exchange SPE (SAX) with 3 mL volume.

4. The sample was diluted (1 : 2) with 3 to 5 mL of water and 15 µL of 0.1% 2-mercaptoethanol (pH 7) was added to the sample before addition to the column. The sample was added slowly to the column and washed with two 1.5-mL rinses of conditioning buffer (0.01 mM phosphate buffer at pH 7.0 with 0.1% v/v 2-mercaptoethanol).

5. Thus, the column was used in its phosphate form for anion exchange. Folic acid was eluted with 2.5 mL of 0.1 M sodium acetate containing 10% w/v sodium chloride and 1% w/v ascorbic acid.

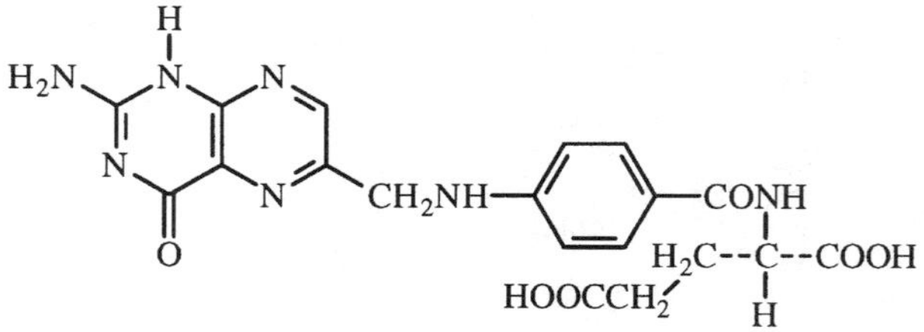

Figure 9.13. Structure of folic acid.

SUGGESTED READING

Casanova, J. A. 1996. Use of solid-phase extraction disks for analysis of moderately polar and nonpolar pesticides in high-moisture foods, *J. AOAC Int.*, **79**: 936–940.

Consalter, A. and Guzzo, V. 1991. Multiresidue analytical method for organophosphate pesticides in vegetables, *Fresenius' J. Anal. Chem.*, **339**: 390–394.

Coulibaly, K. and Jeon, I. J. 1996. An overview of solid-phase extraction of food flavor compounds and chemical residues, *Food Rev. Int.*, **12**: 131–151.

Gere, D. R., Randall L. G., and Callahan, D. J. 1995. *SFE: Food Applications and Principles, Instrumental Methods in Food Analysis*, J. Pare, Ed. Elsevier, Amsterdam.

Gere, D. R., and Derrico, E. M. 1994. SFE theory to practice-first principles and method development, Part I, *LC-GC*, **12**: 352–366.

Lee, S. M., Papathakis, M. L., Feng, H. M. C., Hunter, G. F., and Carr, J. E.. 1991. Multipesticide residue method for fruits and vegetables: California Department of Food and Agriculture, *Fresenius' J. Anal. Chem.*, **339**: 376–383.

Majors, R. E. 1996. The changing role of extraction in preparation of solid samples, *LC-GC*, **14**: 88–96.

Marble, L. K. and Delfino, J. J. 1988. Extraction and solid phase cleanup methods for pesticides in sediment and fish, *Am Lab.*, November, 23–32.

McDonald, P. D. and Bouvier, E. S. P. 1995. *Solid Phase Extraction Applications Guide and Bibliography, a Resource for Sample Preparation Methods Development*, 6th ed. Waters, Milford, MA.

Mindrup, R. F. 1995. Solid phase microextraction simplifies preparation of forensic, pharmaceutical, and food and beverage samples, Supelco, *Reporter*, **14**: 2–5.

Roch, O. G., Blunden, G., Coker, R. D., and Nawaz, S. 1992. The development and validation of a solid phase extraction/HPLC method for the determination of aflatoxins in groundnut meal, *Chromatographia*, **33**: 208–212.

Smith, R. M. 1995. Supercritical fluid extraction of natural products, *LC-GC*, **13**: 930–939.

Simpson, N. and Van Horne, K. C. 1993. *Varian Sorbent Extraction Technology Handbook*, Eds. Varian Sample Preparation Products, Harbor City, CA.

REFERENCES

Arthur, C. L. and Pawliszyn, J. 1990. Solid-phase microextraction with thermal desorption using fused silica optical fibers, *Anal. Chem.*, **62**: 2145–2148.

Barker, S. A., Long, A. R., Hines II, M. E. 1993. Disruption and fractionation of biological materials materials by matrix solid-phase dispersion, *J. Chromatog.*, **629**: 23–24.

Bevan, C. D. and Marshall, P. S. 1994. The use of supercritical fluids in the isolation of natural products, *Natural Products Rep.*, **11**: 451–453.

Booth, S. L., Davidson, K. W. and Sadowski, J. A. 1994. Evaluation of an HPLC method for the determination of phylloquinone (*vitamin* K₁) in various food matrices, *J. Agricul. Food Chem.*, **42**: 295–300.

Castioni, P., Christen, P., and Veuthey, J. L. 1995. Supercritical fluid extraction of compounds from plant origin, *Analusis*, **23**: 95–106.

Conte, E. D., Shen, C. Y., Perschbacher, P. W., and Miller, D. W. 1996. Determination of geosmin and methylisoborneol in catfish tissue (Ictalurus punctatus) by microwave-assisted distillation—solid phase adsorbent trapping, *J. Agricul. Food Chem.*, **44**: 829–835.

Fish, J. R. and Revesz, R. 1996. Microwave solvent extraction of chlorinated pesticides from soil, *LC-GC*, **14**: 230–234.

Friedman, M., Levin, C. E., and McDonald, G. M. 1994. α-Tomatine determination in tomatoes by HPLC using pulsed amperometric detection, *J. Agricul. Food Chem.*, **42**: 1959–1964.

Gere, D. R. and Derrico, E. M. 1994. SFE theory to practice—First principles and method development, Part I, *LC-GC*, **12**: 352–366.

Holcomb, M., Thompson, Jr., H. C., and Hankins, L. J. 1993. Analysis of fumonisin B1 in Rodent feed by gradient elution HPLC using precolumn derivatization with FMOC and fluorescence detection, *J. Agricul. Food Chem.*, **41**, 764–767.

Majors, R. E. 1995. Sample preparation and handling for environmental and biological analysis, *LC-GC*, **13**: 542–552.

Oka, H., Ikai, Y., Hayakawa, J., Masuda, K., Harada, K., Suzuki, M., Martz, V., and MacNeil, J. D. 1993. Improvement of chemical analysis of antibiotics. 18. Identification of residual tetracyclines in bovine tissues by TLC/FABMS with a sample condensation technique: *J. Agricul. Food Chem.*, **41**: 410–415.

Majors, R. E. 1996. The changing role of extraction in preparation of solid samples, *LC-GC*, **14**: 88–96.

Smith, R. M. 1995. Supercritical fluid extraction of natural products, *LC-GC*, **13**: 930–939.

Vahteristo, L. T., Ollilainen, V., Koivistoinen, P. E., and Varo, P. 1996. Improvements in the analysis of reduced folate monoglutamates and folic acid in food by high-performance liquid chromatography, *J. Agricul. Food Chem.*, **44**: 477–482.

Yang, X. and Peppard, T. 1994. Solid-phase microextraction for flavor analysis, *J. Agricul. Food Chem.*, **42**: 1925–1930.

Zhou, Y., Riesen, R., and Gilpin, C. S., 1996. Comparison of Amberlite XAD-2/Freon 11 extraction with liquid/liquid extraction for the determination of wine flavor components, *J. Agricul. Food Chem.*, **44**: 818–822.

CHAPTER

10

AUTOMATION OF SOLID-PHASE EXTRACTION

Automation of solid-phase extraction (SPE) is an important contribution to sample processing. If many samples are to be analyzed in the laboratory, automation provides precise and accurate methods when compared to manual methods. This chapter will discuss three major ideas on automation. First is why automate SPE. Second are the types of automation hardware, and third, how to automate an SPE method is discussed. Finally several case studies will be examined that show the automation process taken from a manual method.

10.1. WHY AUTOMATE SOLID-PHASE EXTRACTION?

Automation of a manual SPE method can provide many benefits, which include health and safety, improved results, and cost savings. Automating a manual SPE method removes the analysts from extended contact with biological samples, such as biological fluids (i.e., blood, serum, and urine) that may contain human immunodeficiency virus (HIV) and hepatitis viruses, or environmental samples that may contain hazardous substances. Because automated workstations are mechanical, they can work in environments where humans might not or cannot work. For example, they include hostile work environments, noisy production locations, or a refrigerated room.

The use of automation results in improved precision, by reducing sample-to-sample variation, and it reduces the number of operator errors compared to manual methods. Operator burnout from the repetitive motions of SPE is also removed. Just-in-time analysis may be used, and automated SPE methods may be linked directly with liquid chromatography or gas chromatography (GC) for totally automated analysis. Furthermore, overnight runs are possible to make maximum use of time for sample production. In summary, automation reduces cost by freeing-up personnel from the use of vacuum boxes and tedious pipetting and column conditioning and elution. An automated SPE instrument allows for 24-hr operation, with or without supervision.

243

10.2. AUTOMATION EQUIPMENT

There are many types of automation equipment for SPE. They include semi-automated instruments, which are instruments where some intervention is required; workstations, which carry out the entire SPE operation without intervention, including on-line analysis by GC and high-pressure liquid chromatography (HPLC); and customized SPE, which are robotic systems that are capable of many activities besides SPE and are custom designed for the user.

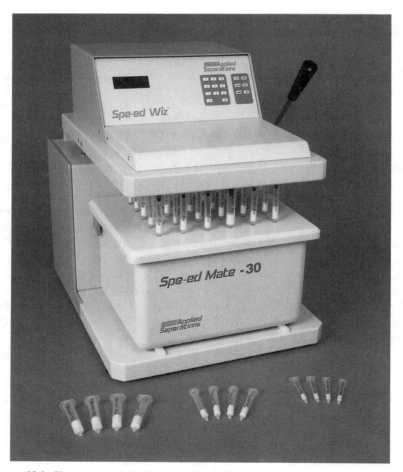

Figure 10.1. Photograph of the Spe-ed Wiz. (Published with the permission of Applied Separations.)

10.2.1. Semiautomated Solid-Phase Extraction

Applied Separation's instrument, the Spe-ed Wiz, is an example of a semi-automated SPE system (Fig. 10.1), where some manual intervention is required. In this system, the analyst must manually pour the samples onto the SPE column, and add the elution receiver or test tube. The Spe-ed Wiz will condition the cartridge and prepare it for sample addition. The sample is manually added to the test tube and the Spe-ed Wiz will aspirate the sample, or the lid can be closed to load the samples. After the sample is loaded on the columns (syringe type), the columns are rinsed, and elution tubes must be manually placed in the elution rack. The elution is automated.

The instrument can work on a variety of column sizes, as well as glass- or plastic-column barrels. The instrument can be programmed with or without a personal computer (PC). The PC adds the capability of saving methods and allows one the control of the cartridge flow rates. The general features of the Spe-ed Wiz are shown in Table 10.1.

Table 10.1. Features of the Spe-ed Wiz

Feature	Spe-ed Wiz
General	
Sample capacity	30 samples
Number of reagents	12 reagents
Internal standard line	No
Dilution	No
Filtration	No
Mixing	No
Temperature controlled racks	No
Evaporation	No
Derivatization	No
Bar code reader	Yes
Batch or serial	Batch
Multiple methods	Runs one method at a time.
SPE	
SPE column sizes	1, 3, 5, 12, 35, and 60 mL
Syringe-barrel columns	Yes
Individual flow rates	Yes with computer system 6030
Sample load volumes	Column dependent
Elution container	Your choice
Max. elution fractions	Unlimited

(contd.)

Table 10.1. (*continued*)

Feature	Spe-ed Wiz
Glass or plastic columns	Both
Dry column with gas	Yes, with air under vacuum
Positive pressure or vacuum	Both

Writing Methods

Documentation dedicated computer	Optional computer with RS-232 interface allows unlimited method storage and the ability to control flow rates.

Analytical Analysis

HPLC injection	No
GC injection	No
UV-Vis interface	No
Vial filling	No
Fully documented report	No

Dimensions and Pricing

Price (U. S. Dollars 1996)	$16,000

Published with the permission of Applied Separations.
Applied Separations
P.O. Box 20032
Lehigh Valley, PA 18002-0032
Phone: 215-770-0900

10.2.2.   Workstations

The second type of automation is the workstation. They are instruments that can perform multiple functions using software tools and dedicated hardware to perform a set of predefined operations, which are related to or part of the process of solid-phase extraction. They differ from laboratory robots that are capable of many laboratory functions. There are seven instruments that will be discussed in this section (Table 10.2–10.8), which are dedicated to solid-phase extraction. They range in price from approximately $20,000 to $50,000 (1996 dollars) for workstations and from $50,000 and up for laboratory robots.

10.2.2.1.   ASPEC XL

This unit is made by Gilson Medical Electronics, Inc. (Fig. 10.2, Table 10.2). The workstation for SPE allows you to write a program that will condi-

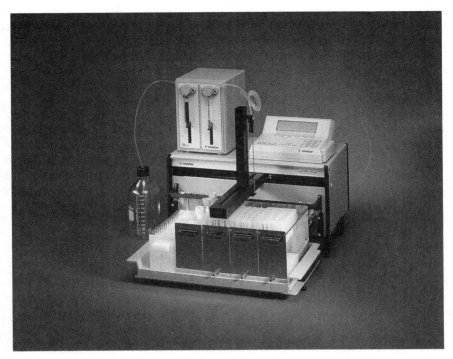

Figure 10.2. ASPEC XL SPE workstation. (Photograph published with the permission of Gilson, Inc.)

Table 10.2. Features of the ASPEC XL by Gilson, Inc.

Feature	ASPEC XL
	General
Sample capacity	108 1-mL, 60 3-mL, 45–60 6-mL columns
Number of reagents	20 reagents
Internal standard addition	Included
Dilution	Volumetric
Filtration	No
Mixing	Cycling or bubble with nitrogen
Temperature-controlled racks	Cooler available
Evaporation	Yes, additional cost

(*contd.*)

Table 10.2. (*continued*)

Feature	ASPEC XL
Derivatization	Yes
Bar code reader	No
Batch or serial	Batch or serial
Multiple methods	Many, requires computer control, not keypad

SPE

SPE column sizes	1, 3, 5, 12, 35, and 60 mL
Syringe-barrel columns	Yes
Individual flow rates	Yes
Sample load volumes	Dependent on container, generally 5–10 mL
Elution container	Multiple tube sizes available
Max. elution fractions	6–8, configuration dependent
Glass or plastic columns	Plastic
Dry column with gas	Yes
Multicolumn SPE	Yes
Positive pressure or vacuum	Positive pressure

Writing Methods

Documentation	Yes
Dedicated computer	No
Computer requirements	Key pad or IBM-compatible computer
Writing method	"Task Automated Software"
Windows based	Yes, when using computer.

Analytical Analysis

HPLC injection	Yes, 1 or 2 Rheodyne 7010, or Rheodyne 7413
GC injection	Yes
UV-Vis interface	No
Vial filling	Will work with 2-mL vials or WISP vials (Waters)
Fully documented report	Yes with Gilson HPLC

Dimensions and Pricing

Price (U.S. Dollars 1996)	$21-25,000

Published with the permission of Gilson, Inc.
Gilson, Inc.
3000 W. Beltline Highway,
P.O. Box 620027
Middleton, WI 53562-0027
Phone: 1-608-836-1551 and 1-800-445-7661

tion, load, rinse, and elute the SPE columns. There is also a provision for on-line introduction of the sample to an HPLC or GC. A dedicated computer is not required to run the instrument, and the instrument can be programmed through a keypad controller. The instrument operates in a batch or serial mode. It can work with 1-, 3-, or 6-mL plastic or 8-mL glass SPE columns, Whatman filters packaged in a syringe barrel, and with disposable sealing caps to maintain the positive pressure for the SPE steps. The individual flow rates for condition, load, rinse, and elution are able to be selected. The workstation may also be used to perform volumetric dilutions, solvent or reagent addition, and sample mixing.

10.2.2.2. Microlab SPE from Hamilton Company

The microlab approach is a solution-handling robot that has a rectangular shape (Fig. 10.3, Table 10.3). A dedicated PC is required to run the instrument. The FLEXPREP software contains preprogrammed routines or users can write

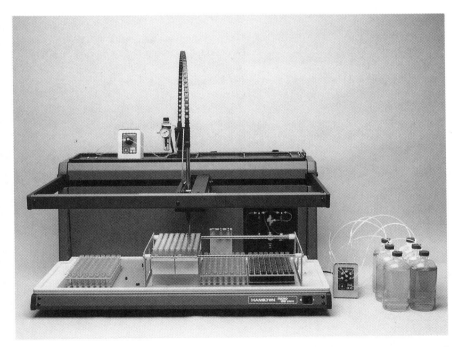

Figure 10.3. The Hamilton Microlab SPE unit. (Photograph published with the permission of Hamilton Company.)

Table 10.3. Features of the Hamilton Microlab SPE. Published with permission of Hamilton Company

Feature	Microlab SPE
	General
Sample capacity	96 samples,
Number of reagents	1–12 reagents
Internal standard line	Available
Dilution	Volumetric
Filtration	Yes
Mixing	Cycling
Temperature-controlled racks	No
Evaporation	Yes
Derivatization	No
Bar code reader	Yes
Batch or serial	Batch or serial
	SPE
SPE column sizes	1, 3, or 6 mL
Syringe-barrel columns	Yes
Individual flow rates	Gas pressure selected, same for all
Sample load volumes	Dependent on container, generally 5–10 mL
Elution container	Multiple tube sizes available
Max. Elution fractions	1–13, configuration dependent
Glass or plastic columns	Plastic or glass
Dry column with gas	Yes
Multicolumn SPE	Yes
Positive pressure or vacuum	Positive pressure
	Writing Methods
Documentation	No
Dedicated computer	No
Computer requirements	IBM-compatible computer
Writing method	FLEXPREP software, preprogrammed or write your own
	Analytical Analysis
HPLC injection	Yes, 1 Rheodyne 7010
GC injection	No
UV-Vis interface	No
Vial filling	Yes
Fully documented report	No

(contd.)

Table 10.3. (*continued*)

Feature	Microlab SPE

Dimensions and Pricing

Price (U.S. Dollars 1996)	$25,000

Publised with the permisssion of Hamilton Company.
Hamilton Company
P.O. Box 10030
Reno, Nevada 89520
Phone: 1-800-648-5950

their own routines. The instrument operates in serial or batch mode working with 1-, 3-, or 6-mL plastic SPE columns. Flow rates are selected by regulating the gas pressure. The workstation may be used to perform volumetric dilutions and sample mixing. Both filtration and HPLC injection options are available.

10.2.2.3. PrepStation by Hewlett-Packard

The Hewlett-Packard (HP) 7686 PrepStation System workstation allows you to write a program that will condition, load, rinse, and elute your SPE columns (Fig. 10.4, Table 10.4). A dedicated DOS-based ChemStation computer system by HP is required to run the instrument. The software uses a Menu Selectable Approach. The instrument operates in a serial mode and works with HP's disposable SPE cartridges. Individual flow rates for condition, load, rinse, and elute are selected. The workstation may be used to perform volumetric dilutions, sample mixing, bar-code reading, evaporation, and derivatization. When used with an HP 1050 HPLC or HP 5890 GC direct injections are possible. The instrument is also available in a stand-alone

Table 10.4. Features of the PrepStation System (Published with permission of Hewlett-Packard Company)

Feature	PrepStation System
	General
Sample capacity	1002 -mL vials
Number of reagents	8 reagent reservoirs
Internal standard line	Yes
Dilution	Volumetric

(contd.)

Table 10.4. (*continued*)

Feature	PrepStation System
Filtration	Yes with HP filter
Mixing	Yes
Temperature controlled racks	Cooler available
Evaporation	Yes, heat or nitrogen
Derivatization	Yes
Bar code reader	Yes
Batch or serial	Serial with limited batch
Multiple methods	Yes
	SPE
SPE column sizes	HP cartridge with 100 or 300 mg of packing
Syringe-barrel columns	HP disposable cartridges
Individual flow rates	Yes
Sample load volumes	Loads from vials or reagent line
Elution container	2-mL vial
Max. elution fractions	Multiple
Glass or plastic columns	Plastic
Dry column with gas	Yes
Multicolumn SPE	Yes
Positive pressure or vacuum	Positive pressure
	Writing Methods
Documentation	Yes
Dedicated computer	Yes
Computer requirements	DOS-based HP computer
Writing method	"Menu Selectable Approach"
Windows based	Yes.
	Analytical Analysis
HPLC injection	Yes, on-line to HP's HPLC Model 1050
GC injection	Yes, on-line to HP's 5890 GC
UV-Vis interface	No
Vial filling	Yes
Fully documented report	Yes with special software, "Bench Supervisor"
	Dimensions and Pricing
Price (U.S. Dollars 1996)	$15–20,000

Published with the permission of Hewlett-Packard Company.
Hewlett-Packard Company
3495 Deer Creek Road
Palo Alto, CA 94304
Phone: 800-334-3110 × 204

Figure 10.4. Photograph of the PrepStation by Hewlett-Packard Company. (Published with the permission of Hewlett-Packard Company.)

version that is separate from the HPLC and the GC. The system uses its own solid-phase extraction cartridges rather than the conventional cartridges available from SPE manufacturers.

10.2.2.4. PROSPEKT (Jones Chromatography)

The PROSPEKT is a fully automated SPE system with on-line HPLC injection that can process up to 96 samples automatically (Fig. 10.5, Table 10.5). Samples are automatically loaded onto disposable cartridges, purged with one

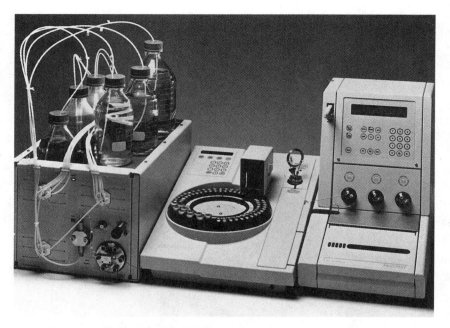

Figure 10.5. PROSPEKT automated solid-phase extraction system. (Photograph published with the permission of Spark Holland and Jones Chromatography.)

Table 10.5. Features of the PROSPEKT System

Feature	PROSPEKT System
	General
Sample capacity	96 samples of 2 mL; larger volumes possible
Number of reagents	Up to 16 reagent reservoirs
Internal standard line	Yes with autosampler
Dilution	Yes with autosampler
Filtration	No
Mixing	Yes with autosampler
Temperature-controlled racks	Optional
Evaporation	No
Derivatization	No
Bar code reader	No
Batch or serial	Serial
Multiple methods	Yes, up to 99

(contd.)

Table 10.5. (*continued*)

Feature	PROSPEKT System
	SPE
SPE column sizes	One size with 20 or 45 mg of packing
Syringe-barrel columns	Proprietary cartridges with standard sorbents
Individual flow rates	Yes
Sample load volumes	Loads from vials or reagent line
Elution container	On-line HPLC or off-line options
Max. elution fractions	Unlimited
Glass or plastic columns	PVDF copolymer with stainless steel frits
Dry column with gas	Yes
Multicolumn SPE	Yes
Positive pressure or vacuum	Positive pressure
	Writing Methods
Documentation	Yes
Dedicated computer	Not required
Computer requirements	N/A
Writing method	Through PROSPEKT module
Windows based	No
	Analytical Analysis
HPLC injection	Yes, on-line HPLC
GC injection	Yes, on-line to GC
UV-Vis interface	Yes
Vial filling	Yes
Fully documented report	No
	Dimensions and Pricing
Price (U.S. Dollars 1996)	$20–50,000 module dependent

Published with the permission of Spark Holland and Jones Chromatography.
Jones Chromatography
P.O. Box 280329
Lakewood, CO 80228
Phone: 800-874-6244

or more reagents for sample clean-up and are directly eluted on line to the HPLC system. The complete PROSPEKT system consists of three modules that can be used in several configurations. The heart of the system is the PROSPEKT module that contains the cartridge transport and sealing mechanism and a microprocessor to control the other modules. The solvent delivery unit selects and delivers various solvents (up to 16) to the cartridge at constant flow rates and high pressures. The PROSPEKT uses its own type of SPE cartridge that fits the automated unit. Packing materials are comparable to HPLC bonded phases with particle sizes from 10 to 40 μm.

A recent advance using the PROSPEKT is the interface of the PROSPEKT with a mass spectrometry–mass spectrometry (MS–MS) system without the use of HPLC. The concept is that the small particle size SPE column (10 μm) can be used for chromatographic separation that is adequate for on-line MS–MS for quantifying drugs at the picogram level (Bowers et al., 1997). The authors found that a sample as small as 200 μL could be used to identify drugs with the on-line SPE MS–MS system with detection limits as low as 50 pg/mL. The advantage is that small samples could be run during the developmental stages of drugs, when only small masses of analyte are available.

10.2.2.5. AutoTrace Workstation

The AutoTrace Workstation was originally designed by Zymark and is now sold by Tekmar (Fig. 10.6, Table 10.6). It is a dedicated SPE workstation designed for large-volume water samples, from 10 mL to 2 L. A dedicated computer is not required, and the instrument will run from a diskette on which the method is saved. The software is menu driven for ease of use. The instrument operates in batch mode, up to six samples at a time. The workstation was designed for large-volume work and will work with 1-, 3-, and 6-mL plastic SPE columns, or 6- and 8-mL glass columns, or with 47-mm extraction disks. All steps are performed using positive pressure. Individual flow rates of condition, load, rinse, and elution are selected. The AutoTrace workstation is being applied to water samples that are in compliance with the Safe Drinking Water Act, Resource Conservation and Recovery Act (RCRA), and National Pollution Discharge Elimination Systems for compliance monitoring and is a good automation system for environmental work.

Table 10.6. Features of the Tekmar AutoTrace System

Feature	Tekmar AutoTrace System
	General
Sample capacity	6 samples per run
Number of reagents	5 reagent reservoirs
Internal standard line	Yes with autosampler

(contd.)

Table 10.6. (*continued*)

Feature	Tekmar AutoTrace System
Dilution	Yes with autosampler
Filtration	No
Mixing	No
Temperature-controlled racks	No
Evaporation	Yes
Derivatization	No
Bar code reader	No
Batch or serial	Batch
Multiple methods	No

SPE

SPE column sizes	1, 3, 6 mL plastic, 6, 8 mL glass, 47-mm disk
Syringe-barrel columns	Yes or disk
Individual flow rates	Yes
Sample load volumes	10 mL to 2 L
Elution container	Test tubes or GC vials
Max. elution fractions	2
Glass or plastic columns	Glass or plastic
Dry column with gas	Yes
Multicolumn SPE	No
Positive pressure or vacuum	Positive pressure

Writing Methods

Documentation	Yes
Dedicated computer	Not required
Computer requirements	MS-DOS system
Writing method	Pull-down menus
Windows based	No

Analytical Analysis

HPLC injection	No
GC injection	No
UV-Vis interface	No
Vial filling	WISP (Waters), GC vial

Dimensions and Pricing

Price (U.S. Dollars 1996)	$20–30,000

Published with the permission of Tekmar-Dohrmann.
Tekmar-Dohrmann
P.O. Box 429576
Cincinnati, Ohio 45242-9576
Phone: 800-543-4461

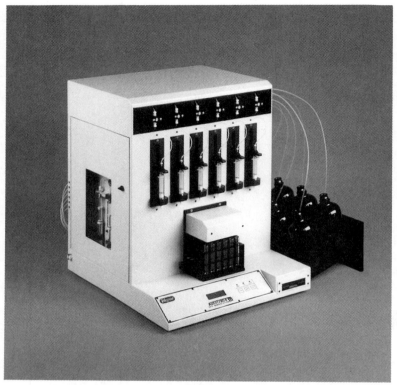

Figure 10.6. Photograph of the AutoTrace workstation. (AutoTrace is a registered trademark of Zymark Corporation. Tekmar-Dohrmann uses this trademark under full license of Zymark Corporation. Photograph published with the permission of Tekmar-Dohrmann.)

10.2.2.6. BenchMate Workstation by Zymark

The BenchMate workstation for SPE allows one to write programs that will condition, load, rinse, and elute the SPE columns (Fig. 10.7, Table 10.7). A dedicated computer is not required and the instrument will operate from a diskette on which the method has been stored. Software is menu driven and the instrument operates in a serial mode. The workstation was designed for samples up to 10 mL. The BenchMate workstation was designed to work with 1- and 3-mL plastic syringe-barrel SPE columns. All steps are performed using positive pressure and the flow rates are selected for condition, load, rinse, and elution. The workstation may be used to perform volumetric or gravimetric dilutions, sample mixing, and filtration. The BenchMate work-

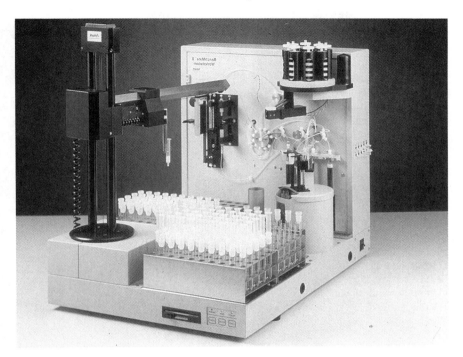

Figure 10.7. BenchMate workstation for SPE by Zymark Corporation. (Photograph published with the permission of Zymark Corporation.)

Table 10.7. Features of the BenchMate Workstation

Feature	BenchMate System
	General
Sample capacity	100–200 samples per run
Number of reagents	6 reagent reservoirs
Internal standard line	Yes
Dilution	Yes, volumetric or gravimetric
Filtration	Yes
Mixing	Vortex or cycling
Temperature-controlled racks	Yes, 4–60 °C
Evaporation	No
Derivatization	Yes, without heat
Bar code reader	For input only
Batch or serial	Serial
Multiple methods	Yes

(*contd.*)

Table 10.7. (*continued*)

Feature	BenchMate System
	SPE
SPE column sizes	1 or 3 mL
Syringe-barrel columns	Yes
Individual flow rates	Yes
Sample load volumes	0.5–9.8 mL
Elution container	Test tubes
Max. elution fractions	2
Glass or plastic columns	Plastic
Dry column with gas	Yes
Multicolumn SPE	Requires multiple methods
Positive pressure or vacuum	Positive pressure
	Writing Methods
Documentation	Yes, gravimetric
Dedicated computer	Not required
Computer requirements	MS-DOS system
Writing method	Pull-down menus
Windows based	No
	Analytical Analysis
HPLC injection	Yes, 1 Rheodyne
GC injection	No
UV-Vis interface	Yes
Vial filling	WISP (Waters), GC vial, or 96 deep well
	Dimensions and Pricing
Price (U.S. Dollars 1996)	$25–35,000

Published with the permission of Zymark Corporation.
Zymark Corporation
Zymark Center
Hopkinton, MA 01748
Phone: 508-435-9500

station offers direct HPLC injections, or loading 4-mL WISP vials, GC vials or deep-well microtiter plates or microtubes.

10.2.2.7. *RapidTrace SPE Workstation by Zymark*

The RapidTrace SPE workstation for SPE allows one to write a procedure that will condition, load, rinse, and elute SPE columns (Fig. 10.8, Table 10.8). By

Figure 10.8. RapidTrace SPE workstation from Zymark. (Photograph published with the permission of Zymark Corporation.)

Table 10.8. Features of the RapidTrace SPE Workstation

Feature	Zymark System
General	
Sample capacity	100 samples per run for workstation of 10 modules
Number of reagents	8 reagent reservoirs
Internal standard line	Not dedicated
Dilution	Yes
Filtration	No
Mixing	Yes, cycling
Temperature-controlled racks	Yes
Evaporation	No
Derivatization	No
Bar code reader	No
Batch or serial	Batch and serial simultaneously
Multiple methods	Yes
SPE	
SPE column sizes	1 or 3 mL
Syringe-barrel columns	Yes
Individual flow rates	Yes
Sample load volumes	0.1–9 mL
Elution container	Test tubes, vials
Max. elution fractions	10

(contd.)

Table 10.8. (*continued*)

Feature	Zymark System
Glass or plastic columns	Plastic
Dry column with gas	Yes
Multicolumn SPE	Yes with multiple methods
Positive pressure or vacuum	Positive pressure

Writing Methods

Documentation	Yes
Dedicated computer	Yes
Computer requirements	Windows 3.1 or higher
Writing method	Windows-based menus
Windows based	Yes

Analytical Analysis

HPLC injection	No
GC injection	No
UV-Vis interface	No
Vial filling	No

Dimensions and Pricing

Price (U.S. Dollars 1996)	$15–75,000 dependent on the number of modules

Published with the permission of Zymark Corporation.
Zymark Corporation
Zymark Center
Hopkinton, MA 01748
Phone: 508-435-9500

processing up to 10 samples simultaneously, the workstation provides fast sample turn-around-time and throughput, from 30–60 samples per hour.

Software is Windows based and designed for ease of use. The software was designed to facilitate a structured methods development process by allowing parameters such as column types, reagent concentration, and flow rates to be incrementally varied and tested. The workstation was designed for samples up to 9 mL. The RapidTrace was designed to work with 1- and 3-mL plastic syringe-barrel SPE columns. All steps are performed using positive pressure, the flow rates are selected for condition, load, rinse, and elution.

10.2.2.8. Solid-Phase Microextraction

With this equipment, the automation of solid-phase microextraction (SPME) is taken directly from sample bottle to gas chromatography. Literally, the water sample is placed in the autotray of the GC and the microfibers are immersed into the sample directly, with agitation of the sample, but not heating. After equilibration of the microfiber, the fiber is withdrawn and injected automatically into the GC (Berg, 1993). The method is quite simple and appears robust. The cost is also quite inexpensive at approximately $15,000 for the unit and software. The instrument is available for most brands of gas chromatographs making the unit quite attractive for automated GC/MS

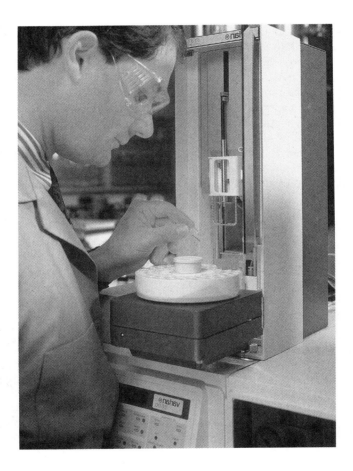

Figure 10.9. Automated solid-phase microextraction (SPME). (Photograph published with the permission of © Varian Associates, Inc.)

analysis of compounds amenable to this method. The unit is shown in Figure 10.9. The details of SPME are discussed in Chapter 12.

10.2.3.   Total Automation

At least two different configurations to completely automated systems are available for use in the laboratory for solid-phase extraction coupled to instrument operation. The first is the rectilinear configuration with an articulated arm, which is marketed by Sagian (Fig. 10.10). This robot is capable of handling samples, weighing, bar-code reading, and interfacing directly with instruments such as supercritical fluid extraction and GC/MS. The concept of the bench and how one does laboratory work in this configuration of the robot is presented by Majors and Holden (1993).

A second approach is the cylindrical robotic system (Fig. 10.11), which is marketed by Zymark. In this system, the robot is most often in the center of the bench and may move around in a circle carrying out the tasks asked of it. They include capping, weighing, centrifuging, filtering, and evaporation. Obviously, SPE is possible with both types of robotic systems. Robotic systems automating the entire method are more costly, often in excess of $100,000 per unit. But for continuous processing of a specific method, they could be cost effective. Furthermore, these units could be interfaced to other less expensive automated SPE units and work in conjunction with them for the entire analysis.

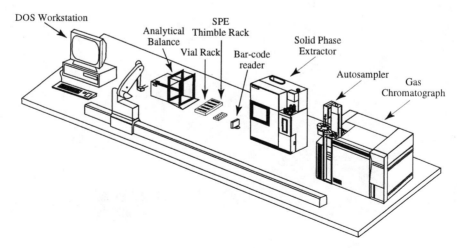

Figure 10.10. Schematic of a rectilinear robotic application. [Reprinted with permission from *LC-GC* magazine, from Majors and Holden (1993)].

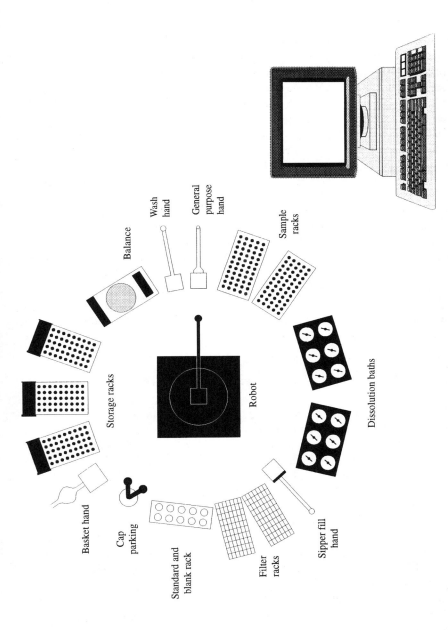

Figure 10.11. Schematic of a cylindrical robotic application. [Reprinted with permission of *LC-GC* magazine, from Majors and Holden (1993)].

10.3.　AUTOMATING A MANUAL SOLID-PHASE EXTRACTION METHOD

In automating a manual SPE method, the goal is to provide a rugged automated SPE method that is free from interferences or carryover and is optimized for recovery and throughput. Before beginning to automate a manual SPE method, one must consider these ideas. First, does the manual SPE method work? A well-characterized, fully developed manual method is essential for a starting point. The reproducibility in the liquid-handling system can be used to improve the accuracy, precision, and day-to-day ruggedness of a SPE method. It is a good practice to validate a working manual method before automating.

Second, work in the sample matrix. Choose a sample matrix for automating the manual SPE method that is free of analyte. Thus, all the interferences and other problems associated with the matrix will be addressed immediately. Third, spike the clean sample matrix to a known concentration so that recoveries may be calculated. Samples should be spiked to a midlevel concentration range. Finally, run samples in triplicate. Triplicates will give the percent relative standard deviation and are some indication of the reproducibility of your automated procedure. Samples that show a low relative standard deviation typically prove that the method is a well-written automated procedure. High relative standard deviations suggest problems with the matrix and recovery of the sample.

10.4.　PROCESS OF AUTOMATION

The process of automation involves a five-step procedure (Fig. 10.12). First is the initial experiment, the goal of which is to get some recovery from the SPE sorbent. Second is to minimize interferences from the matrix. Third is to optimize recovery of the method, trying to achieve at least 85% recovery in the final method. Fourth is to reduce carryover between samples. Finally, the last step is to optimize throughput for the fastest possible production rate.

10.4.1.　The Initial Experiment

The goal of the initial experiment is to get some recovery from the SPE method. It is important to work with your sample in a clean matrix (contains no analyte except for the spike) and to spike the concentration at midlevel. The recovery of the initial experiment may be higher, lower, or the same as is achieved in the manual method. Additional experiments are performed later to maximize recovery and throughput.

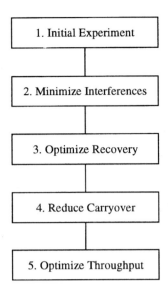

Figure 10.12. The process of SPE method automation. [After Jordan (1993), published with permission.]

When working with a vacuum box to perform SPE, the pressure is set to give an approximate flow rate. With an automated SPE workstation, individual flow rates are selected for each step of a SPE method: condition, load, rinse, and elute. Thus, the flow rates must be selected on an automated method. Approximate flow rates may be measured with the vacuum box and transferred to an automated method.

Visual observations are important when running the initial experiment. One should watch each step of the automated method to be sure that it is being performed the same as the manual method. If the initial experiment shows no recovery, one should check for obvious problems first. Was the correct column used? Are the reagents in the correct lines for the workstation? Is the sample spiked to correct level in the matrix? After obtaining recoveries comparable to the manual method, the next step is to minimize interferences.

10.4.2. Minimize Interferences

Interferences must be minimized prior to further method optimization to give accurate quantitation of the analyte. Interferences are removed from the SPE column by washing the SPE sorbent with solvent, without removing the analyte. The parameters in an automated procedure that affect rinsing are rinse

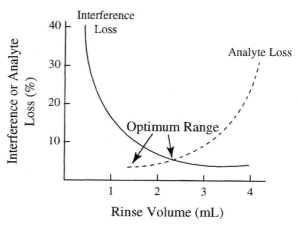

Figure 10.13. Optimizing rinsing of the SPE column. [After Jordan (1993), published with permission.]

volume and flow rate. It is important to change only one variable at a time when optimizing an automated procedure.

Figure 10.13 shows the optimum rinse volume to remove interferences without removing the analyte. An insufficient rinse volume may lead to interferences in the final elution. Therefore, removal of interferences are important. The goal is the smallest rinse volume that adequately removes interferences while leaving the majority of the analyte on the sorbent (> 90%). Biological samples, such as urine and blood, usually require rinses, while environmental samples, such as water, do not. Soil and plant materials often require a rinse.

If the rinse flow rate is too fast, interferences may not be adequately removed. Diffusion between the packing material and the rinse reagent may be enhanced at slower flow rates. Verify that the rinse flow rate is set correctly by decreasing it and determine if the interferences decrease. Automated systems for SPE have many reagent lines available, which gives one the opportunity to use additional reagents or to test these reagents during methods development. These reagents should be strong enough to remove interferences, but not remove the analyte.

10.4.3. Optimize Recovery

Recovery of the analyte is optimized after interferences have been minimized. Automated systems for SPE should yield recoveries equivalent to or better than the manual method. The considerations for optimized recovery include load and elute flow rates and column drying. First, only one variable

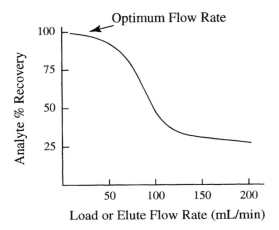

Figure 10.14. Effect of flow rate on the recovery of the analyte from SPE sorbent. [After Jordan (1993), published with permission.]

should be changed in any single experiment in order to evaluate the importance of the change. Second, make large changes in the variable at first to determine if there is an effect that may be optimized. The goal is to obtain the greatest recovery possible for the analyte. Flow rates should be run as fast as possible while still maintaining good recovery, because this will affect the amount of throughput that the method can accommodate.

Figure 10.14 shows the effect of load and elute flow rates on percent recovery. As the flow rate is increased to greater and greater amounts, the percent recovery is markedly decreased because equilibrium sorption and desorption are not occurring. In general, the sorption process is rapid so the flow rate may be fast. However, for elution, the flow rate should be lowered to allow the solvent time to diffuse into the sorbent and make adequate contact for desorption. The smallest amount of desorption solvent possible is an advantage later when the solvent may have to be evaporated for analysis. Likewise if elution is taking too much solvent, one should consider a stronger solvent for elution in order to minimize elution volume.

10.4.4. Column Drying

Drying of the SPE column is necessary when two immiscible reagents are used in sequence. Inadequate drying can cause low recovery because water prevents the organic solvent from adequately solubilizing the analyte. Thus, drying is meant to remove water. Correct gas pressure is important. If the gas pressure is too low, the column-drying step may not be effective and longer

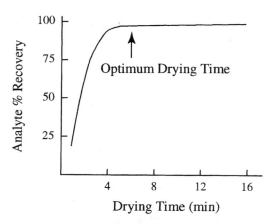

Figure 10.15.   Effect of column drying on percent recovery of the analyte. [After Jordan (1993), published with permission.]

drying times are required. Figure 10.15 shows the effect of drying time on percent recovery.

One should evaluate the percent recovery by observing changes in recovery with drying time. The automated workstation is efficient at running these types of changes, since a different method may be programmed with increasing drying times. One should remember to make large changes in order to see the effect of the variable. The goal is to determine the shortest drying time for the highest percent recovery.

10.4.5.   Reduce Carryover

Carryover of analyte between samples should be checked once the recovery has been optimized. Carryover may be evaluated with five samples: two blanks, followed by the highest sample that is expected to be found, followed by two blanks. The first two blanks set the baseline, the high sample or standard then tests the system, and the last two blanks evaluate the amount of carryover that is possible into the following sample.

If carryover is detected, cleaning or rinse steps need to be inserted in the method. The most commonly used technique is to add steps to clean the fluid path between samples. Explore the option of washing the fluid path with a cleaning reagent between samples. Remember, multiple low-volume washes are more effective than a single high-volume wash for cleaning a fluid path. Again, Figure 10.16 shows the concept of the least wash volume for the maximum removal of carryover.

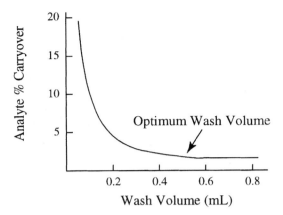

Figure 10.16. Optimizing wash volume to reduce carryover in SPE. [After Jordan (1993), published with permission.]

In choosing an optimum wash volume, select the smallest wash volume that will adequately remove carryover. Choose a wash reagent that dissolves the analyte and that will remove anything that may be remaining from the sample matrix. Miscibility is also an important factor when selecting a wash reagent. For example, did the fluid path last contain hexane for an elution reagent, and if water is the next reagent in the path, then a wash with methanol is required prior to the water reagent.

10.4.6. Optimize Throughput

Optimize sample throughput only after recovery and carryover requirements have been met. There are many parameters that may be adjusted to increase throughput. The first important thing to do is to determine your throughput goal. If an automated workstation makes direct HPLC injections, your sample throughput may be limited by the analytical run time, which then becomes the throughput time. After setting a goal, the times of each step are determined and minimized. The final step is to recheck recovery and carryover.

Figure 10.17 shows the time of the entire method as a series of bar graphs that give the total time of the method. When determining which steps to optimize for throughput, work first on the steps that may yield the biggest return, that is, those steps that require the most time. As changes are made, the method should be rechecked to verify that the recovery of analyte and the low carryover of analyte between samples has not been jeopardized.

The sample loading and elution rates are often more critical to optimizing recovery than are the condition and rinse flow rates. If the load flow rate is too

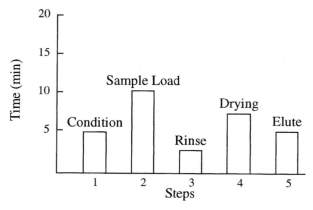

Figure 10.17. Method step times in an automated method. [After Jordan (1993), published with permission.]

fast, the compounds may not adequately sorb to the column. Refer to earlier experiments about flow rate and sorption, elution, and rinsing to optimize these rates. It is always a good idea not to push the limits of flow rate but to be somewhat conservative and maintain good recovery.

Finally, the automated method should give as good if not better recoveries and precision than the original manual method. Commonly, the automated methods are not as rapid as the manual method. This is because of the serial nature of the automated workstations, whereas manual methods are typically done by vacuum box, which is a batch method. The automated method is superior to manual methods in the area of precision because any repetitive steps of the method are reproducibly delivered and the robot will not make the errors prone to humans.

10.5.   EXAMPLES OF AUTOMATED SOLID-PHASE EXTRACTION

10.5.1.   Phenylurea Herbicides in Water

Phenylurea herbicides are widely used for weed control around the world. This method works directly on a filtered river water sample. This fully auto-mated method was accomplished on the PROSPEKT automated system and was interfaced directly to an HPLC. Figure 10.18 shows the configuration of the system. It consists of the solvent delivery unit (SDU), which prepares the sorbent, pumps the sample, and elutes the cartridge. The system also includes Marathon autosampler, which coordinates the 10-mL water samples, and the

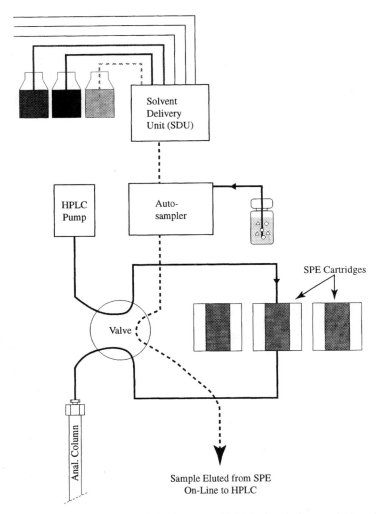

Figure 10.18. Diagram of the PROSPEKT system. (Published with the permission of Spark Holland.)

PROSPEKT unit itself, which runs the automated operation of the SPE and interfaces with the HPLC for sample analysis.

Phenylurea Automated Method

1. A 40-μm C-8 cartridge was used on the PROSPEKT, which was conditioned with 0.5 mL of methanol and 2 mL of deionized water.

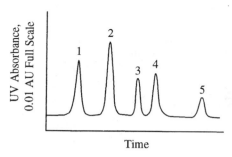

Figure 10.19. Chromatographic separation of phenylurea herbicides at 10 μg/L (Published with the permission of Spark Holland).

2. Load 10 mL of river water on the sorbent via the solvent selector valve.
3. Wash sorbent with 2 mL of deionized water.
4. Elute the sample with 2.4 mL mobile phase of the liquid chromatograph pump (45% methanol in 0.02 M phosphate buffer at pH 7).
5. Detection is in ultraviolet (UV) at 254 nm (Fig. 10.19).

10.5.2.   Triazine Herbicides in Water

Triazine herbicides represent another class of herbicides that are found widely in groundwater and surface water. This method uses the automated analysis of 10-mL water samples by SPE followed by analysis directly by GC/MS (Brinkman, 1995). The method uses innovative technology to interface the SDU of the PROSPEKT with a precolumn in the GC. The GC is modified such that the sample may be injected onto a precolumn, which is essentially a retention gap column of uncoated deactivated fused-silica capillary that is several meters in length, with the analytical column off-line (see Fig. 10.20). Then the GC may be turned on and analyze the sample automatically. Because the retention-gap column is uncoated, there is refocusing of the analyte on the retaining precolumn, even with large injection volumes of up to 100 μL.

Automated Method for Triazines by GC/MS

1. A 10-mL sample is processed through the SDU and the analytes are concentrated on the SPE column, which has been preconditioned by the SDU.
2. The SPE column must be dried thoroughly before the analytes are eluted from the SPE sorbent. This is accomplished with nitrogen for a 15-min drying time.

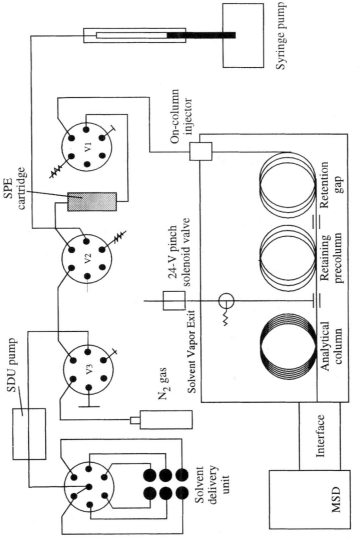

Figure 10.20. Automated SPE-GC-MSD. [Brinkman (1995), published with permission.]

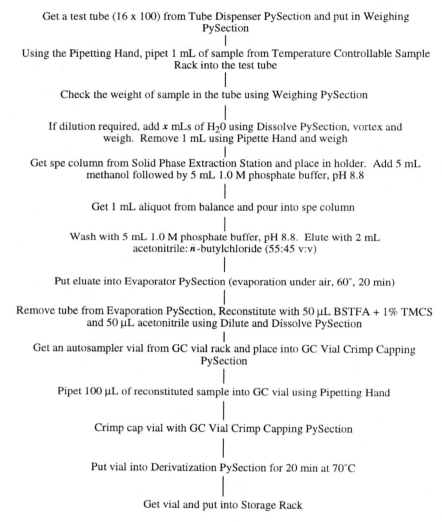

Get a test tube (16 x 100) from Tube Dispenser PySection and put in Weighing PySection

|

Using the Pipetting Hand, pipet 1 mL of sample from Temperature Controllable Sample Rack into the test tube

|

Check the weight of sample in the tube using Weighing PySection

|

If dilution required, add x mLs of H_2O using Dissolve PySection, vortex and weigh. Remove 1 mL using Pipette Hand and weigh

|

Get spe column from Solid Phase Extraction Station and place in holder. Add 5 mL methanol followed by 5 mL 1.0 M phosphate buffer, pH 8.8

|

Get 1 mL aliquot from balance and pour into spe column

|

Wash with 5 mL 1.0 M phosphate buffer, pH 8.8. Elute with 2 mL acetonitrile: n-butylchloride (55:45 v:v)

|

Put eluate into Evaporator PySection (evaporation under air, 60°, 20 min)

|

Remove tube from Evaporation PySection, Reconstitute with 50 μL BSTFA + 1% TMCS and 50 μL acetonitrile using Dilute and Dissolve PySection

|

Get an autosampler vial from GC vial rack and place into GC Vial Crimp Capping PySection

|

Pipet 100 μL of reconstituted sample into GC vial using Pipetting Hand

|

Crimp cap vial with GC Vial Crimp Capping PySection

|

Put vial into Derivatization PySection for 20 min at 70°C

|

Get vial and put into Storage Rack

Figure 10.21. Extraction procedure for cocaine and benzoylecgonine in urine. (Reproduced from the *Journal of Analytical Toxicology* by permission of Preston Publications, Division of Preston Industries, Inc.)

3. Desorption requires 50 to 150 μL of solvent, such as ethyl acetate at a flow rate of approximately 50 μL/min. During the elution, the analytes are quantitatively desorbed and transferred to the retention-gap column, with concurrent evaporation of solvent.

4. After transfer of sample, the GC may be operated in any mode, such as mass spectrometry or another type of GC detector.

10.5.3. Cocaine in Urine

Cocaine use is a major drug problem in the United States, and the analysis of this compound and its metabolite, benzoylecgonine, is a routine analysis problem for many drug-testing laboratories. Confirmation must be done by GC/MS. Large laboratories may analyze hundreds of samples per day by GC/MS. Turn around times are typically short, from 24 to 48 hr for analysis. Thus, the use of a totally automated analysis is financially feasible. The method described here (Taylor and Le, 1991) uses the Zymark robotic system for the entire sample preparation procedure from measuring the sample volume, through SPE, through derivatization to finally placing the vial in the storage rack of the GC/MS. The GC/MS must be loaded and activated by a user, which itself is also automated.

The robot used is the circular approach as discussed earlier. There are 14 steps in the procedure as outlined in Figure 10.21. Beginning with the pipetting of sample, addition of internal standard, SPE, elution, evaporation, derivatization, reconstitution, addition to a vial, sealing the vial, and loading the storage rack for the GC/MS. The quality of the GC/MS analysis was good, and no spurious peaks were found in the selected ion chromatograms when using the automated system.

10.6. CONCLUSIONS

The use of automation results in fast, easy, and reliable methods for SPE. There are a number of instruments available for totally automated SPE, as well as automated methods development of SPE, which span the spectrum in cost. In environmental applications, the choices of automation are limited, with only the Autotrace by Tekmar available for large-volume samples (1 L). In the area of on-line SPE–HPLC, nearly all of the workstations for SPE have the option for direct injection into the HPLC. For on-line SPE–GC/MS, there are several instruments and methods available (a modified PROSPEKT, Varian automated SPME, and the Hewlett-Packard solid-phase extraction workstation). For total automation including addition of internal standards, derivatization, sample handling, SPE, and on-line HPLC or GC/MS, there is the ability to interface several robotic systems. For example, the interfacing of the Zymark circular total robotic system with the linear system of Hewlett-Packard could accomplish the task of total sample handling because it uses the features of both systems. Such a melding of systems would be quite expensive, and suitable perhaps, in cases where harsh conditions prevail and human intervention might be dangerous.

SUGGESTED READING

Application Briefs of Zymark for Benchmate, 1991.

Application Briefs of Jones Chromatography for Prospekt, 1995.

Berg, J. R. 1993. Practical use of automated solid phase microextraction, *Am. Lab.* November, 18–24.

Brinkmann, U. A.Th. 1995. On-line monitoring of aquatic samples, *Environ. Sci. Tech.*, **29**: 79A–84A.

Brinkman, U. A. Th. and Vreuls, R. J. J. 1996. Solid-phase extraction for on-line sample treatment in capillary gas chromatography, *LC-GC*, **14**: 581–585.

Dimson, P., Brocato, S, and Majors, R. E. 1986. Automating solid-phase extraction for HPLC sample preparation, *Am. Lab.*, October, 82–94.

van der Hoff, G. R., Gort, S. M., Baumann, R. A., van Zoonen, P., and Brinkman, U. A. Th. 1991. Clean-up of some organochlorine and pyrethroid insecticides by automated solid-phase extraction cartridges coupled to capillary, *GC-ECD, J. High Resolution Chromatog.*, **14**: 465–470.

Johnson, E. L., Reynold, D. L., Wright, D. S., and Pachla, L. A. 1988. Biological sample preparation and data reduction concepts in pharmaceutical analysis, *J. Chromatog. Sci.*, **26**: 372–379.

Jordan, L. 1993. Automating a solid-phase extraction method, *LC-GC*, **11**: 634–638.

Jordan, L. 1993. *Handling Biological Samples with the BenchMate™ Workstation.* Zymark Publication, Hopkinton, MA.

Jordan, L. and Goffredo, M. E. 1993. *Automating Solid Phase Extraction Methods.* Zymark Publication, Hopkinton, MA.

Majors, R. E. and Holden, B. D. 1993. Laboratory robotics and its role in sample preparation, *LC-GC*, **11**: 488–496.

Majors, R. E. and Fogelman, K. D. 1993. The integration of automated sample preparation with analysis in gas chromatography, *Am. Lab.*, February, 40W–40FF.

Majors, R. E. 1995. Trends in sample preparation and automation—What the experts are saying, *LC-GC*, **13**: 742–748.

Pico, Y., Vreuls, J. J., Ghijsen, R. T., and Brinkman, U. A.Th. 1994. Drying agents for water-free introduction of desorption solvent into a GC after on-line SPE of aqueous samples, *Chromatographia*, **38**: 461–469.

REFERENCES

Berg, J. R. 1993. Practical use of automated solid phase microextraction; *Am. Lab.* November, 18–24.

Bowers, G. D., Clegg, C. P., Hughes, S. C., Harker, A. J., and Lambert, S. 1997. Automated SPE and tandem MS without HPLC columns for quantifying drugs at the picogram level, *LC-GC*, **15**: 48–53.

Brinkman, U. A. Th. 1995, On-line monitoring of aquatic samples, *Environ. Sci. Tech.,* **29**: 79A–84A.

Jordan, L. 1993. Automating a solid-phase extraction method, *LC-GC*, **11**: 634–638.

Majors, R. E. and Holden, B. D. 1993. Laboratory robotics and its role in sample preparation, *LC-GC*, **11**: 488–496.

Taylor, R. W. and Le, S. D. 1991. Robotic method for the analysis of cocaine and benzoylecgonine in urine, *J. Anal. Toxicol.*, **15**: 276–278.

CHAPTER

11

SOLID-PHASE EXTRACTION DISKS

11.1. INTRODUCTION

Solid-phase extraction (SPE) disks (particle-loaded membranes) were recently introduced by 3 M as the Empore disk (1989), by Ansys, Inc. as the SPEC disc, by Alltech Associates, as the Novo-Clean disk, and by J. T. Baker as the Speedisk. Recently (1997) several other manufacturers have introduced disk SPE (Restek and Whatman), and more disks will surely be available in the near future. The disks are used much like filter papers in a filtration apparatus. The concept of the disk is identical to conventional cartridges used in SPE. The disks may be used for the solid-phase extraction of analytes from liquid samples, to clean-up a sample matrix prior to analysis, or to concentrate an analyte to meet the sensitivity range of an analytical instrument. In addition, the disks may be used to remove impurities while letting the analytes pass through the disk. Essentially, the disks may be used for all the same isolation chemistry that has been discussed throughout this book. The major difference lies in the format and particle size of the packing material, not in the chemistry of the packing material. The disks are available in free disks, a syringe barrel format, and most recently in a 96-well microtiter plate (Fig. 11.1).

The disks are currently available in the same phases available for cartridges. The disks also come in a variety of sizes from small disks of 4 mm (2–4 mg of sorbent) to 90 mm with over a gram of sorbent. In general, all of the disks contain approximately 8 to 30 µm particles imbedded in some type of matrix. It is this small particle size that makes the disks effective for rapid sorption of analytes. The large surface area represented by the small diameter particles makes the disks an effective format for rapid sorption and effective removal of analytes. This is especially useful for environmental samples where one liter of water is processed. Rapid mass transfer also means that channeling is reduced and small volumes of conditioning and elution solvents may be used. This result also lends the disk to microscale work, and the placement of disks in syringe barrels as small as 1 mL allows for trace quantities of analyte and solvent to be easily used in SPE applications. In fact, a new application is the placement of 4-mm disks in pipette tips (see Fig. 11.1). This application was developed by Ansys, Inc. (1997) for handling small samples. More discussion is given later in Chapter 12 on handling small samples by

281

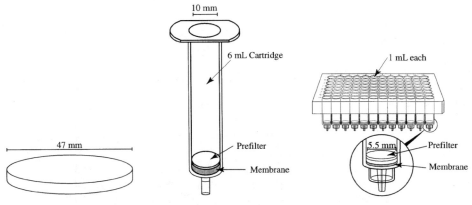

Figure 11.1. Format of disks used in SPE, including free disks, syringe-barrel format (syringe disk cartridge), and in 96-well microtiter plate format. (Published with permission of 3M.)

SPE. This chapter will discuss these aspects of disk extraction with emphasis on the types of disk formats that are currently available, both free disks and disks in syringe barrels, how to use an SPE disk, examples of disk methods, automation of disks, and new technology with disks.

11.2.   "FREE" DISK FORMAT

11.2.1.   Empore Disk

The disk is constructed in several formats by the different vendors. One of the popular environmental extraction disks, called the Empore Extraction Disk (3M), consists of 8 to 12 µm particles of sorbent imbedded into an inert matrix of polytetrafluoroethylene (PTFE) fibrils (see Fig. 1.7). The disks consist of 90% sorbent particles and 10% PTFE by weight, with a bed height of approximately 0.5 mm.

The disks are available in membrane format alone or in syringe-barrel format (Figs. 1.7 and 11.1). Recently, 3M introduced a new format that consists of disks in a 96-well plate (Fig. 11.1), similar to the 96-well plates used in immunoassay. Table 11.1 gives the variety of sizes, formats, and packings that are available currently for SPE disks. The disks are conditioned and used in a similar fashion to the syringe-barrel columns. The 47- and 90-mm disks are ideally suited to sample preparation of large-volume samples (1-2 L). However, care must be taken in filtration of water samples before using the Empore disk in order to get the maximum flow rates. The disks will filter a sample quite effectively trapping particles greater than 0.45 µm as well

Table 11.1. Types of Disks Available for SPE

Sorbent	Structure	Sizes
Free Disks		
C-18 disk (3M)	8–10 μm, Teflon fibrils, 18%C, 60 Å	25, 47, 90 mm
C-8 disk (3M)	8–10 μm, Teflon fibrils, 60 Å	25, 47, 90 mm
CH disk (3M)	8–10 μm, Teflon fibrils, 60 Å	25, 47, 90 mm
SDB-XCDisk (3M)	8–10 μm, Teflon fibrils, 80 Å	25, 47, 90 mm
SDB-RPSDisk (3M)	8–10 μm, Teflon fibrils, 80 Å	25, 47, 90 mm
Oil and grease disk (3M)	8–10 μm, Teflon fibrils, 80 Å	25, 47, 90 mm
Anion-SR disk (3M)	8–10 μm, Teflon fibrils, 80 Å	25, 47, 90 mm
Cation-SR disk (3M)	8–10 μm, Teflon fibrils, 80 Å	25, 47, 90 mm
C-18 AR disk (Toxi-Lab)	30 μm, 70-Å pores diameter, 7 μm particles, glass fiber filter	47, 90 mm
C-8 AR disk (Toxi-Lab)	30 μm, 70-Å pores diameter, 7 μm particles, glass fiber filter	47 mm
C-18 Speedisk (J. T. Baker)	$-(CH_2)_{17}CH_3$, 17.2%C, endcapped, trifunctional, 40 μm, 60-Å pores, irregular silica	50 mm
C-18 Speedisk XF (J. T. Baker)	$-(CH_2)_{17}CH_3$, 17.2%C, endcapped, trifunctional, 40 μm, 60-Å pores, irregular silica	50 mm
C-8 Speedisk (J. T. Baker)	$-(CH_2)_7CH_3$, 14%C, endcapped, trifunctional, 40 μm, 60-Å pores, irregular silica	50 mm
Speedisk DVB (J. T. Baker)	Styrene–divinylbenzene copolymer, neutral sorbent	50 mm
Speedisk SAX (J. T. Baker)	Strong anion exchanger, Styrene–divinylbenzene copolymer	50 mm
Oil and grease (J. T. Baker)	C-18, endcapped, trifunctional, 40 μm, 60-Å pores, irregular silica	50 mm
Resprep TM-C8	$-(CH_2)_7CH_3$, glass-fiber SPE disk	47 mm
Syringe Cartridge Disks		
Reversed-phase sorbents: C18, C-8, C-2, SDB, C-18/OH (3M)	8–10 μm, Teflon fibrils, 60 Å	1 mL (4 mm, 4 mg) 3 mL (7 mm, 12 mg) 6 mL (10 mm, 24 mg)
Normal phase sorbents silica (3 M)	8–10 μm, Teflon fibrils, 60 Å	1 mL (4 mm, 4 mg) 3 mL (7 mm, 12 mg) 6 mL (10 mm, 24 mg)

(*contd.*)

Table 11.1. (*continued*)

Sorbent	Structure	Sizes
Ion-exchange sorbents: strong cation and anion, (3M)	8–10 µm, Teflon fibrils, styrene–divinylbenzene	1 mL (4 mm, 4 mg) 3 mL (7 mm, 12 mg) 6 mL (10 mm, 24 mg)
Reversed phase sorbents: C-18, C-18AR, C-8, Phenyl, C-2 (Toxi-Lab)	30 µm, 70 Å pores diameter 7 µm particles	3 mL (9 mm, 15 mg) 10 mL (12 mm, 35 mg)
Normal phase sorbents: CN, NH₂, primary and secondary amine, silica, (Toxi-Lab)	30 µm, 70-Å pores diameter 7 µm particles	3 mL (9 mm, 15 mg) 10 mL (12 mm, 35 mg)
Ion-exchange sorbents: anion and cation exchange (Toxi-Lab)	Styrene–divinylbenzene	3 mL (9 mm, 15 mg) 10 mL (12 mm, 35 mg)
Mixed-mode sorbents: nonpolar/strong cation slightlypolar/strong cation (Toxi-Lab)	Chemistry not given	3 mL (9 mm, 15 mg) 10 mL (12 mm, 35 mg)

as smaller colloidal particles inside the pores of the disk. Thus, if a glass-fiber filter is used to filter a water sample, which is typical for most water samples, then a river or lake sample may still contain fine particulate matter that will be trapped on the membrane and eventually will decrease flow rates of the disk.

Furthermore, the concept of the disk is that the whole water sample may be applied to the disk without any filtration. The result is that fine particulate matter will plug the disk. This trapping of particulates will reduce the flow rate of the Empore disk. Thus, the disks have rapid flow rates generally only on tapwater and groundwater samples. There are filter aids that are sold in conjunction with the Empore disk. These filter aids are 40-µm silica glass beads (Filter Aid 400) that are added to the sample to prevent a fine filter cake from building up on the filter and slowing the flow rate. The filter aids create a larger pore size and faster flow rates than on the disk itself. The recommendation is to add approximately enough Filter Aid 400 to have a depth of 1 cm on the disk.

11.2.2. SPEC DISC

The fiber-glass disc made by Ansys, Inc. has the bonded silica particles (10–30 µm) woven into the fibers of the fiber-glass disc. This gives an even more porous filter than the Empore disk, which results in a somewhat more rapid flow rate for difficult to filter samples that contain particulate matter and algae. The SPEC disc is twice as thick as the Empore disk (1 vs. 0.5 mm, respectively). The result is that the SPEC disc may occasionally leak in the conventional filtration apparatus used for filtering water samples because of the extra thickness of the disc. Ansys, Inc. does have another type version of a filtration device for the discs that will not leak. Teflon tape may be wrapped around a conventional filtration apparatus when using the SPEC disc for a tight seal. Figures 8.14 and 8.15 show examples of the SPEC "free" discs, which consist of sample reservoir, filter, and the inlet of the SPE disc.

Ansys, Inc. has the mixed-mode discs available for the drug-analysis field with many solid phases available in disc format, including reversed phase, normal phase, and ion exchange (Table 11.1). The methodology for drug analysis also uses derivatization on the disc itself, which is an innovative and important technology to explore, especially with gas chromatography/mass spectrometry (GC/MS) analysis. One advantage of the syringe-barrel disk is that it is easy to automate for drug analysis using workstations that are designed for syringe-barrel sorbents. However, there are only two systems currently for free disks, the Tekmar Autotrace and the automated disk workstation of Horizon Technology, which will be discussed later in this chapter.

11.2.3. Speedisk

J. T. Baker makes a disk for environmental samples in a cartridge holder that consists of free bonded-silica packing material that is approximately 10 μm in diameter and is held between two glass-fiber filters with a screen to hold the glass-fiber filters in place (Fig. 11.2). The design of the Speedisk allows the disk to be used in a standard vacuum manifold or in a filtration flask. The Speedisk is available in C-18, C-18XF, C-8, strong anion exchange (SAX), styrene–divinylbenzene, and disks for oil and grease. All the Speedisk come in a standard size of 47 mm. Of the three types of disks (Empore, SPEC disc, and Speedisk), the Speedisk has the fastest flow rate for turbid samples based on manufacturer's test and on personal observations for turbid samples (unpublished data). Apparently the glass-fiber filter can remove the particulate matter and yet maintain a fast flow rate through the 10-μm C-18 sorbent (based on manufacturer's literature and with experiments in the laboratory with turbid samples). However, this disk comparison was based on disks being used without any Filter Aid 400, which would speed up filtration. Generally speaking, the use of disks on water samples that contain particulate matter may greatly slow up the process of sample processing if the sample volume is greater than 500 mL. The Empore disk is being redesigned with a larger particle size

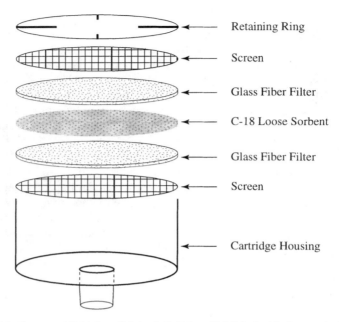

Retaining Ring

Screen

Glass Fiber Filter

C-18 Loose Sorbent

Glass Fiber Filter

Screen

Cartridge Housing

Figure 11.2. Structure of the Speedisk by J. T. Baker. (Published with the permission of J. T. Baker, Inc.)

(approximately 30 μm), which has considerably faster flow rates and are currently available (1997, personal communication with 3M). No comparisons of the various disks are available at this time, and other manufacturers are currently preparing disks for SPE applications.

11.2.4.   Novo-Clean Disk

A third disk manufacturer is Alltech Associates, which originally made only cation- and anion-exchange disks. Now the disks are available in C-8 and C-18 sorbents (1996). The disks are a Teflon membrane impregnated with ion-exchange media for specific removal of ionic interferences, especially in ion chromatography (Saari-Nordhaus and Anderson, 1995; Saari-Nordhaus et al., 1994). Medical-grade polypropylene is used for the housing of the membrane (Fig. 11.3). The disks come in 13- and 25-mm diameters. The disks may be purchased without plastic holders for use in standard-size stainless filter holders with Luer-lock fittings for use on syringes. They were originally designed for use in ion chromatography but may have applications for any use of ion-exchange media in solid-phase extraction.

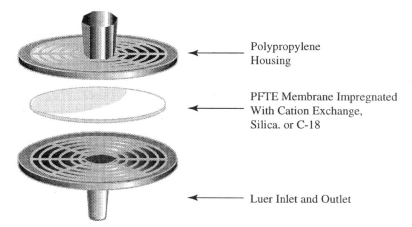

Polypropylene
Housing

PFTE Membrane Impregnated
With Cation Exchange,
Silica. or C-18

Luer Inlet and Outlet

Figure 11.3. Design of the Novo-Clean disk. (Published with the permission of Alltech Associates, Inc.)

11.2.5. MemSep

Waters also makes an ion-exchange disk called MemSep, which is a stacking of cellulose disks that are compressed into a continuous bed of 12,000-Å through-pores. These pores are functionalized with ion-exchange groups suitable for many types of biological samples using ion exchange.

11.3. SYRINGE DISK FORMAT

There are two manufacturers that currently produce the disks in syringe format, 3M and Ansys, Inc. Table 11.1 lists the sizes of syringe formats that are available, and Figure 1.7 shows the construction of the Empore syringe with disk. The syringe format consists of a standard polyethylene syringe that is fitted with a Teflon 20-μm frit, Empore disk, and a prefilter of glass fiber. The glass-fiber filter is designed to keep flow rates as rapid as possible by removing particulate matter before it can plug the disk. The syringe-barrel format lends itself to small sample and eluent volumes because rapid mass transfer occurs. For example, the 4-mm disk in syringe format requires only 100 μL of elution solvent and the 7-mm disk (syringe format) uses only 250 μL. This arrangement allows for microscale work using the disk. Similar to the construction of the Empore disk, the SPEC disc is a glass-fiber filter rather than Teflon as the matrix but has a similar construction in syringe format.

Advantages of the syringe format are that there are a larger variety of sorbents available for the syringe disk made by Ansys, Inc., than there are for the free disk format; also, the syringe disk format may be used in any of the automation systems mentioned in Chapter 10. The syringe disk cartridges range in size from 1 mL (4-mm disk, 4 mg), 3 mL (7-mm disk, 12 mg), 6 mL (11-mm disk, 24 mg), to 10 mL (12-mm disk, 35 mg). The types of sorbents range across the entire spectrum of packing materials (Table 11.1), including mixed-mode sorbents. These milligram masses of sorbent are adequate for small samples, that is, the approximate volume of the syringe, from 1- to 10-mL sample volumes. A 1-mL syringe has 4 mg of C-18 packing, and for a volume of 1 mL has a mass-to-volume ratio or 1 : 4. This ratio compares favorably with the 47-mm disk that is typically used on 1-L samples, which can process 1000 mL of sample with 500 mg of sorbent for a ratio of 2 : 1. Thus, as much as 8 mL could be used on the smallest disk with the same volume-to-weight ratio. Thus, in spite of the smallest disk sizes, there is more than adequate capacity for most applications involving drug analysis from body fluids, which typically use 1- to 5-mL samples.

An advantage to the small masses of sorbent in the disks is that small volumes of solvent may be used to elute the disk. For example, the 4-mm disk requires only 100 μL of solvent. The use of small volumes is quite useful for many analyses, where there are micro amounts of sample and sampling handling must be minimized.

11.4. HOW TO USE A SOLID-PHASE EXTRACTION DISK

11.4.1. Manifold and Hardware for Disk Extraction

The manifold is constructed of stainless steel, and each station is controlled by a valve that allows extraction or venting to the atmosphere in the off position. There is also specially designed glassware to go with the manifold (Fig. 11.4). Figure 11.4 shows how the manifold is constructed, and also shows how the manifold is connected to the vacuum system and how the elution is performed.

Manual methods using 1-, 3-, and 6-station manifolds are also available (Fig. 11.4). Using the manifold, up to six extractions can be completed simultaneously and multiple manifolds can be managed by a single operator. Extraction time is dependent on the amount of particulate matter in the sample. Typically drinking waters take approximately 30 min per batch for a 1-L sample. For lake and river water, it is necessary to add Filter Aid 400 in order to keep extraction time from taking hours for a 1-L sample.

11.4.1.1. *Condition the Disk*

The procedure for using an SPE disk with reversed phase is similar to that with an SPE cartridge. The method described here is identical for free disks or for syringe-cartridge disks. First, condition or wet the disk with methanol. This step activates the sorbent so that the bonded phase is prepared for sample

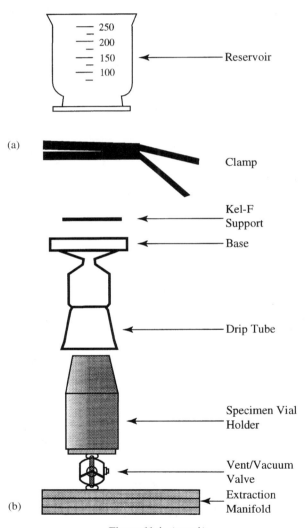

Figure 11.4. (*contd.*)

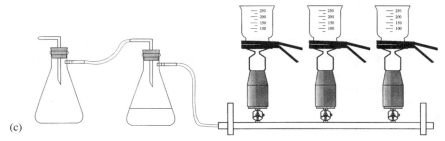

(c)

Figure 11.4 (*continued*). Manifold assembly and manifold stations for disk extractions. (Published with permission of 3M.)

addition. Place the disk into a holder as shown in Figures 11.4 and 11.5 and secure the disk. After addition of approximately 15 mL of methanol for a 47-mm disk, place a vacuum of ~20 in of mercury on the disk and aspirate the methanol through the disk. This step prepares the disk for sample application, and the C-18 sorbent is fully opened so that an effective reversed-phase mechanism is operational. Wash away the residual methanol with 10 mL of distilled water, being careful not to let the disk run dry but to leave it partially wetted. Drying of the disk will deactivate the C-18, which will result in poor recoveries.

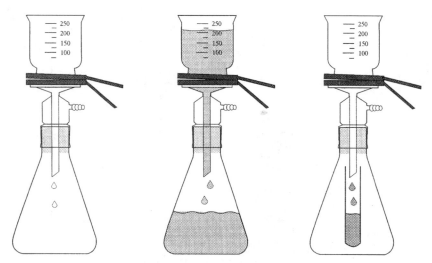

Figure 11.5. Procedure for using Empore disk, SPEC-47, Novo-Clean disk, or the Speedisk for analysis of water samples using a reversed-phase sorbent.

11.4.1.2. Extract the Sample

Carefully transfer the sample into the holder of the disk and apply vacuum to aspirate the sample through the disk. If the disk should become dry before addition of the sample, recondition the disk and apply the sample. A vacuum of 12 in. of mercury provides an appropriate flow rate for sample addition. If the sample contains particulates and a large volume is to be filtered, that is, > 100 mL for a 47-mm disk, the addition of Filter Aid 400 beads (3M) is recommended. Filter Aid 400 is nonporous, inert, and resistant to leaching. It is placed on top of the disk to a depth of about 1 cm. It acts as a depth filter to prevent the suspended solids from plugging the disk. The Filter Aid 400 are silica beads of 40-μm diameter. Again do not allow the disk to go dry after addition of the sample. Because of the high efficiency of the small particle size of the C-18 (10 μm), the sample recovery is unaffected by flow rate and the sample should be past as fast as the extraction disk will allow.

11.4.1.3. Wash the Disk

Add distilled water or buffer to the filtration device and aspirate the water through the disk with vacuum. This step washes interferences from the disk and retains the analytes, which are bound by a reversed-phase mechanism of sorption.

11.4.1.4. Elute the Analyte

Elute the analyte from the disk with the appropriate solvent. Typically, ethyl acetate will work quite well for removing solutes bound by a reversed-phase mechanism. The typical generic method, described in Chapter 3, will work well. For a 47-mm disk, 10 mL of elution solvent in two 5-mL aliquots will give good elution recovery from a C-18 disk for most compounds. A slower flow rate for elution will give better recoveries, and it is a good idea to soak the disk for several minutes with elution solvent. This step allows the solvent to wet the disk well and to give effective elution recovery.

11.5. DISK AUTOMATION

The automation equipment available for the disks include the Tekmar system (Chapter 10, Fig. 10.6) and the Horizon Technology system, which are specially designed for 47-mm disks and environmental analysis of water samples (Fig. 11.6). The modular system consists of a controller and up to eight satellite extraction stations (Johnson, 1996). The extractor can control

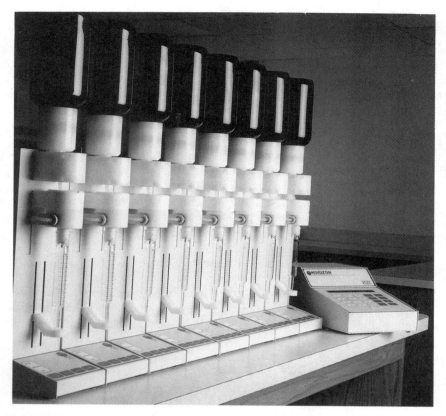

Figure 11.6. Photograph of automated disk extraction with Horizon Technology, called the SPE-DEX. (Published with the permission of Horizon Technology.)

four solvents for wetting the SPE disk, and another two for eluting the analytes from the disk. The extractor also controls the extraction of the sample from the container and rinses the container with solvent to improve recovery, as Environmental Protection Agency (EPA) methods often prescribe. The disk is eluted for GC or high-pressure liquid chromatography (HPLC) analysis.

Johnson (1996) reports on the recovery of this system according to the EPA Method 525 and its analytes. This work showed that the automated disk method gave comparable results to a manual disk method. The extractor uses a Teflon body construction with a minimum internal volume of 20 μL to minimize carryover between samples. The instrument is priced at $6900 for controller and extractor, additional extractors cost $4000 each (1996 dollars).

11.6. EXAMPLES OF DISK METHODS

11.6.1. Polynuclear Aromatic Hydocarbons from Water

The extraction of polynuclear aromatic hydrocarbons (PAHs) from water may effectively be accomplished with C-18 and a reversed-phase mechanism. The method is straightforward and consists of isolation of the analytes from a 1-L sample, followed by elution with ethyl acetate and with methylene chloride. The combination of solvents is needed for the most hydrophobic of the PAHs. The disk is cleaned with the elution solvents before addition of the sample.

SPE Conditions Used

Sample	Groundwater sample, 1 L.
Solutes	PAHs.
Sorbent	47-mm disk Empore C-18. Condition with 5 mL methanol, 5 mL of methylene chloride/ethyl acetate. Let stand on disk for 3 min to soak into disk. Draw remaining solvent through the disk. Dry under vacuum for 1 min. Next add 5 mL methanol and let stand for 3 min to condition disk. Add sample directly to the trace amounts of methanol that are left standing on the disk.
Eluent	Elute with 5 mL ethyl acetate, then 5 mL of methylene chloride. Combine extracts and evaporate for GC/MS analysis.
Reference	Bakerbond Application Note EMP-002 Extraction of PAHs from water. Published with permission.

11.6.2. Extraction of Polychlorinated Biphenyls from Water

The extraction of Polychlorinated Biphenyls (PCBs) from water may effectively be accomplished with C-18 and a reversed-phase mechanism. The method is similar to the PAH method and consists of isolation of the analytes from a 1-L sample, followed by elution with ethyl acetate followed by methylene chloride. The combination of solvents is needed for the hydrophobic nature of PCBs. The disk is cleaned with the elution solvents before addition of the sample.

SPE Conditions Used

Sample	Ground water sample, 1 L.
Solutes	PCBs.
Sorbent	47-mm disk Empore C-18. Condition with 5 mL methanol, 5 mL of methylene chloride/ethyl acetate. Let stand on disk for

<table>
<tr><td></td><td>3 min to soak into disk. Draw remaining solvent through the disk. Dry under vacuum for 1 min. Next add 5 mL methanol and let stand for 3 min to condition disk. Add sample directly to the trace amounts of methanol that are left standing on the disk.</td></tr>
<tr><td>Eluent</td><td>Elute with 5 mL ethyl acetate, then 5 mL of methylene chloride. Combine extracts and evaporate for GC/MS analysis.</td></tr>
<tr><td>Reference</td><td>Bakerbond Application Note EMP-003 Extraction of PCBs from water. Published with permission.</td></tr>
</table>

11.6.3.   EPA Method 552.1: Haloacetic Acids and Dalapon in Drinking Water

This method extracts the chlorinated organic acids formed during the chlorination of drinking water. The method uses a 100-mL sample with an isolation using anion exchange. The method consists of adjusting the pH to 5.0 ± 0.5 with 50% H_2SO_4/water. A surrogate standard is added to check recovery. Next, the sample is added to the strong anion-exchange disk, where it is sorbed by anion exchange. The sorbent is eluted with a mixture of 10% sulfuric acid and 90% methanol. The sample is methylated and analyzed by gas chromatography.

SPE Conditions Used

Sample	Drinking water sample, 100 mL.
Solutes	Haloacetic acids and Dalapon.
Sorbent	47-mm disk Empore SAX. Condition with 10 mL methanol, and let soak for 30 sec. Aspirate all but leave disk wet. Add 10 mL of 1N HCl/methanol and soak for 30 sec. Aspirate all but leave disk wet. Wash disk with 2×10 mL distilled water washes. Condition the disk with 10 mL of $1.0N$ NaOH and let soak for 30 sec. Wash disk with 2×10 mL of distilled water washes. Disk is ready for sample addition.
Eluent	Elute with 4 mL of 10% H_2SO_4/methanol. Let eluent soak for 30 sec before aspiration.
Reference	Varian Empore disk extraction summary. Published with permission.

11.6.4.   References for Further Environmental Methods Development

There are a number of recent papers that deal with the application of disks to the extraction of contaminated waters. For example, Chiron and co-workers

(1993) compared disk extraction to liquid extraction for monitoring selected pesticides in different environmental waters, including both river water and groundwater. Using the Empore disk, they obtained detection limits for compounds in the 0.01 to 0.5 µg/L level. Compounds included triazines and carbamates with detection by liquid chromatography/mass spectrometry (LC/MS).

Kwakman and co-workers (1992) used disk extraction to isolate organophosphorus pesticides on-line with gas chromatography for rapid analysis. The method is on-line for sorption and elution using ethyl acetate as a desorbing solvent and a retention time gap approach for the GC analysis. The detection limit was 10 to 30 ng/L in tapwater and 50 to 100 ng/L in river water. Barcelo and co-workers (1993) and Molina and co-workers (1994) also found that organophosphorus pesticides and triazines could be effectively isolated on disks for further analysis by HPLC/MS using thermospray and electrospray.

Klaffenbach and Holland (1993) analyzed sulfourea herbicides in soil water extracts and in water samples using disk extraction. They found that the thermally unstable sulfourea herbicides could be extracted onto the C-18 disks eluted with ethyl acetate and derivatized with diazomethane for GC/MS confirmation. The method also worked well for aqueous soil extracts that could also be passed through the C-18 disks for rapid concentration, derivatization, and analysis.

11.7. THEORY OF DISK OPERATION

Because the disks use the same packing material that has been described throughout the book, the theory of operation of the disks is identical for all of the methods described. The difference with the disks lies in the use of smaller particle size and in faster flow rates for large-volume samples. Figure 11.7 shows the relative difference in particle size for 40 to 60-µm particles and for the 10-µm particle sizes that are used in disks. For the same bed height as a cartridge, the disk has a much more tortuous path of flow, which means that there is considerably more surface area available, and the kinetics of sorption will be substantially quicker. This result is shown in the Empore trade literature, which shows that dye analytes are tightly bound near the surface of the disk.

In fact, work by Fernando and co-workers (1993) investigated the breakthrough of the 3M disk with respect to flow rate and predicted the capacity or breakthrough of solutes on the membranes. They found that the membranes have typically 4 to 9 theoretical plates (see Chapter 4 for discussion of plate theory) and that flow rates of 10 to 100 mL/min are possible with the 47-mm disk. The capacity of the disks was related to the capacity factor of the solute

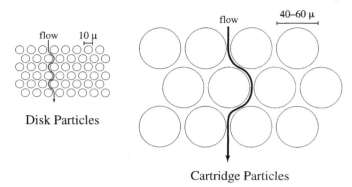

Figure 11.7. Relative difference in particle size between a disk and standard SPE cartridge.

for the membrane in a methanol/water eluent (30:70). This result shows that the height equivalent of a theoretical plate is equal to:

$$0.5 \text{ mm}/5 \text{ plates} = 0.1 \text{ mm per plate} \qquad (11.1)$$

Thus, a typical Empore disk has a plate height of 0.1 mm while a typical cartridge sorbent has a bed height of 1 cm and approximately 20 theoretical plates, which is equal to:

$$10 \text{ mm}/20 \text{ plates} = 0.5 \text{ mm per plate} \qquad (11.2)$$

This result indicates that the typical disk of 10 μm is approximately 5 times more efficient or has 5 times more effective plates per unit length than does the best syringe cartridge.

The next consideration with the disks is the area of the disk versus the area of the syringe barrel with the same mass of packing material. Consider a 47-mm disk that contains 500 mg of packing and is 0.5 mm in thickness versus a cartridge that contains 500 mg of packing material and is 9 mm in diameter and 12 mm in length. The disk has approximately 25 times more surface area than the cartridge and the cartridge is approximately 25 times longer than the disk. Thus, the disk could be considered to be a very wide version of the cartridge! The difference, however, is that for the same bed height the disk has at least 5 times more plates or column efficiency than the cartridge has, which, of course, is due to the smaller-diameter particle size (10 vs 40–60 μm). Thus, the disk is a new and improved version of a cartridge and represents nothing more than a decrease in particle size with a very small bed height so that back pressures are low. The low back pressures allow for fast flow rates, and the large surface areas of the small particle size give nearly equilibrium sorption at fast flow rates. The innovation lies in preparing the disks in formats that may be used with only vacuum systems, and now at least three different formats have been invented with the Empore disk, SPEC disc, and the Speedisk. For the latest information, the reader should check the web sites of the different vendors, which are given in the Appendix.

11.8. NEW TECHNOLOGY WITH DISKS

11.8.1. Disk Derivatization

A feature of the high surface area of the disk has been used by Field and Monohan (1995) in the isolation and subsequent on-disk derivatization of dacthal metabolites in groundwater. The method that they have pioneered involves the ion-exchange isolation of dacthal metabolites onto a strong anion-exchange disk (Fig. 11.8). The method appears to be quite useful for environmental samples and involves sorbing the analytes onto a disk and then derivatization of the analyte on the disk itself. The method involves the use of ion-exchange disks to isolate organic acids from water and the subsequent derivatization of the acids on the disk using methyl or ethyl iodide and acetonitrile. The derivatization is done with the disk in an autosampler vial so that the method allows for the direct analysis of the analyte onto the GC. The method calls for the isolation of the organic acid, in this case a metabolite of dacthal, from 100 mL of water onto a strong anion-exchange disk that is 13 mm in diameter. The disk is dried under vacuum and placed in an autosampler vial with 1 mL of acetonitrile and 140 μm of ethyl-iodide derivatization reagent for 1 hr at 100 °C. The kinetics of the derivatization are enhanced on the solid phase, and the derivatization goes to completion in a less than 1 hr. The method is quite effective for trace levels of dacthal metabolite in ground-water, with detection limits of 0.02 μg/L (Field and Monohan, 1995).

Another method that Field and colleagues have developed (Krueger and Field, 1995) involves the elution of the disk in the autosampler vial of the gas chromatograph, followed by an on-column derivatization. The concept here is to elute directly in the autosampler vial so that there is no loss of analyte and a transfer step is not involved. In this method, the Empore disk is cut to a smaller diameter with a cork-boring device (13 mm) in order to fit these small disks into the autosampler vials. The method is clever in that the cost of disks is much reduced by buying them in larger sizes and "making your own". The small disk is now free and can be placed in the autosampler vial for subsequent elution by tetramethylammonium hydrogen sulfate in chloroform. The tetra-methylammonium hydrogen sulfate is the on-column derivatization reagent

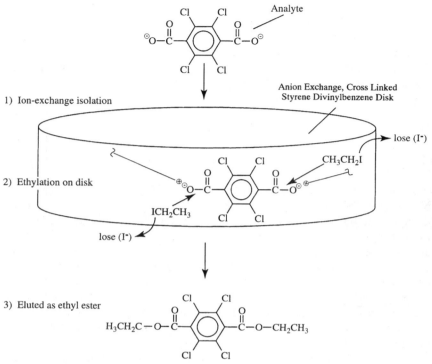

Figure 11.8. Structure of dacthal and the on-disk derivatization method. [After Krueger and Field (1995).]

for linear alkylbenzenesulfonates, which are anionic compounds. When derivatized they can be analyzed by GC/FID (flame ionization detector) or GC/MS.

11.8.2. Sample Preservation on Disk

The preservation of samples for environmental analysis is an important topic, and field methods are always an area for new methods development. Work by Martinez and Barcelo (1996) found that pesticides could be preserved on disk by passing the sample through the disk in the field, then immediately storing the disk in the frozen state until elution could be done in the laboratory. Ferrer and Barcelo (1997) found a similar result for the preservation of pesticides on cartridges. They found that samples could be stored safely for several weeks without losses of a number of pesticides. The advantage of this method is that samples are nearly always filtered in the field during environmental studies; thus, sample processing may easily be linked at this stage with, for example, a two-stage filtering device that first removed particulates and then through a free disk for organic compound isolation by reversed phase. The disk then can be stored on ice until elution in the laboratory on the following day, or could in fact, be eluted directly into an autosampler vial for analysis. The use of disks will continue to expand with more new and interesting applications.

11.8.3. Pipette-Tip Extraction Disk

Ansys, Inc. has recently introduced a solid-phase extraction pipette tip, called the SPEC · PLUS · PT, which is shown in Figure 11.9. The disk is placed in a pipette tip with the filter at the bottom of the column instead of at the top. The disk is conditioned in the normal way and sample is drawn up through the pipette tip and sorbed to the disk. The disk may be washed and then eluted with solvent. The elution consists of a back elution with the solvent pushed through the disk. The disk will be useful for small samples. It is fitted with a 4-mm disk with various phases. The method is a flexible, simple, and an unique approach to small sample handling by SPE.

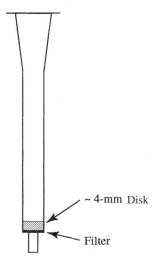

~ 4-mm Disk

Filter

Figure 11.9. Solid-phase extraction pipette tip, called the SPEC · PLUS · PT. (Published with the permission of Ansys, Inc.)

SUGGESTED READING

3M Empore Application Notes: Available upon request (see Appendix).

Speedisk Brochure, J. T. Baker: Available upon request (see Appendix).

SPEC disc Brochure, Ansys, Inc: Available upon request (see Appendix).

Novo-Clean Brochure, Alltech Associated, Inc. Deerfield, IL: Available upon request (see Appendix).

Allanson, J. P., Biddlecombe, R. A., Jones, A. E. and Pleasance, S. 1996. The use of automated solid-phase extraction in the 96-well format for high-throughput bioanalysis using lipid chromatography coupled to tandem mass-spectrometry, *Rapid Commun. Mass Spectrom.*, **10**: 811–816.

Barnabas, I. J., Dean, J. R., Hitchen, S. M., and Owen, S. P. 1994. Selective extraction of organochlorine and organophosphorus pesticides using a combined solid phase extraction-supercritical fluid extraction approach, *Anal. Chim. Acta*, **291**: 261–267.

Blevins, D. D. and Schultheis, S. K. 1994. Comparison of extraction disk and packed-bed cartridge technology in SPE, *LC-GC*, **12**: 12–16.

Blevins, D. D. and Henry, M. P. 1995. Pharmaceutical applications of extraction disk technology, *Am. Lab.*, May, 32–35.

Brouwer, E. R., Lingeman, H., Brinkman, and U. A. Th. 1990. Use of membrane extraction disks for on-line trace enrichment of organic compounds from aqueous samples, *Chromatographia*, **29**: 415–418.

Dirksen, T. A., Price, S. M., and Mary, S. J. St. 1993. Solid-phase disk extraction of particulate-containing water samples, *Am. Lab.*, December, 24–27.

Hagen, D., Markell, C. and Schmitt, G. 1990. Membrane approach to solid-phase extractions, *Anal. Chim. Acta*, **236**: 157–164.

Hearne, G. M. and Hall, D. O. 1993. Advances in solid-phase extraction technology, *Am. Lab.*, January, 28H–28M.

Larrivee, M. L. and Poole, C. F. 1994. Solvation parameter model for the prediction of breakthrough volumes in solid-phase extraction with particle-loaded membranes, *Anal. Chem.*, **66**: 139–146.

Lensmeyer, G. L., Wiebe, D. A. and Darcey, B. A. 1991. Application of a novel form of solid-phase sorbent (Empore™ Membrane) to the isolation of tricyclic antidepressant drugs from blood, *J. Chromatog. Sci.*, **29**: 444–449.

Markell, C., Hagen, D. F., and Bunnelle, V. A. 1991. New technologies in solid-phase extraction, *LC-GC*, **9**: 332–337.

Markell, C. and Hagen, D. F. 1996. Solid-phase extraction basics for water analysis in *Principles of Environmental Sampling*, 2nd ed., Keith, L. H., Ed. ACS Professional Reference Book, American Chemical Society, Washington, D.C., Chapter 17.

Senseman, S. A., Lavy, T. L., Mattice, J. D., Myers, B. M. and Skulman, B. W. 1993. Stability of various pesticides on membranous solid-phase extraction media, *Environ. Sci. Tech.*, **27**: 516–519.

Snyder, J. L.; Grob, R. L., McNally, M. E., and Oostdyk, T. S. 1994. A different approach—using solid-phase disks and cartridges to extract organochlorine and organophosphate pesticides from soils, *LC-GC*, **12**: 230–242.

Viana, E., Redondo, M. J., Font, G., and Molto, J. C., 1996. Disks versus columns in the solid-phase extraction of pesticides from water, *J. Chromatog.*, **733**: 267–274.

Wells, D. A., Lensmeyer, G. L., and Wiebe, D. A. 1995. Particle-loaded membranes as an alternative to traditional packed-column sorbents for drug extraction: In-depth comparative study, *J. Chromatog. Sci.*, **33**: 386–392.

REFERENCES

Barcelo, D., Durand, G., Bouvot, V., and Nielen, M. 1993. Use of extraction disks for trace enrichment of various pesticides from river water and simulated seawater samples followed by liquid chromatography-rapid-scanning UV-visible and thermospray-mass spectrometry detection, *Environ. Sci. Tech.*, **27**: 271–277.

Chiron, S., Alba, A. F., and Barcelo, D. 1993. Comparison of on-line solid-phase disk extraction to liquid-liquid extraction for monitoring selected pesticides in environmental waters, *Environ. Sci. Tech.*, **27**: 2352–2359.

Fernando, W. P. N., Larrivee, M. L. and Poole, C. F., 1993. Investigation of the kinetic properties of particle-loaded membranes for the solid-phase extraction by forced flow planar chromatography, *Anal. Chem.*, **65**: 588–595.

Ferrer, I, and Barcelo, D. 1997. Stability of pesticides stored on polymeric solid-phase extraction cartridges, *J. Chromatog.*, **778**: 161–170.

Field, J. A. and Monohan, K. 1995. In-vial derivatization and Empore disk elution for the quantitative determination of the carboxylic acid metabolites of dacthal in groundwater, *Anal. Chem.*, **67**: 3357–3362.

Johnson, R. 1996. Performance-based validation of automated disk extraction, *Am. Environ. Lab.*, May, 22–24.

Klaffenbach, P. and Holland, P. T. 1993. Analysis of sulfonylurea herbicides by gas-liquid chromatography. 2. Determination of chlorsulfuron and metsulfuron-methyl in soil and water samples, *J. Agricul. Food Chem.*, **41**: 396–401.

Krueger, C. J. and Field, J. A. 1995. In-vial C_{18} Empore disk elution coupled with injection port derivatization for the quantitative determination of linear alkylbenzenesulfonates by GC-FID, *Anal. Chem.*, **67**: 3363–3366.

Kwakman, P. J. M., Vreuls, J. J., Brinkman, U. A. Th. and Ghijsen, R. T. 1992. Determination of organophosphorus pesticides in aqueous samples by on-line membrane disk extraction and capillary gas chromatography: *Chromatographia*, **34**: 41–47.

Martinez, E.; Barcelo, D., 1996. Chromatographia, v. 42, pp. 72–76.

Molina, C., Honing, M. and Barcelo, D. 1994. Determination of organophosphorus pesticides in water by solid-phase extraction followed by liquid chromatography/high-flow pneumatically assisted electrospray mass spectrometry, *Anal. Chem.*, **66**: 4444–4449.

Saari-Nordhaus, R., Nair, L. M., and Anderson, Jr., J. M. 1994. Elimination of matrix interferences in ion chromatography by the use of solid-phase extraction disks, *J. Chromatog. A*, **671**, 159–163.

Saari-Nordhaus, R. and Anderson, Jr., J. M. 1995. Membrane-based solid-phase extraction as a sample clean-up technique for anion analysis by capillary electrophoresis, *J. Chromatog. A*, **706**: 563–569.

CHAPTER

12

NEW TECHNOLOGY IN SOLID-PHASE EXTRACTION

There are a number of new technologies that have appeared in solid-phase extraction (SPE) and SPE-related products. Although several of these products are not solid-phase extraction by the simple definition in Chapter 1, they do, however, fit into the general category of extraction and sample preparation using solid phases. Because of their innovation and application to classical SPE, these methods are included in this chapter. They include:

1. Solid-phase microextraction (SPME), which involves the use of a microfiber immersed into the sample for sorption, followed by direct desorption in the inlet of the gas chromatograph.
2. There is a new SPE method for solid materials, such as food, called solid-phase matrix dispersion.
3. There are many new phases being brought to the marketplace, including the use of graphitized carbon for polar compounds and internal reversed-phase sorbents for clean-up of complex biological materials, affinity chromatography for pesticides, and molecular recognition SPE.
4. Recently, isolation and derivatization have been combined using both silica sorbents modified with specific derivatization agents, derivatization on ion-exchange disks, and semipermeable membrane devices.

These different innovations are discussed in this chapter with pertinent references.

12.1. SOLID-PHASE MICROEXTRACTION

12.1.1. Design

Solid-phase microextraction was invented at the University of Waterloo (Ontario, Canada) by Pawliszyn and associates (Belardi and Pawliszyn, 1989) and is being sold by Supelco. The method involves the equilibrium sorption of

303

analytes onto a small microfiber, which is made of a fused-silica optical fiber coated with a hydrophobic polymer (Fig. 12.1). Analytes either in the air or in a water sample come into equilibrium with the fiber according to their affinity for the solid phase. The microfiber, which is incorporated into a gas chromatography (GC) syringe, is directly injected into the GC either manually or by an automated system (Fig. 10.9). Because the SPME fiber is heated by the inlet, the injection is directly to the GC without solvent.

The method is different from conventional SPE in that SPE isolates the majority of the analyte from a sample (> 90%) but only injects about 1 to 2% of the sample onto the GC. Solid-phase microextraction isolates a much smaller quantity of analyte (2–20%), but all of that sample is injected into the GC. The extraction efficiency of the fiber is a combination of extraction time, the thickness of the stationary phase, and the magnitude of the partition coefficient for the stationary phase.

The SPME unit consists of two elements: a length of fused-silica fiber bonded to a stainless steel plunger and a holder that looks like a modified microliter syringe (Fig. 12.1). The fused-silica fiber can, by using the plunger,

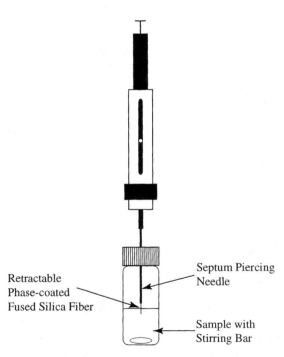

Figure 12.1. Schematic of the SPME device. (Published with the permission of Supelco, Inc.)

be drawn into the septum-piercing needle for protection. To use the unit, the analyst draws the coated fiber into the needle, passes the needle through the septum that seals the sample vial, and depresses the plunger, lowering the fiber into the sample. Organic analytes sorb to the phase coated on the fused-silica fiber, aided by vigorous stirring of the sample. The sorption reaches equilibrium in 2 to 30 min.

After sample sorption, the fused-silica fiber is withdrawn into the needle, and the needle is removed from the sample vial and introduced into the GC injector, where the sorbed analytes are thermally desorbed and analyzed. The fiber assembly is reusable. Extraction may be accomplished either by direct sampling in the water or by sampling the headspace above the sample. Because the SPME can attain detection limits of 15 ng/L for both volatile and nonvolatile compounds, the technique can be used for the U.S. Environmental Protection Agency (EPA) Methods and the Ontario Municipal/Industrial Strategy for Abatement (MISA) Program (Arthur et al., 1992).

Several fiber coatings have been investigated. The most successful phase has been polydimethylsiloxane (PDMS), which is available commercially from several optical fiber manufacturers or from Supelco in thicknesses ranging from 7 to 100 μm (Arthur et al., 1992; Supelco Catalog, see Appendix). A second coating that is proving useful is a polyacrylate polymer for phenols and chlorinated phenols (Bucholz and Pawliszyn, 1993).

12.1.2. Theory

The theory for sorption onto the fibers has been described in a study by Louch and co-workers (1992). According to the authors, SPME does not exhaustively extract the solute onto the fiber in most cases because there is only one theoretical plate or one sorption step. Rather, an equilibrium is developed between the aqueous concentration and the sorbed concentration. The number of moles of analyte on the fiber, n_s, is linearly related to the concentration in the aqueous phase by the following equation:

$$n_s = KV_1C_{2^0} \qquad (12.1)$$

where K is the distribution constant of an analyte partitioning between the stationary and aqueous phases, V_1 is the volume of the stationary phase, and C_{2^0} is the initial analyte concentration in the water (Louch et al., 1992). Time is not considered in this relationship, nor is the volume of sample considered, rather it is the equilibrium condition.

Proceeding to a finite volume, the term V_2 is added (Louch et al., 1992), which is the volume of the sample (12.2):

$$n_s = KV_1V_2C_{2^0} / KV_1 + V_2 \qquad (12.2)$$

This equation is simply a rearrangement of the distribution constant equation that relates the mass sorbed to the stationary phase divided by the mass in the solution phase. The authors note that the amount of analyte sorbed by the coating is proportional to the initial analyte concentration in both Eqs. (12.1) and (12.2). However, the additional term of KV_1 is now present in the denominator of Eq. (12.2). This term decreases the amount of solute sorbed (n_s) when this term is comparable in size to V_2. When it is much smaller than V_2, then only the volume of sample is important. As KV_1 becomes much greater than V_2, then the terms KV_1 in numerator and denominator cancel, and one is left with the conclusion that the majority of the original analyte, C_2^0, is sorbed. Thus, the extraction is quantitative at this point. In practice, the authors have found that for 90% of the sample to be sorbed into the coating, the distribution coefficient must be about an order of magnitude greater than the phase ratio, V_2/V_1. For this to occur, the K must be approximately 1000, which is equivalent to compounds with an octanol–water partition coefficient (K_{ow}) of approximately the same value, or log K_{ow} of 3.

Thus, for many of the compounds that are extracted by this method, the equilibrium state does not quantitatively remove the analyte from solution. Rather from 2 to 20% of the analyte may be removed (Louch et al., 1992), although the application may be quantitative if standards are treated in an identical way. Another important point brought out in this study is that at high K values (>1000) the time to equilibrium is much longer than at lower K values. This effect is the result of the slow diffusion of the nonpolar analytes into the coated phase (Louch et al., 1992). Thus, for compounds with high K values (>1000) a thin film is used, ~ 7 μm (Supelco) in order to attain quick equilibrium conditions. In the case of headspace analysis where sorption occurs from the gaseous state rather than the liquid state, the rate of equilibrium is much faster because of faster diffusion rates around the fiber and the lack of the water boundary layer. Thus, equilibrium for headspace analysis is approximately 10 times faster than the liquid-state sorption.

12.1.3. Examples of Solid-Phase Microextraction

To reiterate, the choice of coating thickness is important in the analysis of organic analytes. For example, thick coatings (100 μm) are useful for volatile organic compounds because they have low values of K; thus, more capacity is obtained for the isolation. An example method is EPA Method 624 for volatiles (Supelco). The sample may be analyzed either in the headspace or in the solution. When the headspace method is used, salt is added to the sample to the saturation point. This step helps to drive the compounds into the vapor phase.

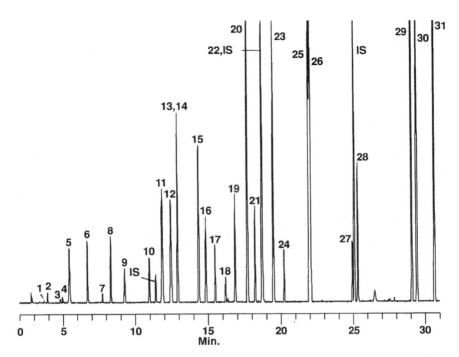

Figure 12.2. EPA Method 624 analytes at 50 µg/L by SPME using a 100-µm polydimethylsilox-ane coating. (Published with the permission of Superco, Inc.)

An example of headspace analysis is Figure 12.2, which shows the chromatogram for volatile sampling according to EPA Method 624 for a 50-µg/L solution of standard compounds on a 100-µm polydimethylsiloxane coating. The most volatile compounds, such as compounds 1 to 4 in Table 12.1 (chloromethane, vinyl chloride, bromomethane, and chloromethane), gave the least response because they had the lowest recovery in the stationary phase. Compounds, such as 21 to 34 in Table 12.1 (e.g., toluene through 1,2-dichlorobenzene), gave excellent response because of good overall recoveries. The sample volume for this analysis was 1.5 mL. In this method, Supelco has reported that the use of a 0.75-mm interior diameter inlet sleeve improved GC column performance much more than the standard splitless sleeve.

Semivolatile compounds also may be extracted by the 100-µm PDMS coating, and for most compounds works well. However, because of the high boiling polynuclear aromatic hydrocarbons and higher K_{ow} (PAHs, $K > 1000$), the extraction time has to be greatly increased for sufficient extraction

Table 12.1. Volatile Organic Compounds (VOC) Determined by SPME in Figure 12.2

1. Chloromethane	17. Bromodichloromethane
2. Vinyl chloride	18. 2-Chloroethylvinyl ether
3. Bromomethane	19. cis-1,3-Dichloropropane
4. Chloroethane	20. Toluene
5. Trichlorofluoromethane	21. trans-1,3-Dichloropropane
6. 1,1-Dichloroethene	22. 1,1,2-Trichloroethane
7. Methylene chloride	IS. 1-Bromo-2-chloropropane
8. 1,2-Dichloroethene	23. Tetrachloroethene
9. 1,1-Dichloroethene	24. Dibromomonochloromethane
10. Chloroform	25. Chlorobenzene
IS. Bromochloromethane (IS)	26. Ethylbenzene
11. Tetrachloroethane	27. Bromoform
12. Carbon tetrachloride	IS. Dichlorobutane
13. Benzene	28. 1,1,2,2-Tetrachloroethane
14. 1,2-Dichloroethane	29. 1,3-Dichlorobenzene
15. Trichloroethene	30. 1,4-Dichlorobenzene
16. 1,2-Dichloropropane	31. 1,2-Dichlorobenzene

Published with the permission of Supelco, Inc. IS = Internal Standard.

(>30 min). If the coating is reduced to 7 μm PDMS, the extraction time is greatly reduced. This extremely thin coating worked best for the PAHs but did not work as well, of course, for the lower boiling compounds. Thus, film thickness is an important parameter to select for good recovery of compounds.

Film chemistry can play a role as well. For example, Supelco has a specialized coating, polyacrylate, which is especially useful for the isolation of phenols and an entire series of chlorinated phenols (Fig. 12.3, Table 12.2). It comes in a film thickness of 85 μm. Furthermore, Supelco has published a series of application notes on the use of SPME as well as a bibliography of important papers. Finally, automation is available for SPME by Varian, which markets a system that will work on any GC or gas chromatography/mass spectrometry (GC/MS) (Fig. 10.9).

The future for SPME looks interesting. There are many applications in environmental chemistry, forensic chemistry, pharmaceutical, food, beverage, and flavor analysis, with more applications being published continually (see Supelco Highlights for further reading). Recently, Supelco has published a note on the interface of SPME and high-pressure liquid chromatography (HPLC) (Supelco Spring Highlights, 1996). The interface consists of a six-port valve and a desorption chamber that replaces the injection loop in the HPLC. In this system, the elution of the fiber is by solvent desorption rather than by heating. Finally, there is a book on SPME Pawliszyn (1997) recently available.

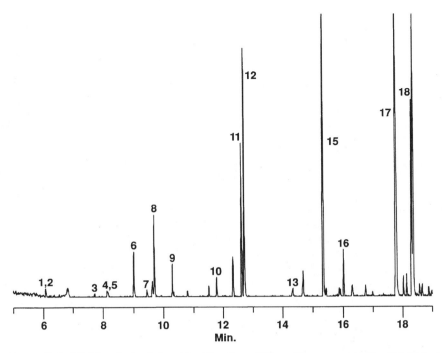

Figure 12.3. SPME analysis of phenols at 25 µg/L at pH 2 on a 97-µm acrylate fiber. (Published with the permission of Supelco, Inc.)

12.2. MATRIX SOLID-PHASE DISPERSION

Matrix solid-phase dispersion (MSPD) was developed by researchers at Louisiana State University's School of Veterinary Medicine in order to isolate, identify, and quantify veterinary drug residues in livestock (Barker and Hawley, 1992). Compared to traditional methods, MSPD reduces solvent use by 98% and turnaround time by as much as 90%. The method involves the mechanical blending of a sample matrix with bulk C-18 sorbent. The C-18 hydrophobic phase has the ability to incorporate the lipids in meat and other food products into its matrix. Mechanical shearing forces initially disrupt the sample structure and disperse the sample over the surface of the C-18 bonded silica. The process causes the sample and polymer phase to become semidry, which then allow the material to be packed into a column (see Fig. 9.4).

Matrix solid-phase dispersion gives a major reduction in scale over classical approaches in that the latter uses 10 to 50 g of tissue and is typically extracted with 100 to 500 mL of solvent with repeated processing in many

Table 12.2. Phenols Determined by SPME in Figure 12.3

Internal Standard is 2-Fluorophenol
1. Phenol
2. 2-Chlorophenol
3. 2-Methylphenol
4. 3- & 4-Methylphenol
5. 2-Nitrophenol
6. 2,4-Dimethylphenol
7. 2,4-Dichlorophenol
8. 2,6-Dichlorophenol
9. 4-Chloro-3-methylphenol
10. 2,4,5-Trichlorophenol
11. 2,4,6-Trichlorophenol
12. 2,4-Dinitrophenol
13. 4-Nitrophenol
14. 2,3,4,6-Tetrachlorophenol
15. 2-Methyl-4,6-Dinitrophenol
IS 2,4,6-Tribromophenol (Internal Standard)
16. Pentachlorophenol
17. Dinoseb
18. Unknown

Published with the permission of Supelco, Inc.

cases. Furthermore, the extracts require extensive clean-up before chromato-graphic analysis (see discussion in Chapter 9). With MSPD the method consists of adding approximately a half of gram of sample to 2 g of C-18 pack-ing and grinding the sample until the pulverized material (milk, fat, liver, kid-ney, muscle, etc.) is incorporated into the packing material. The packing material is then transferred by funnel into a syringe-barrel column and plugged with a filter paper disk. The column head is covered with a second disk, and the contents are compressed by a plunger to a volume of ~ 4.5 mL. The column may then be eluted with a solvent or series of solvents for the analyte of interest (Barker et al., 1993).

A series of drugs have been extracted by MSPD from meat, tissue, and milk (Barker et al., 1993). Typically, the methods call for washing of the sorbent with hexane, and elution with acetonitrile. The extraction methods for the various drug classes gave recoveries of greater than 60% (Barker et al., 1993). In most cases, no clean-up steps were required. Limits of detection were at the action level of the regulatory agencies. The extracts generally gave clean chromatograms for either HPLC or GC (Barker et al., 1993).

The efficiency of the MSPD is related to the dispersal of a sample over a larger theoretical surface area of approximately 1000 m^2 (2 g of packing material at ~ 500 m^2/g). Barker and co-workers (1993) measured the thickness of the 0.5-g sample with electron microscopy and found that it was approximately 100 Å thick. Therefore, micellular lipids are disrupted and cells unfolded within the tissue. They hypothesize that the C-18 phase behaves as a "detergent" with the meat to disrupt and free the drugs from the cellular material.

Although there are direct similarities between MSPD and SPE, the MSPD differs in that it appears to be a mixture of interactions, including partitioning, adsorption, and ion pairing, which makes this an effective method of sorption and extraction. It is possible to elute fractions that contain neutral lipids (hexane), phospholipids (dichloromethane), fatty acids and sterols (acetonitrile), a mixture of phospholipids, amino acids, inositols, mono-, disaccharides, and citric acid (methanol), and finally nucleotides and protein (water).

12.3. NEW SOLID PHASES

There are a host of new solid phases being brought to the market. Some of the most interesting currently include graphitized carbon, functionalized styrene–divinylbenzene copolymers, restricted access solid-phase reversed phase, affinity chromatography with antibodies, and molecular imprinted polymers. These five types of sorbents have different applications from environmental to biological sample preparation.

12.3.1. Graphitized Carbon

Graphitized carbon (Supelco, Appendix) has made a comeback in the environmental field along with high surface-area styrene–divinylbenzene copolymers. Both of these phases have the ability to isolate polar molecules with high water solubility (>1000 mg/L). Solutes such as these are the most difficult to isolate because they have low affinity for most reversed-phase sorbents (Fig. 2.9); however, the increased surface area of the graphitized carbon increases the possibility of a reversed-phase interaction. Furthermore, the graphitized carbon also has a somewhat positively charged surface that also sorbs by an ion-exchange-like mechanism. The third mechanism available on graphitized carbon is the π–π electron interaction, which can be thought of as a flow of electrons between the graphite sheets of carbon. Thus, aromatic analytes may be tightly held on carbon by this interaction. Because of the "mysterious" nature of graphitized carbon, it has received a reputation for being difficult to elute. Supelco, which has done the most to market this sorbent, recommends a mixture of methylene chloride and methanol (80 : 20, respectively) as the best eluent.

Table 12.3. Physical Characteristics of Carbon Sorbents Available from Supelco

Sorbent	Surface Area (m^2/g)	Porosity	pH
ENVI-Carb	100	none	9.5
ENVI-Carb X	240	0.15	9.7
Carboxen 1000	1200	0.85	8.4
Carboxen 1002	1100	0.94	10.7

Published with the permission of Supelco, Inc. (1997).

Supelco has a publication on the use of various forms of graphitized carbon and a comparison of various sorbents for environmental compounds (Nolan and Betz, 1997). Table 12.3 shows the various types of carbon with surface areas that vary from 100 to 1200 m^2/g. The alkaline pH of carbon suggests that its surface contains oxides, which give a slight positive charge to the surface and create a small anion-exchange capacity. The work of Nolan and Betz (1997) shows that the maximum capacity for a suite of environmentally important compounds (such as phenols and pesticides) follows surface area, with the largest capacity being the Carboxen sorbents with over 1000 m^2/g of surface area.

Research carried out by Di Corcia has found that graphitized carbon works well for the isolation of pesticides from water (Di Corcia and Marchetti, 1991; Borra et al., 1986). They designed a multiresidue method for both polar and nonpolar pesticides in water for subsequent analysis by liquid chromatography. Low detection limits of 0.003 to 0.007 µg/L were reported for a number of compounds including polar compounds, such as oxamyl and methomyl (solubility of 28,000 and 58,000 mg/L, respectively). Generally speaking, the Carbopak B sorbent is an excellent sorbent for these polar compounds and nonpolar compounds as well. The problem appears in the elution of these compounds from the sorbent. There may be difficulty in desorption, the probable cause being a mixed-mode activity or specific sorption of the molecule due to incorporation of positively-charged oxygen into the graphitized-carbon structure. Di Corcia and Marchetti (1991) recommend back elution of the Carbopak B sorbents to improve recovery.

12.3.2. Functionalized Styrene–Divinylbenzene

There are several new polymeric sorbents with surface areas of 800 to 1100 m^2/g that could be quite useful for isolating polar organic compounds from water. They include the Isolute styrene–divinylbenzene (SDB) from

International Sorbent Technology, SBD-1 from J. T. Baker, and the Oasis HLB poly(divinylbenzene-*co*-*N*-vinylpyrrolidone) polymer of Waters. All of these polymers contain some trace of hydrophilic character to improve their wetting characteristics for good mass transfer, but yet the polymers still have high capacities for polar organic compounds.

Both the SDB sorbent made by International Sorbent Technology (IST) and the SDB-1 by J. T. Baker are specifically designed for polar compounds in both environmental and drug applications. They are high-surface-area polymers (> 1000 m^2/g) with a light sulfonation that increases their ability to imbibe water and gives the polymer good mass-transfer characteristics (personal communication with IST). These sorbents have nearly twice the surface area of a typical C-18 bonded phase and are nearly 90% carbon versus the 18% carbon loading of C-18 bonded silica. Therefore, the high-surface-area SDB sorbents will have a high capacity for polar solutes. They differ from the graphitized carbon in that they do not contain the positively charged oxygen sites that give anion-exchange capacity with carbon. Also the SDB polymers have a milder form of π–π interaction than the graphitized carbon, which increases the recovery of analytes during elution when compared with graphitized carbon. For these reasons, the SDB has been a popular new sorbent for polar compounds in water.

Table 12.4 shows the characteristics of several of the SDB sorbents and the older styrene–divinylbenzae copolymers (XAD) as a comparison. The *XAD* resins have generally a somewhat lower surface area, approximately half that of the newer SDBs. This lower surface area generally means that the sorbent will have somewhat less capacity for polar solutes. For example, Pichon

Table 12.4. Physical Characteristics of Polymeric Sorbents

Sorbent	Chemistry	Surface Area (m2/g)	Pore Diameter (Å)
XAD-1 (Rohm and Haas)	Styrene–divinylbenzene	100	200
XAD-2 (Supelco)	Styrene–divinylbenzene	330	90
XAD-4 (Supelco)	Styrene–divinylbenzene	750	50
PRP-1 (Hamilton)	Styrene–divinylbenzene	415	—
SDB (International Sorbent Technology)	Styrene–divinylbenzene	1100	—
SDB-1 (J. T. Baker)	Styrene–divinylbenzene	1060	—
Oasis (Waters)	[poly(divinylbenzene-*co*-*N*-vinylpyrrolidone)]	830	82

Reference for XAD resins is Aiken and co-workers (1992); and Pichon and co-workers (1996) and manufacturers' literature for the other sorbents.

co-workers (1996) found that polar herbicide metabolites had about a 10-fold lower capacity factor in water, k'_w, on the PRP-1 with a surface area of 415 m^2/g compared to the SDB-1 with a surface area of 1060 m^2/g. Furthermore, they found that acidic pesticides could be effectively removed from water without pH adjustment when they were sorbed on the SDB-1 for a 500-mL sample of water using a 200-mg cartridge. This result shows the highly effective nature of the high-surface-area SDB polymers.

Another feature of the high-surface-area SDB polymers is that they exclude humic substances because of the small pore size of the sorbent. Table 12.4 shows that as the surface area of the polymer goes up the effective pore size is decreasing. The effect of the decreasing pore size for XAD resins was shown by Aiken and co-workers (1979) to have an effect on the loss of humic substances from XAD resin. Likewise, Pichon and co-workers (1996) show that humic substances are poorly retained on the high-surface-area SDB polymers at neutral pH, which indicates a type of size exclusion during sorption. However, Pichon and co-workers (1996) found that if the pH is lowered to pH 3, the humic substances are removed and interfere with the analysis of the pesticides when using HPLC. Thus, one advantage of the polymer packings for acidic compounds is the removal of the majority of the humic peak in the HPLC chromatogram.

The Oasis HLB sorbent by Waters is a high-surface-area copolymer of [poly(divinylbenzene-co-N-vinylpyrrolidone)] (Fig. 12.4) that has the property of conditioning with only water. Methanol is not needed! This is the only

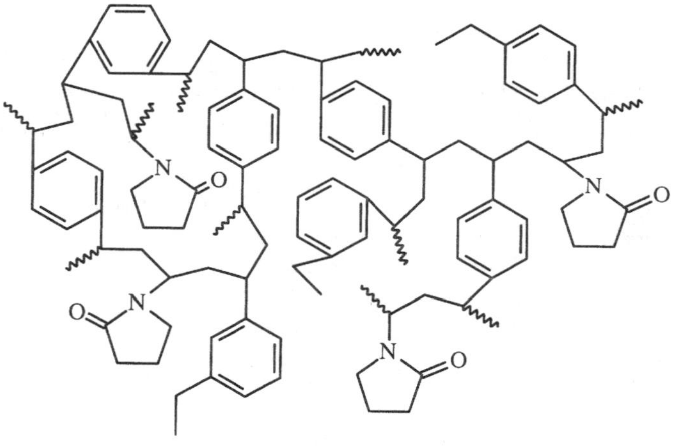

Figure 12.4. Structure of the Oasis HLB sorbent [poly(divinylbenzene-co-N-vinylpyrroli-done)]. (Published with the permission of Waters.)

such polymer currently available. The usefulness of this property has been found in the analysis of drugs in urine and blood. Commonly, many laboratories run hundreds of urine samples per day through SPE cartridges using a vacuum manifold. In the course of running many samples, it often happens that the cartridge may dry while the analyst is not present. If the sorbent is then used, the recovery of analyte will be low. In only 5 min of drying the sorbent totally loses its conditioning and will function poorly.

The cyclic amide structure of the Oasis HLB has the ability to wet with water even after drying and will still sorb effectively. Figure 12.5 shows the recovery of five drugs that were isolated on both the Oasis HLB and on C-18 after different drying times during the conditioning of the sorbents. In all cases, the Oasis rewetted and effectively removed the analytes with > 95% recovery. On the other hand, the C-18 sorbent lost its effectiveness rapidly and after 5 min had recoveries of less than 50%. Thus, the Oasis HLB has an effective capability in that it can rewet with only water. No data are available to compare the Oasis with other SDB sorbents for its effectiveness for polar compounds.

12.3.3. Restricted-Access Sorbents

There are various types of restricted-access packings that have been developed for sample clean-up. Reviews by Unger (1991) and Haginaka (1991) discuss internal-surface reversed phases, shielded hydrophobic phases, semipermeable surfaces, dual-zone phases, and mixed functional phases. The majority of these packings have been developed to purify drugs from body fluids.

The restricted-access solid-phase extraction may work in dual mode when it combines size exclusion and reversed-phase partitioning to isolate and clean up complex matrices such as serum, blood, and plasma (Haginaka, 1991). In this case, it is an internal-surfaced reversed phase (ISRP), as shown in Figure 12.6. The solid phase is a porous silica particle with a pore diameter of 80 Å. The internal surface of the pores are coated with reversed-phase moieties, while the exterior of the particle is coated with a polar group. Interferences such as proteins cannot pass through the small pores, do not interact with the polar exterior of the particles, and pass through the sorbent nonretained. Smaller analytes of interest enter the interior of the particle where they are sorbed by the reversed-phase sorbent (Fig. 12.6).

If a surface barrier is created based on hydrophilicity rather than molecular size, a different type of exclusion is carried out. This type of sorbent is called a semipermeable surface, or SPS. An example is the coating of a nonionic surfactant, such as Tween (a fatty-acid ester of bis-polyethyleneoxide modified sorbitol), onto a reversed-phase silica (Fig. 12.7). In this case, small

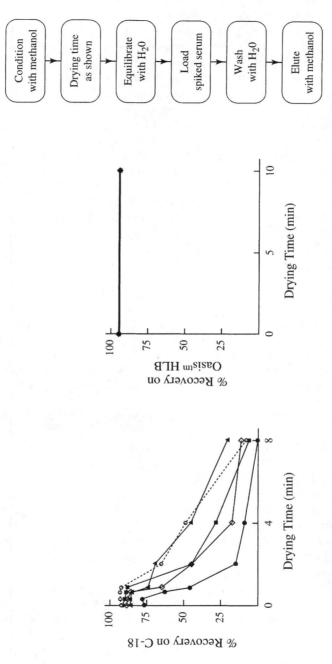

Figure 12.5. Percent recovery versus cartridge drying time for pharmaceutical compounds in porcine serum using C-18 and Oasis HLB sorbents. (Published with the permission of Waters.)

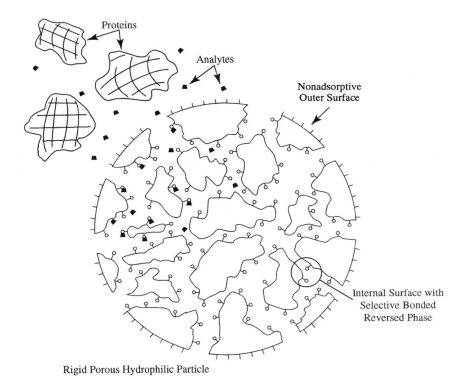

Proteins

Analytes

Nonadsorptive
Outer Surface

Internal Surface with
Selective Bonded
Reversed Phase

Rigid Porous Hydrophilic Particle

Figure 12.6. Restricted-access solid-phase extraction. [Reproduced with permission from Haginaka (1991).]

solutes are retained by a combination of hydrogen bonding at the surface and by reversed phase at the inner surface (Fig. 12.7). An advantage of the SPS surface is that the surfaces may be controlled and varied including both the density and nature of the outer and inner surfaces.

Another type of restricted-access packing is the shielded hydrophobic packings (SHP). This sorbent consists of hydrophobic regions on both the outside and inside of the sorbent, which are embedded in a hydrophilic surface (Fig. 12.8). The stationary phase functions such that a hydrophilic polymer layer acts as a shielding barrier for the lipophilic groups. This packing is different than the SPS in that the shielding covers both the external and internal surfaces. In this type of packing, the pore size of the silica is not critical because the shielding covers all the surfaces.

Another example of restricted-access sorbents is the dual-zone phase (DZP). This consists of an outer zone enriched in a hydrophilic moiety, a

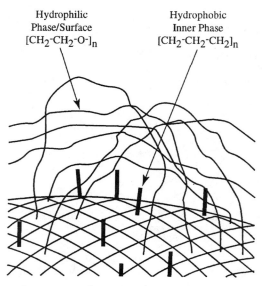

Figure 12.7. Schematic representation of a semipermeable surface packing material. [Reproduced with permission from Haginaka (1991).]

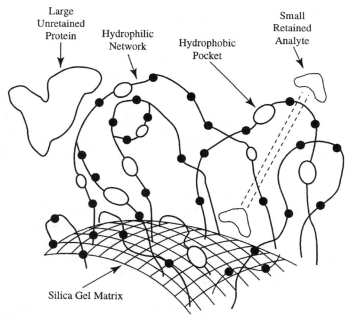

Figure 12.8. Schematic representation of a shielded hydrophobic phase (SHP). [Reproduced with permission from Haginaka (1991).]

perfluorobutylethylenesilyl group, and an internal zone enriched with hydrophobic ODS groups (Haginaka, 1991). Lastly, there are mixed functional groups phases (MFP). These were discussed in detail in Chapter 8 and in the preceding section on styrene–divinylbenzene modified phases.

12.3.4. Affinity Solid-Phase Extraction

Affinity SPE is analogous to the use of antibodies in enzyme immunoassay to bind specific molecules. However, instead of attaching the antibodies to microtiter-plate walls, the antibodies are bound to silica or other types of solid substrates. This type of chromatography is called affinity chromatography, and there are at least two recent books published on the topic (Turkova, 1993; Kline, 1993). The analyte of interest preferentially binds to the site as a key to the keyhole model (Fig. 12.9). Normally, the specific binding site that is attached to the solid phase is a ligand immobilized on a silica support. The analyte is eluted off by disrupting the binding site, usually by elution with acid or buffer.

This separation technique is widely used in the separation of specific proteins and biological molecules. But it has had much less use on environmental samples until recently with work on small molecules, such as atrazine (Thomas et al., 1994). The advantage of this technique is that it is highly specific for the compounds of interest or class of compounds, which may be bound from a complex matrix that would be difficult to assay directly.

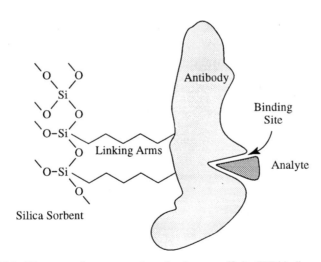

Figure 12.9. Diagrammatic representation of an immunoaffinity SPE binding an analyte.

The binding process involves the use of diol-bonded silica onto which the anti-atrazine antibodies are attached using a Schiff-base method (Thomas et al., 1994). After the preparation of the affinity SPE column, an on-line system was developed so that the isolation and analysis was completed in a single step (Fig. 12.10). The method calls for the isolation of a class of triazines commonly found in water. They included atrazine and simazine, and several metabolites (deethylatrazine, deisopropylatrazine, and hydroxatrazine). The method gave very clean chromatograms because of the specific isolation of the triazines. Recent similar examples are the work of Rule and co-workers (1994) for carbofuran and Pichon and others (1995a,b) for atrazine metabolites, for phenylureas (Pichon and others 1995b), and pesticides in general (Pichon et al., 1995b; Pichon et al., 1996). Affinity SPE promises to be a useful technique for difficult to isolate molecules, or for analytes in complex matrices, but it is not yet offered as a commercial sorbent phase.

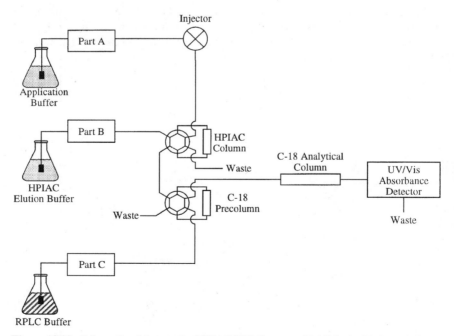

Figure 12.10. Schematic of the atrazine HPIAC/HPLC system. [Published with the permission of Thomas and co-workers (1994).]

12.3.5. Molecular Recognition and Molecular-Imprinting Solid-Phase Extraction

This new technology uses the concept of affinity chromatography but does not use antibodies. Rather an imprint is left or a "hole" is synthesized that will fit the analyte of interest. A general overview on molecular recognition and SPE is given by Izatt and co-workers (1994) with special emphasis on inorganic ions. They describe molecular recognition as an approach where a host polymer is designed that is capable of a high degree of molecular recognition for specific ions or groups of ions called guests, even if the conditions of isolation are difficult because of matrix interferences. The host molecules are attached to a silica support or silica gel. The system of ion-binding molecules plus support is called AnaLig (IBC Advanced Technologies, Inc.). AnaLig systems are used in either the fixed-bed column or particle-loaded membrane (disk). The term molecular recognition technology (MRT) is used to describe the process of designing ligands and creating SPE sorbents. These SPE sorbents also have the capability to be used in industrial applications as well as in environmental or biological applications (Izatt et al., 1994).

An application of the AnaLig system is shown in Figure 12.11 for the isolation of Pb^{2+} from a solution containing other metal ions and acid. A macrocyclic compound, called a cryptand (Izatt et al., 1994), is attached to the silica gel. This ligand is capable of selective binding of trace levels of lead even in complex matrices. The lead is sorbed in the cavity of the ligand and held until a solution containing the complexing agent EDTA is used to displace it. The cryptands may be synthesized such that the cavity size varies for different sized cations. Thus, there may be a selectivity chosen for different metal ions.

Another example of new sorbents is the molecular imprinted polymers (MIP) from the work of Siemann and co-workers (1996). They synthesized a methacrylic acid–ethylene glycol dimethacrylate copolymer with atrazine as an imprint molecule. Imprint synthesis entails polymerization around an imprint species with monomers that are selected for their ability to form specific and definable interactions with the imprint molecule. The atrazine is chemically removed from the polymer leaving holes or cavities. The "cavities" are formed in the polymer matrix whose size and shape are complementary to that of the imprint molecule (Siemann et al., 1996). These recognition sites enable the polymer to rebind the imprint species selectively from a mixture of closely related compounds, in many instances with binding affinities approaching those demonstrated by antigen–antibody systems.

The number of sites associated with this polymer was 7.7 μm/g of dry polymer. This work represented a pilot study on the use of MIP sorbents for environmental molecules and promises to be a useful tool in the near future for

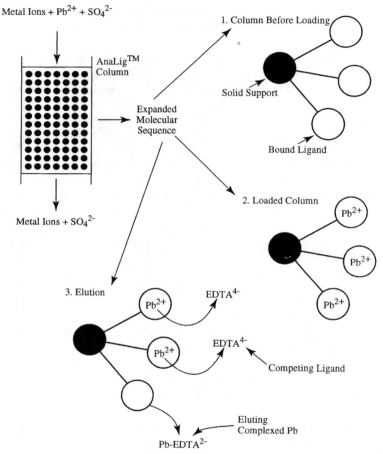

Figure 12.11. Schematic illustrating the separation of Pb^{2+} from a matrix of other ions using an AnaLig SPE column. [Published with the permission of Izatt and others (1994).]

difficult molecules or samples of complex matrices. Advantages of MIP sorbents is that they have high mechanical, thermal, and solvent stability, but they are still in the experimental stage for SPE.

12.3.6.   Layered Sorbents

Raisglid and Burke (1997) recently reported on the concept of layering sorbents for effective SPE. One example included layering an amino sorbent over a C-18 in order to remove interferences from humic substances when sorbing pesticides from water samples. The concept is that the amino sorbent will bind the humic molecules through their phenolic groups and remove them. The pesticides do not effectively bind to the polar amino phase and pass through into the C-18 phase where they are sorbed by reversed-phase interaction. They then may be eluted with an organic solvent and the humic interferences remain bound to the amino sorbent.

A second example is the layering of C-2 over C-18 for sorbing a mixture of hydrophobic and moderately hydrophobic pesticides. The concept is that the most hydrophobic pesticides sorb to the C-2 and are quantitatively eluted. While the more polar pesticides pass through the C-2 and sorb on the C-18. The concept of layering sorbents is an interesting approach that could be exploited for other applications of SPE.

12.4. SMALL VOLUME SOLID-PHASE EXTRACTION

Small-volume SPE is an emerging area with new products appearing at the Pittsburgh Conference in Atlanta, Georgia (1997). There are a number of companies making cartridges of 1-mL volume that contain only 50 to 100 mg of packing material and can be eluted with 100 to 200 μL of solvent. Furthermore, there are several makers of extraction disk cartridges of the 1-mL volume that may also be eluted with volumes as small as 100 μL. There is the pipette-tip disk that has been introduced by Ansys, Inc. (Fig. 11.9), which is specially designed to handle small volumes of sample, from as little as 50 μL to 1 mL. There is also the introduction of the 96-well microtiter plate in both disk form and in cartridge form that works well with 1-mL samples. Thus, there is a set of new products specially designed for small-volume work. Finally, there is the automated micro-SPE work discussed in Chapter 10 using the PROSPEKT where as little as 200 μL of sample may be analyzed directly using SPE–MS–MS on concentrations as low as 50 pg/mL (Bowers et al., 1997).

12.5. DERIVATIZATION ON SOLID-PHASE EXTRACTION SORBENTS

Most recently derivatization on SPE has been tried with good success. A review study by Zhou and co-workers (1992) outlines recent examples of silica based, solid-phase reagents (SPR) for derivatizations in chromatography. Solid-phase reagents can be made on many types of matrices, including silica, alumina, and organic polymers. Silica is most suitable because of the large surface area available for modification.

They explain three types of immobilization procedures: (1) physical adsorption of reagent onto the solid matrix, (2) covalent attachment of a reactive tag through a chemical bond with silanols on the silica surface, and (3) covalent attachment of a reactive tag to an organic polymer coating on the silica surface.

With physical adsorption of the reagent to the silica surface, the SPE column becomes the derivatization agent, and sample may be passed through the column where derivatization takes place. In the second approach, the reagent is attached directly to the surface of the silica and can bind the analyte directly. An example of this mechanism was given earlier in Chapter 7 on the isolation of aldehydes from air using SPE (see Fig. 7.9). The reaction involves a hydrazone derivative that is chemically attached to a silica sorbent. Air is drawn through the cartridge and the aldehyde chemically reacts with the hydrazone to form a derivative that may be later eluted and quantified with HPLC. This system is offered by Waters as part of its SPE product line.

Another method that has been recently published and appears to be quite useful for environmental samples involves sorbing the analytes onto a disk and then derivatization of the analyte on the disk itself (Field and Monohan, 1995). The method involves the use of ion-exchange disks to isolate organic acids from water and the subsequent derivatization of the acids on the disk using methyl iodide and acetonitrile. The derivatization is done on the disk in a GC vial so that the method allows for the direct analysis of the analyte on the GC. The method calls for the isolation of the organic acid, in this case a herbicide, dacthal, from 100 mL of water on an strong anion-exchange disk that is 13 mm in diameter. The disk is dried under vacuum and placed in an autosampler vial of the GC with 1 mL of acetonitrile and 140 μL of ethyl-iodide derivatization reagent for 1 hr at 100 °C. The kinetics of the derivatization are enhanced on the solid phase and the derivatization goes to completion in a short time. The method is quite effective for trace levels of a dacthal metabolite in groundwater, with detection limits of 0.02 μg/L (Field and Monohan, 1995).

12.6.   SEMIPERMEABLE MEMBRANE DEVICE

The semipermeable membrane device (SPMD) is a thin-walled, layflat polyethylene tubing that contains approximately 1 g of a lipid, triolein (600 molecular weight), that cannot diffuse through the polyethylene bag. The SPMD was invented by Huckins and co-workers (1993). The principle of the SPMD is that it is a passive sampler for nonionic organic substances in water or air.

Figure 12.12 shows an expanded view of the polyethylene bag that contains the high-molecular-weight lipid. It is based on the principle of a liquid extraction of flowing water in a stream, lake, or groundwater, and the diffusion of the analytes through the pores of the polyethylene bag. Contaminant organic compounds, such as PAHs, polychlorinated biphenyls (PCBs), and organochlorine pesticides, have very low water solubility and high octanol–water partition coefficients. These classes of compounds are difficult to analyze by conven-

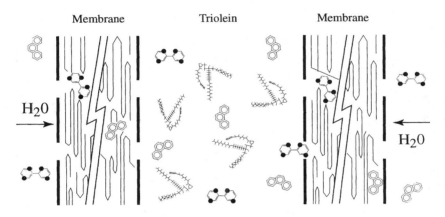

Compound	Symbol	Dimensions (Å)	
		Length	Breadth
Triolein		27	28
Phenanthrene		11.6	7.9
2,2',5,5'-TCB		11.6	8.6

Figure 12.12. Example of a deployed semipermeable membrane device (SPMD). [Published with the permission of Huckins and others (1993).]

tional SPE because they will sorb onto various surfaces, including glassware, tubing, and so forth, which makes the quantitation difficult. Furthermore, it is only convenient to extract approximately 1-L of sample for most SPE situations, even those that use disk technology. This volume is inadequate for quantitation of these most insoluble of compounds at the levels at which they occur in water.

The SPMD allows for the collection of upward of 100 L of water, which greatly improves the ability to detect these insoluble compounds in the water column. Furthermore, the bioconcentration of these substances is an important environmental chemistry question, and the SPMD. allows for direct comparisons between the sampler and the lipid tissue of fish and other organisms (Huckins et al., 1993).

The SPMD allows for the diffusion of nonionic organic compounds, but not water, salts, and ionized organic molecules (humic substances, organic acids, etc.). After deployment, which may last several weeks, the SPMD is at "equilibrium" with its surroundings. It may be returned to the laboratory for dialysis and gel-permeation clean-up of the lipid. The solutes are solubilized in methylene chloride and hexane for analysis by gas chromatography/electron capture detector (GC/ECD) and GC/MS. This method will give detection limits for PCBs, PAHs, and organochlorines in the range of picograms per liter.

12.7. FUTURE OF SOLID-PHASE EXTRACTION

The future of SPE is quite exciting and will continue to involve miniaturization of SPE methods and more on-line use of both liquid and gas chromatography. As instruments such as GC/MS and HPLC/MS become more sensitive, smaller sample sizes may be used and on-line methods will become routine. Sampling handling will be minimized with automated systems that fit directly on the gas chromatograph, so that analysis proceeds directly from the water sample or biological sample to the instrument.

New phases will continue to be introduced to take full advantage of specific interactions. The commercialization of immunoaffinity SPE will no doubt occur, especially for compounds that are polar and difficult to recover by the available sorbents of today. Polymers will be refined to enhance the recovery of polar compounds and more new phases will be introduced. Finally, other uses of SPE will be developed and optimized, such as derivatization on disks, and products will no doubt be introduced that make this process routine. At last sampling handling and solid-phase extraction will reach the level of sophistication that its relatives in liquid chromatography have reached, and perhaps go beyond.

SUGGESTED READING

Arthur, C. L., Killam, L., Buchholz, K. D., and Pawliszyn, J. 1992. Automation and optimization of solid-phase microextraction, *Anal. Chem.*, **64**: 1960–1966.

Arthur, C. L., Killam, L., Motlagh, S., Lim, M., Potter, D., and Pawliszyn, J. 1992. Analysis of substituted benzene compounds in groundwater using SPME, *Environ. Sci. Tech.*, **26**: 979–983.

Arthur, C. L., Potter, D. W., Buchholz, K. D., Motlagh, S., and Pawliszyn, J. 1992. Solid-phase microextraction for the direct analysis of water: Theory and practice, *LC-GC*, **10**: 656–661.

Barker, S. A, Long, A. R., and Hines II, M. E. 1993. Disruption and fractionation of biological materials by matrix solid-phase dispersion, *J. Chromatog.*, **629**: 23–34.

Bouvier, E. S. P., Martin, D. M., Iraneta, P. C., Capparella, M., Cheng, Y. F., and Phillips, D. J. 1997. A novel polymeric reversed-phase sorbent for solid-phase extraction, *LC-GC*, **15**: 152–158.

Buchholz, K. D. and Pawliszyn, J. 1994. Optimization of solid-phase microextraction conditions for determination of phenols, *Anal. Chem.*, **66**: 160–167.

Buszewski, B., Kawka, S., and Wolski, T. 1993. Polyamide PA-6 and silica gel as a mixed packing material for isolating dabsyl amino acids using solid-phase extraction, *LC-GC*, **11**: 366–369.

Gao, C. X., Krull, I. S., and Trainor, T. 1990. Determination of aliphatic amines in air by on-line solid-phase derivatization with HPLC-UV/FL, *J. Chromatog. Sci.*, **28**: 102–108.

George, G. D.and Stewart, J. T. 1990. The use of internal surface reversed-phase packing for the solid phase extraction of drugs from serum, *J. Liquid Chromatog.*, **13**: 3861–3889.

Hawthorne, S., Miller, D., Pawliszyn J., and Arthur, C. L., 1992. Solventless determination of caffeine in beverages using solid phase microextraction with fused silica fibers, *J. Chromatog.*, **603**: 185–191.

Hearne, G. M. and Hall, D. O. 1993. Advances in solid-phase extraction technology, *Am. Lab.*, January, 28H.

Horack, J. and Majors, R. E. 1993. Perspectives from the leading edge in solid-phase extraction, *LC-GC*, **11**: 74–90.

Irth, H. and Brinkman, U. A. Th. 1990. Ligand exchange for trace enrichment and selective detection of ionic compounds, *Trends Anal. Chem.*, **9**: 235–241.

Johnson, R. 1996, Performance-based validation of automated disk extraction, *Am. Environ. Lab.*, **8**: 22–24.

Kimata, K., Hosoya, K., Tanaka, N., and Araki, T. 1991. Effect of stationary phase structure on retention and selectivity of restricted-access reversed-phase packing materials, *J. Chromatog.*, **558**: 19–30.

Kleinmann, I., Plicka, J., Smidl, P. and Vins, I. 1994. Hydrophobic interaction chromatography of proteins on HEMA-based sorbents, *Am. Lab.* April, 34H–34L.

Liu, H., Cooper, L. M., Raynie, D. E., Pinkston, J. D., and Wehmeyer, K. R. 1992. Combined supercritical fluid extraction/solid-phase extraction with octadecylsilane cartridges as a sample preparation technique for the ultratrace analysis of a drug metabolite in plasma, *Anal. Chem.*, **64**: 802–806.

Louch, D., Motlagh, S., and Pawliszyn, J. 1992. Dynamics of organic compound extraction from water using liquid-coated fused silica fibers, *Anal. Chem.*, **64**: 1187–1199.

Majors, R. E. 1993. Perspectives from the leading edge in solid-phase extraction, *LC-GC*, **11**: 74–90.

Majors, R. E. 1995. New approaches to sample preparation, *LC-GC*, **13**: 82–94.

Martin, P., Morgan, E. D., and Wilson, I. D. 1996. An investigation of the properties of a shielded phase for the solid-phase extraction of acidic and basic compounds, J. Pharmaceut. *Biomed. Anal.*, **14**: 419–427.

McDonnelland, T., Rosenfeld, J., and Rais-Firouz, A. 1993. Solid-phase sample preparation of natural waters with reversed-phase disks, *J. Chromatog.*, **629**: 41–53.

Onnerfjord, P., Barcelo, D., Emneus, J., and Gorton, L., and Markovarga, G. 1996, Online solid-phase extraction in liquid chromatography using restricted access precolumns for the analysis of s-triazines in humic-containing waters, *J. Chromatog.*, **737**: 35–45.

Pichon, V. and Hennion, M. C. 1995. On-line preconcentration and LC analysis of phenylurea pesticides in environmental water using a silica-based immunosorbent, *Anal. Chim. Acta*, **311**: 429–436.

Pichon, V. and Hennion, M. C. 1996. Selective trace enrichment using immunosorbents for multiresidue analysis of pesticides, *J. Chromatog.*, **725**: 107–119.

Urano, T., Trucksess, M. W., and Page, S. W. 1993. Automated affinity liquid chromatography system for on-line isolation, separation, and quantitation of aflatoxins in methanol-water extracts of corn and peanuts: *J. Agricul. Food Chem.*, **41**: 1982–1985.

Zhang, Z. and Pawliszyn J. 1993. Headspace solid phase microextraction, *Anal. Chem.* **65**: 1843–1852.

REFERENCES

Aiken, G. R., Thurman, E. M., Malcolm, R. L., and Walton, H. F. 1979. Comparison of XAD macroporous resins for the concentration of fulvic acid from aqueous solution, *Anal. Chem.*, **51**: 1799–1803.

Aiken, G. R., McKnight, D. M., Thorn, K. A. Thurman, E. M. 1992. Isolation of hydrophilic organic acids from water using nonionic macroporous resins, *Org. Geochem.*, **18**: 567–573.

Arthur, C. L., Potter, D. W., Buchholz, K. D., Motlagh, S., and Pawliszyn, J. 1992. Solid-phase microextraction for the direct analysis of water: Theory and practice, *LC-GC*, **10**: 656–661.

Barker, S. A.and Hawley, R. 1992. Efficient biological analysis using MSPD, *Am. Lab.*, October, 42–43.

Barker, S. A., Long, A. R., and Hines II, M. E. 1993. Disruption and fractionation of biological materials by matrix solid-phase dispersion: *J. Chromatog.*, **629**: 23–34.

Belardi, R. and Pawliszyn, J. 1989. *J. Water Pollution Res. Canada*, **24**: 179.

Borra, C., Di Corcia, A., Marchetti, M., and Samperi, R. 1986. Evaluation of graphitized carbon black cartridges for organic trace enrichment from water: Applications for pollutant phenols, *Anal. Chem.*, **58**: 2048–2052.

Bowers, G. D., Clegg, C. P., Hughes, S. C., Harker, A. J. and Lambert, S. 1997. Automated SPE and tandem MS without HPLC columns for quantifying drugs at the picogram level, *LC-GC*, **15**: 48–53.

Buchholz, K. and Pawliszyn, J. 1993. Determination of phenols by solid-phase microextraction and gas chromatographic analysis, *Environ. Sci. Tech.*, **27**: 2844–2848.

Di Corcia, A. and Marchetti, M. 1991. Multiresidue method for pesticides in drinking water using a graphitized carbon black cartridge extraction and liquid chromatographic analysis, *Anal. Chem.*, **63**: 580–585.

Di Corcia, A. and Marchetti, M. 1992. Method development for monitoring pesticides in environmental waters: Liquid-solid extraction followed by liquid chromatography, *Environ. Sci. Tech.*, **26**: 66–74.

Fernando, W. P. N., Larrivee, M. L., and Poole, C. F. 1993. Investigation of the kinetic properties of particle-loaded membranes for the solid-phase extraction by forced flow planar chromatography, *Anal. Chem.*, **65**: 588–595.

Field, J. A. and Monohan, K. 1995. In-vial derivatization and Empore disk elution for the quantitative determination of the carboxylic acid metabolites of dacthal in groundwater, *Anal. Chem.*, **67**: 3357–3362.

Haginaka, J. 1991. Drug determination in serum by liquid chromatography with restricted access stationary phases, *Trends Anal. Chem.*, **10**: 17–22.

Huckins, J. N., Manuweer, G. K., Petty, J. D., Mackay, D., and Lebo, J. A. 1993. Lipid-containing semipermeable membrane devices for monitoring organic compounds in water, *Environ. Sci. Tech.*, **27**: 2489–2496.

Izatt, R. M., Bradshaw, J. S., Bruening, R. L. and Bruening, M. L. 1994. Solid phase extraction of ions of analytical interest using molecular recognition technology, *Am. Lab.*, December, 28C–28M.

Kline, T., Ed. 1993. *Handbook of Affinity Chromatography*, Marcel Dekker, New York.

Louch, D., Motlagh, S., and Pawliszyn, J. 1992. Dynamics of organic compound extraction from water using liquid-coated fused silica fibers, *Anal. Chem.*, **64**: 1187–1199.

Nolan, L. and Betz, W. R. 1997. Extraction of polar analytes from aqueous media using carbogenic extraction: 1997 Pittsburgh Conference, Atlanta, Georgia, Supelco Manufacturers Handout.

Pichon, V. and Hennion, M. C., 1995. Comparison of sorbents for the solid-phase extraction of the highly polar degradation products of atrazine (including ammeline, ammelide, and cyanuric acid), *J. Chromatog.*, **711**: 257–267.

Pichon, V., Chen, L., Hennion, M. C., Daniel, R., Martel, A., Le Goffic, F., Abian, J. and Barcelo, D. 1995. Preparation and evaluation of the immunosorbents for selective trace enrichment of phenylurea and triazine pesticides in environmental waters, *Anal. Chem.*, **67**: 2451–2460.

Pichon, V., Cau Dit Coumes, C., Chen, L., Guenu, S. and Hennion, M.-C. 1996. Simple removal of humic and fulvic acid interferences using polymeric sorbents for the simultaneous solid-phase extraction of polar acidic, neutral and basic pesticides, *J. Chromatog.*, **737**: 25–33.

Raisglid, M. and Burke, M. F. 1997. Solid phase extraction using layered, mixed and single sorbent phases, Pittsburg Conference Abstract number 653, Atlanta, Georgia.

Rule, G. S., Mordehai, A. V., and Hennion, J. 1994. Determination of carbofuran by on-line immunoaffinity chromatography with coupled-column liquid chromatography/mass spectrometry, *Anal. Chem.*, **66**: 230–235.

Siemann, M., Andersson, L. I., and Mosbach, K. 1996. Selective recognition of the herbicide atrazine by noncovalent molecularly imprinted polymers, *J. Agricul. Food Chem.* **44**: 141–145.

Supelco Spring Highlights, 1996. Supelco Inc., Bellefonte, PA.

Thomas, D. H., Beck-Westermeyer, M.and Hage, D. S. 1994. Determination of atrazine in water using tandem high-performance immunoaffinity chromatography and reversed-phase liquid chromatography, *Anal. Chem.*, **66**: 3823–3829.

Turkova, J. 1993. *Bioaffinity Chromatography*, 2nd ed. Elsevier, Amsterdam.

Unger, K. K. 1991. Packing materials with surface barriers, controlled distribution and topography of ligands for sample clean-up and analysis of biologically active solutes in HPLC, *Chromatographia*, **31**: 507–511.

Zhoul, F. X., Thorne, J. M., and Krull, I. S. 1992. Silica based, solid phase reagents for derivatizations in chromatography, *Trends Anal. Chem.*, **11**: 80–85.

APPENDIX

SOLID-PHASE EXTRACTION ON THE INTERNET
AND PRODUCT GUIDE

A1. WEB SITES

Over the past year the Internet has swept the world with the concept of linking computers for a worldwide web of information and knowledge. Solid-phase extraction (SPE) is included in this web of information with a new society (March 1997), called SPE Society. The SPE Society has a web site (www.-spesociety.com) that features a comprehensive databank of SPE applications that is continually updated. The society was started to bring together vendors and customers in SPE for more effective use of technology, resources, and ideas on SPE. There is a publication called *S·P·E*, which is a journazine. It is a publication that falls between the level of review found in a journal and the easy reading and fast turnaround time of a magazine. The focus will be SPE, including technology, users, and its products. The journal is scheduled to begin in July of 1997 and will be published six times per year. The concept of *S·P·E* is a refreshing new idea in sample preparation.

Other information on-line for SPE is a number of vendors who provide information on their products, as well as on various applications of SPE. In fact, the worldwide web is an excellent place to find out the latest information on chromatography and SPE products. Nearly all of the major manufacturers of instrumentation have their hardware in photograph and diagram form on the web. Prices are often available as well. This section of the Appendix discusses the current web sites that were located using the search code "solid phase extraction" as well as looking up the specific vendors cited for their information on web solid-phase extraction.

There are at least 16 web sites that are currently available for browsing on the Internet that deal with solid-phase extraction. Table A.1 lists the 16 sites and their site identification. For SPE products and accessories, they are 3M, Alltech Associates, J&W Scientific, J. T. Baker, Jones Chromatography, Macherey-Nagel, Restek, Supelco, United Chemical Technologies, Varian, Waters, and Whatman. Of these, the most exciting at the time of writing is Alltech Associates, which has much on-line information about chromatography and the practice of solid-phase extraction, including methods and methods development help. The remainder of the web sites appear to be more of a

Table A.1. Web Locations for Information on Solid-Phase Extraction

Company or Organization	Web Site Identification
SPE Products	
3M	www.mmm.com
Alltech Associates	www.alltechweb.com
J&W Scientific	www.jandw.com
J. T. Baker	www.jtbaker.com
Jones Chromatography	www.sciquest.com/jones/
Macherey-Nagel	www.macherey-nagel.com
Restek	www.restekcorp.com
Supelco	www.sigma.sial.com
United Chemical Technologies	www.unitedchem.com
World Wide Monitoring	
Varian	www.varian.com
Waters	www.waters.com
Whatman	www.whatman.com
SPE Automation	
Gilson	www.gilson.com
Jones	www.sciquest.com/jones/
Tekmar-Dohrman	www.tekmar.com
Zymark	www.zymark.com

catalog service, new-product centers, and information about products. Many of the home pages are partially complete at this time but they will surely continue to grow.

An example of how to use the web to find an SPE product is the Varian home page. With a search program like YAHOO hit the "open" button and then type: www.varian.com. This operation will take you to the home page of Varian. Then, using the search program at the Varian home page, type "spe," and it will bring up all of the pages that are available on SPE products, including a catalog. If now the SPE catalog button is pushed, one has an on-line catalog of information with the following address: http://www.varian.com/inst/sppcat/solphase.html. This example provides a model of the type of information that is readily accessible via the Internet.

Waters may be accessed through www.waters.com, which brings up its home page. If the products information is opened followed by chromatography chemistry products, then sample preparation products, one has found the information available on SPE. For example, Waters' extensive applications bibliography is cited in its sample preparation products.

The 3M site can be quickly searched with www.mmm.com, which brings up the home page. There a search may be done across the entire web site using the term "solid phase extraction," which brings up many products and includes information such as references on SPE using disks.

For sites dealing with SPE automation, there are: Gilson, Jones, Tekmar-Dohrman, and Zymark. These sites give information on the types of instruments available by each of the vendors. These are some of the current web sites, and more sites will be made available as the worldwide web becomes a catalog service for many businesses.

A.2.   PRODUCT GUIDE

The product guide (Table A.2) gives the names and addresses of companies producing SPE sorbents, accessories, and automation equipment. Phone numbers and fax numbers are also included for many of the vendors.

Table A.2. Product Guide of SPE Materials and Accessories with Names and Addresses

Company and Address	Products Available for SPE
3M Company Building 220-9E-10 St. Paul, MN 55144-1000 (800) 328-5921 (612) 736-7149 FAX	Empore disks: reversed phase, SDB, ion exchange, oil and grease, chelator Disk cartridges: reversed phase SDBXC, SDB-RPS, silica, ion exchange, accessories for disks, filter beads, manifolds, Applications guide
Alltech Associates Inc. 2051 Waukegan Road Deerfield, IL 60015-1899 USA (847) 948-8600 (847) 948-1078 FAX	Robot-compatible syringes, standard syringe, maxi-clean cartridge, Novo-Clean membranes (ion exchange and reversed phase), many types of syringe filters, reservoirs, inlet and outlet caps, adapters, frits, and filter columns, vacuum manifolds, reversed-phase, normal-phase, and ion-exchange sorbents. Applications bibliography, Bulletin #202, and Catalog

(contd.)

Table A.2. (*continued*)

Company and Address	Products Available for SPE
Ansys 2 Goodyear Irvine, CA 92718-2002 (800) 854-0277 (714)-770-0863 FAX	Filter discs (free and in syringes); reversed phase, normal phase, ion-exchange, and mixed-mode sorbents: application examples available
Applied Separations, Inc. 930 Hamilton Street Allentown, PA 18101-1137 (610) 770-0900 (610) 740-5520 FAX	Automation equipment for SPE Spe-ed Wiz Syringe Cartridges: reversed phase normal phase, ion exchange
J. T. Baker Inc. 222 Red School Lane Phillipsburg, NJ 08865 1-800-JTBaker or (1-800-582-2537) (908) 859-9318 FAX	Robot-compatible syringes, standard syringe, glass columns Teflon frits, Speedisk, reversed phase, normal phase, ion exchange, Special applications: Narc-1 (drugs), Narc-2 (cocaine), spe-500 (EPA), EPA Method 525, Oil and grease vacuum manifold, extraction disk processors Applications bibliography
Gilson, Inc. 3000 W. Beltline Highway, Box 620027 Middleton, WI 53562-0027 (800) 445-7661 (608) 831-4451 FAX	ASPEC XL, automated SPE
Hamilton Company 4970 Energy Way Reno, Nevada 89502 (800) 648-5950 (702) 856-7259 FAX	Microlab SPE, automated SPE
Hewlett-Packard Company 2850, Centerville Road Wilimington, DE 19808 (302) 633-8405 (302)633-8902 FAX	PrepStation System, automated SPE SPE cartridges for automated SPE

(*contd.*)

Table A.2. (*continued*)

Company and Address	Products Available for SPE
Interaction Chromatography 2032 Concourse Drive San Jose, CA 95131 USA (408) 894-9200 (408) 894-0405 FAX	Bifunctional sorbents: SDB-C-18 (Styrene-divinylbenzene-C-18), Vinylpyridine, SDB-SCX
International Sorbent Technology Ltd. 1st House Duffryn Industrial Estate Hengoed Mid Glam, CF82 7RJ England 44-1443-816656 44-1443-816657 FAX	Reversed phase, normal phase, Ion exchange, Applications guide and Handbook, QA/QC Reports
Jones Chromatography P.O. Box 280329 Lakewood, CO 80228-0329 (800) 874-6244 (303) 989-9200 (303) 988-9478 FAX	PROSPEKT, automated SPE IST sorbents SPE accessories
J&W Scientific, Inc. 91 Blue Ravine Road Folsom, CA 95630-4714 (916) 985-7888 (916) 985-1101 FAX	Reversed phase, normal phase ion exchange Alumina-acidic, alumina-neutral alumina-basic, Florisil Vacuum manifold and accessories Applications Guide, QA/QC Reports
Phenomenex, Inc. 2320 W. 205th St. Torrance, CA 90501 (310) 212-0555 (310) 328-7768 FAX	Solid-phase extraction products
Restek Corp. 110 Benner Cir Bellefonte, PA 16823-8812 (800) 356-1688 (814) 353-1309 FAX	Solid-phase extraction disks Solid-phase extraction products
Spark Holland P. de Keyserstr. 8 Emmen, 7825 The Netherlands 31-591-631700 31-591-630035 FAX	PROSPEKT, automated SPE, Micro SPE cartridges for PROSPEKT

(*contd.*)

Table A.2. (*continued*)

Company and Address	Products Available for SPE
Supelco Inc. Supelco Park P.O. Box B Bellefonte, PA 16823-9900 (800) 247-6628 (814) 359-3441 (800) 447-3044 FAX (814) 359-3044 FAX	Reversed phase, normal phase, 　ion exchange, DrugPak-T ENVI-Carb (graphitized carbon) XAD-2, 4, 7, 8, 9, 10, 11, 12, 16, 　XAD-2010, 1180 Vacuum manifold and accessories SPMD devices: polydimethylsiloxane 　and polyacrylate for automation see 　Varian
Tekmar-Dohrmann P.O. Box 429576 Cincinnati, Ohio 45242-9576 (800) 543-4461 (513) 247-7050 FAX	Automation equipment Tekmar AutoTrace System
United Chemical Technologies Worldwide Monitoring 2731 Bartram Road Bristol, PA 19007 (215) 781-9255 (215) 785-1226 FAX	Reversed-phase sorbents, 　normal-phase sorbents 　mixed-mode for drug analysis, Ion-exchange sorbents, 　vacuum manifold and accessories Application Guide for Drug Analysis
Varian Sample Preparation Products 24201 Frampton Avenue Harbor City, California 90710, USA (800) 421-2825 (310) 539-6490 (310) 539-4270 FAX	Reversed phase, normal phase, 　ion exchange, drug columns, 　mixed-mode columns, 　oil and grease vacuum manifold and 　accessories, 96-Well format, Automated SPME liquid/liquid extraction 　cartridges, Applications Guide and 　Sorbent Extraction Technology 　Handbook QA/QC Reports
Waters Chromatography Division 34 Maple Street Milford, MA 01757 USA (508) 478-2000 (508) 872-1990 FAX	Reversed phase, normal phase, 　ion exchange Formats: syringes and cartridges Aldehyde in air sorbent, 　vacuum manifold and accessories Applications bibliography (disk 　format available), Guide to Methods 　Development

(*contd.*)

Table A.2. (*continued*)

Company and Address	Products Available for SPE
Whatman Inc. 9 Bridewell Pl. Clifton, NJ 07014 (201) 773-5800 (201) 472-6949 FAX	Solid-phase extraction disks Solid-phase extraction products

Zymark Corporation
Zymark Center
Hopkinton, MA 01748
(508) 435-9500
(508) 436-3439 FAX
	Automation equipment for SPE,
	  BenchMate
	  The RapidTrace SPE Workstation,
	  Total robotic systems

INDEX